Making Parents

Inside Technology
edited by Wiebe E. Bijker, W. Bernard Carlson, and Trevor Pinch

A list of books in this series appears at the back of the book.

Making Parents

The Ontological Choreography of Reproductive Technologies

Charis Thompson

The MIT Press
Cambridge, Massachusetts
London, England

This book was set in New Baskerville on 3B2 by Asco Typesetters, Hong Kong.

Library of Congress Cataloging-in-Publication Data

Thompson, Charis
Making parents : the ontological choreography of reproductive technologies / Charis Thompson.
p. cm. — (Inside technology)
Includes bibliographical references and index.
ISBN 978-0-262-20156-8 (hc. : alk. paper)—978-0-262-70119-8 (pb. : alk. paper)
1. Human reproductive technology—United States—History. 2. Human reproductive technology—Social aspects—United States. 3. Human reproductive technology—Philosophy. I. Title. II. Series.
RG133.5.T495 2005
616.6′9206—dc22 2004062112

The MIT Press is pleased to keep this title available in print by manufacturing single copies, on demand, via digital printing technology.

For John
and for Thomas, Jessica, and Charlotte

Contents

Acknowledgments

I am grateful to the Universities of Oxford, California at San Diego, École Nationale Superiéure des Mines de Paris, Cornell, Illinois at Urbana-Champaign, Harvard, Michigan, and California at Berkeley for instruction and material and institutional support. Part of this research was supported by grants from the National Science Foundation.

My heartfelt thanks go to the patients, practitioners, research scientists, and others (who I cannot name individually here for reasons of privacy) who have worked and talked with me and extended hospitality to me during my fieldwork.

I acknowledge the following sources, where earlier and substantially different versions or some parts of chapters 1, 2, 3, 5, and 6 appeared: "Ontological Choreography: Agency through Objectification in Infertility Clinics," *Social Studies of Science* 26 (1996):575–610; "Ontological Choreography: Agency for Women Patients in an Infertility Clinic," in M. Berg and A. Mol, eds., *Differences in Medicine: Unraveling Practices, Techniques and Bodies* (Durham, NC: Duke University Press, 1998), 166–201; "Producing Reproduction: Techniques of Normalization and Naturalization in an Infertility Clinic," in S. Franklin and H. Ragone, eds., *Reproducing Reproduction: Kinship, Power, and Technological Innovation* (Philadelphia: University of Pennsylvania Press, 1998), 66–101; "Quit Snivelling Cryo-Baby, We'll Work Out Which One's Your Mama: Kinship in an Infertility Clinic," in R. Davis-Floyd and J. Dumit, eds., *Cyborg Babies: From Techno-Sex to Techno-Tots* (London: Routledge, 1998), 40–66; "Primate Suspect: Some Varieties of Science Studies," in S. Strum and L. Fedigan, eds., *Close Encounters: Primates, Science and Scientists* (Chicago: Chicago University Press, 1999), 329–357; "Strategic Naturalizing: Kinship in an Infertility Clinic," in S. Franklin and S. McKinnon, eds., *Relative Values: Reconfiguring Kinship Studies* (Durham, NC: Duke University Press, 2001), 175–202; "Fertile Ground: Feminists

Theorize Infertility," in M. Inhorn and F. van Balen, eds., *Infertility around the Globe: New Thinking on Childlessness, Gender, and New Reproductive Technologies* (Berkeley: University of California Press, 2002), 52–78.

I thank my extended family—Kristin, Daniel, Matthew, David, Maggie, Liz, Joe, and Freya—and Dr. and Mrs. Lie and Mahie for kinship and heteroclite help throughout. I thank all those who have cared for and taught my children and me over this period.

I thank the many brilliant students, teachers, colleagues, and friends from whom I have benefited in so many ways, especially those working in fields closely related to mine, from whom I learn at every turn. For being able to think about the material and inquiries in this book in particular I thank the following exceptional people: Karen Barad, Gay Becker, Mario Biagioli, Allan Brandt, Judith Butler, John Carson, Adele Clarke, Adrian Cussins, Jerry Doppelt, Amal Fadlalla, Sarah Franklin, Kris Freeark, Peter Galison, Evelynn Hammonds, Donna Haraway, Anne Harrington, Marcia Inhorn, Sheila Jasanoff, Stephanie Kenen, Martha Lampland, Bruno Latour, Irving Leon, John Lie, Margaret Lock, Michael Lynch, Everett Mendelsohn, Chandra Mukerji, Katharine Park, Michael Peletz, Trevor Pinch, Paul Rabinow, Rayna Rapp, Steven Shapin, Brian Smith, Marilyn Strathern, Shirley Strum, David Western, Robert Westman, and Kath Weston. Jennifer Harris was a fine research assistant. Members of the Michigan Institute for Research on Women and Gender group on infertility and adoption were charming and incisive interlocutors. The staff, students, and faculty in the Harvard Department of the History of Science and the Program in Women's Studies provided an intellectually stimulating and convivial environment that I miss greatly. My colleagues, my students, and the staff in the Departments of Women's Studies and Rhetoric as well as in the wider intellectual community at UC Berkeley have been wonderfully welcoming. Judy Lajoie and Yuri Hospodar at Harvard and Sharon Lyons Butler and Althea Grannum-Cummings at Berkeley have been particularly kind and helpful. Sara Meirowitz has been a great editor.

John Lie's erudition, loving kindness, and astonishing efficiency provided the daily example. John's, Thomas's, Jessica's, and Charlotte's lives and my love for them provide the daily fabric.

Making Parents

Introduction

Making Parents: Ontological Choreography, a Biomedical Mode of Reproduction, Methods, Reading This Book, and Where I Stand

Shortly after we arrived in Berkeley, California, in August 2003, we discovered a sprawling CD and DVD store on Telegraph Avenue. It offered something for everyone and deals on all of them. While I was at the cash register, my older daughter read aloud a sign that promised a cornucopia of free goods for anyone who was born as the result of a virgin birth or for anyone who was a clone. As a test-tube baby, she was delighted. The procedure involved in her conception bypasses sexual intercourse, hence the virgin birth. The store manager—alternating between laughter and some apparently genuine concern that he might be required to give her the free goods—told us that he had "Jesus" and not reproductive technologies in mind. "It was meant to be a joke," a second, decidedly nervous sales clerk contributed.

This is a book about making parents through the practice of assisted reproductive and genetic technologies in the United States in the last quarter of a century. During this relatively short period of time, the practices of the field have grown from being theoretical to being commonplace. The first U.S. baby who was born from assisted reproductive technologies in the United States arrived in 1981. In 2001, over 40,000 babies were born from procedures that had been performed in the United States (table I.1). The term *assisted reproductive technologies* (ARTs) refers to the means that are used in noncoital, technically assisted reproduction where gametes are manipulated or embryos are created outside the body. ART clinics are often referred to as infertility clinics, and infertility is the condition that clinicians and patients most commonly aim to overcome. Focusing on infertility gets at the human drama that often characterizes the use of these technologies, but people do not always agree on its definition (a failure to achieve a desired pregnancy after a year of unprotected intercourse is one common definition), and infertility is frequently not treated or treated with little or

Table I.1
Assisted Reproductive Technology Facts and Statistics, United States

Definition The World Health Organization (WHO) defines *infertility* as "the inability of a couple to achieve conception or to bring a pregnancy to term after a year or more of regular, unprotected intercourse."

Infertility numbers It estimates that approximately 8 to 10% of couples worldwide (some 50 to 80 million people) experience infertility. Figures vary regionally according to factors such as prevalence of sexually transmitted diseases, age at attempted conception, and contraceptive-use history.[a]

The American Society for Reproductive Medicine (ASRM) estimates that infertility affects 6.1 million American women and their partners or approximately 10% of the reproductive-age population.

The Vital and Health Statistics Series 23, no. 19 (1995) estimates that 9 million U.S. women have used infertility services, 1.8 million have used fertility drugs, and 2.1 million married couples are infertile.

Based on data from Cycle 5 of the National Survey of Family Growth conducted by the National Center for Health Statistics, which interviewed 10,847 women age 15 to 44 in 1995, the number of women who will experience infertility in 2025 will range from 5.4 to 7.7 million.[b]

The majority of U.S. couples who face infertility first turn to medical solutions if these are available. Decisions to adopt, to foster, to try alternative therapies, or to concentrate on other life goals typically come after a period of medical treatment.[c]

Baby numbers The first in vitro fertilization (IVF) conception in the United States that resulted in a live birth occurred in 1983. In 2000, 35,025 babies were born from assisted reproductive technology (ART) procedures carried out in reporting U.S. clinics; 40,687 were born in 2001.

The Centers for Disease Control and Prevention (CDC) estimates that in 2001, 1% of U.S. babies were born as the result of ARTs.

Clinic numbers There were 421 ART clinics in the United States and Puerto Rico in 2001, and 384 submitted data to the Society for Assisted Reproductive Technology (SART) registry. They began 107,587 cycles of ART involving IVF in 2001.

Of the ART 384 clinics that submitted data for 2001, 84% accepted single women for treatment. Treatment for single women effectively removes marital status and sexual orientation from being barriers to treatment.

Pregnancy rates In 2001, 32.8% of cycles resulted in a pregnancy confirmed by ultrasound, and of those 82% resulted in the birth of at least one live baby for women using their own fresh, nondonor eggs. The national average ART success rate was 27% (compared with under 20% for 1995, the first year for which the CDC reported data).

Table I.1
(continued)

The pregnancy rate for fertile couples with uncontracepted midcycle intercourse is estimated at a maximum of 30% per cycle, with significant numbers of these pregnancies ending in a spontaneous abortion.[d]

There are no statistical differences in the success rates for "Caucasian," "African American," "Asian," and "Hispanic" patients (racial categories drawn from source).[e]

Miscarriage rates The miscarriage rate for IVF pregnancies using fresh nondonor eggs in reported cycles in 2001 was 14% among women younger than 34. The rates increase among women in their mid- to late 30s and continue to increase with age, reaching 30% at age 40 and 41% at age 43.

Fertility rates A woman in her late 30s is about 30% less fertile than she was in her early 20s. After age 40, success rates for ARTs decline precipitously.

Live birth rates per one embryo transfer are 30%; for two embryos, 51.7%; and for three embryos, 46.9%. This translates into a rationale for transferring no more than two embryos and would also prevent the occurrence of higher-order multiple births.

For diagnoses of male-factor, unexplained, or tubal-factor infertility, spontaneous pregnancy rates are very low and are significantly lower than IVF success rates, making IVF a high priority for these diagnoses.[f]

Single and multiple births Of the live births from 2001 cycles, 64.2% were singletons, 32% were twins, and 3.8% were higher order than twins. In 2001 (the results of which were published in December 2003), the SART registry began to separate live births into singleton births and multiple births, to emphasize the optimal nature of singleton pregnancies, and to highlight the problems of neonatal and maternal morbidity and mortality associated with multiple births.

Higher-order multiple births (triplets or more) are higher in states without insurance coverage for infertility treatments.[g]

IVF children have no greater incidence of birth defects than children conceived in the conventional manner.

Donor sperm, donor eggs, and surrogates More than 30,000 babies are born a year from donor sperm.[h]

In 2001, gestational carriers, or surrogates, were involved in 0.7% of ART cycles using fresh nondonor embryos in 2001 (571 cycles).

Donor eggs or embryos were used in slightly more than 11% of all ART cycles carried out in 2001 (12,018 cycles). Few women younger than age 39 used donor eggs; however, the percentage of cycles carried out with donor eggs increased sharply starting at age 39. About 76% of all ART cycles among women older than age 45 used donor eggs.

In 2001, 89% of reporting clinics offered donor-egg programs, 58% offered donor-embryo programs, and 69% offered gestational carriers.

Table I.1
(continued)

Frozen embryos Cryopreservation was offered by 98% of reporting clinics in 2000.

There are an estimated 400,000 frozen embryos in fertility centers in the United States.

Almost 90% of frozen embryos are intended to be used by the couples from whose gametes they were derived—to use if the fresh cycle of treatment does not work or if they want to have another child at a later date. Of the approximately 4% of frozen embryos available for donation, it is estimated that 11,000 are destined for research and 9,000 for other couples. Approximately 9,000 are currently destined to be thawed without use.[i]

"Abandoned" embryos—which are unclaimed after a set period of years and after diligent efforts have been made to contact the couple involved—are not used for research or donation.

Source: Except when indicated otherwise, this table has been adapted from *2001 Assisted Reproductive Technology Success Rates: National Summary and Fertility Clinic Reports,* which can be found at 〈http://www.cdc.gov/reproductivehealth/ART01/index.htm〉. These annual reports are prepared jointly by the Centers for Disease Control and Prevention (CDC), the American Society for Reproductive Medicine (ASRM), the Society for Assisted Reproductive Technology (SART), and RESOLVE, in compliance with the Fertility Clinic Success Rate and Certification Act (FCSRCA) of 1992, section 2[a] of Public Law 102-493 (42 U.S.C. 263(a)-1), which the U.S. Congress passed in 1992 and which requires all clinics performing ART in the United States to annually report their success-rate data to CDC. CDC uses the data to publish an annual report detailing the ART success rates for each of these clinics.
a. Rowe et al. (1993). b. Stephen and Chandra (1998). c. Van Balen et al. (1997). d. Zinaman et al. (1996). e. ASRM (2002b). f. Evers et al. (1998). g. Reynolds et al. (2003). h. Corson and Mechanick-Braverman (1998). i. Hoffman et al. (2003). j. Robertson (1996).

no resort to reproductive technologies. In addition, assisted reproductive technologies are increasingly used for building alternative families and for genetic screening when there is no diagnosis of infertility. Focusing on assisted reproductive technologies captures these additional diagnoses and holds firmly in view the combination of technology and reproduction that is ontologically important but doesn't necessarily evoke emotions and meanings as readily.

Reproductive technologies may aim to produce miracle babies, and yet they are irreducibly mundane. When they work, they make babies and parents. These kinds of apparent contradictions are a signature aspect of assisted reproductive technologies. They are intensely technological, and yet they also make kinship; they are variously commodified, and yet they promise the priceless. They require substantial commitment on the part of the bodies of patients, especially women, and yet involuntary childlessness is usually experienced as a breakdown in gender identity and the life course, and only in some states is it classified as a disease of the body. Assisted reproductive technologies demand as much social as technological innovation to make sense of the biological and social relationships that ARTs forge and deny. In this book, I don't try to reconcile these tensions but instead work out some of the details of moving between, coordinating, and in the process, changing the variously political, social, personal, legal, ethical, bureaucratic, medical, and technical facets of these technologies.

A large number of academic books have been published about assisted reproductive technologies, and many of these books are excellent. I have not tried to provide a corrective to other accounts (excellence hardly lends itself to this) or to produce a definitive account of reproductive technologies (scholarship in the area demonstrates the variety and dynamism of the technologies and patients' experiences of them). Instead, I focus topically, methodologically, and theoretically on a number of things that are particular to my own project. One contribution that I attempt to make is in the use of ethnographic data to raise and address questions typically debated in a much more abstract mien in philosophy or social theory. Many of the questions that I inherited from my background in philosophy of mind, moral and political philosophy, and feminist epistemology proved highly amenable to being addressed with ethnographic data, as I attempt to show in parts II and III through the development of the notions of ontological choreography and of a biomedical mode of reproduction. The particular assisted reproductive technologies on which I concentrate are new enough that I have studied them almost since their inception. This means that the ethnographic narrative has a temporal dimension that approximates a meaningful temporal unit for the technology in question. Thus, there is also an attempt throughout to begin to historicize the ethnographic gaze, which is not common.

Regarding content, I focus not on making babies but on making parents. This focus is motivated by my interest in what it takes to become parents—the biomedical interventions, the legal innovations, and the work that disambiguates the relevant kinship categories—but it also reflects a historical argument about the trend in ARTs over the time that I have worked in the field. When I began to work in

ART clinics, the predominant framing for ARTs in the United States was similar to adoption. That is, access to assisted reproductive technologies, as for adoption, was governed not only by a would-be patient's self-referral based on her desire to have a child but also by judgments made by others about the social appropriateness of such a person as a parent. Historians of adoption in the United States have documented the rise of the idea of the "search for the ideal family" and of the parental screening system that is epitomized today by the "home study."[1] Using a rationale most often referred to as "the best interests of the child," some parents and not others have, in various ways since the 1920s, been considered appropriate to become parents by adoption. The state, through its proxies the public and private adoption agencies, takes an interest in the creation of adoptive kinship only according to tacit criteria of social engineering.

As I describe in chapter 3, through the mid-1990s, access to ARTs tended to mirror the idea that some kinds of patients and not others should have access to becoming parents through the use of ARTs. In more recent years, however, access to ARTs in the United States has come to be framed more within a broad understanding of "reproductive choice" and less in terms of the best interests of the child. That is, it is less common for ART patients to have to prove that they would offer an ideal home to a future child and more common for their infertility to be considered a matter of reproductive privacy and of a right to access to appropriate treatment. At the beginning of the twenty-first century, most women who become pregnant by sexual intercourse have reproductive and kinship rights in this country without having been deemed to be socially fit parents. Exceptions—such as prisoners, women on welfare, drug users, those who refuse medical intervention, and those who are considered to have neglected or abused other children—can be viewed as cases where the woman in question has been robbed of or has herself forfeited the right to certain aspects of fundamental citizenship, effectively making the child-to-be a ward of the state.

During the 1990s and first years after 2000, increasing numbers of clinics treated single women and gay and lesbian parents (see table I.1). Those who have no access to reproductive technologies now are mostly those who cannot afford to pay and whose treatments are not covered by the state or insurance. Homologous to the case for "normal" reproduction, this can be thought of as dividing those with biomedical citizenship from those without it. As with reproduction more generally in this country, the haves and the have-nots of ARTs are increasingly the socioeconomic haves and have-nots of society in general. The normalizing of single, gay, and lesbian parenting in ARTs through the shift from best interests of the child to reproductive privacy has become part of the somewhat increased citizenship rights of these family models in recent years.

It is not prima facie obvious that ARTs should be patterned in the United States more like reproductive choice than like adoption. After all, many different people are involved in making parents and children by ARTs, and in many cases, egg or sperm or embryo donors, or surrogates, are involved in the biological bringing in to being of ART parents and children, even down to involving mandatory adoption by one or both of the intended parents for some procedures in many states. It makes just as much intuitive sense, then, that making ART babies and parents would develop more similarly to adoption, as it does that the techniques would come to resemble other kinds of reproductive choice. In chapter 7, I discuss the ways in which ARTs have become part of reproductive privacy, making parents rather than making babies, by precipitating this shift from the best interests of the child to a parent-centered extension of the idea of reproductive rights.

A consideration of the array of reproductive technologies leads to an interesting kind of double vision, both frames of which need to be kept in focus. First, reproductive technologies are a very small part of a broader picture of pro- and antinatalist sentiment, ideology, practice, and policy. In the United States as elsewhere, reproductive technologies for infertility coexist with countless other pro- and antinatalist phenomena, including ambient contraceptive technologies. Worldwide, reproductive technologies are sometimes part of an explicit state-sponsored pronatalism. In the United States, state orientation to pronatalism is selective at most and tends to be the implicit result of rather than the explicit object of policies that are directly aimed at regulating things like abortion, immigration, and insurance coverage of contraception and infertility treatments (for insurance coverage by state, see table 2.2). In general, though, reproductive and contraceptive technologies, both of which are unevenly available, are directed at overlapping but substantially different populations.[2] The broader agonistic and stratified field of pro- and antinatalisms looks at whose reproduction material, economic, technical, and emotional resources are being invested and thus not just who but what is being reproduced at a given time and place. It also allows one to examine the idea that the technologies might themselves be a site for class and other differentiation and mobility. We are familiar with thinking of technologies as motors of progress that produce or signal global stratification and as objects of consumption that differentiate the rich from the poor. But we are less familiar with the idea that our interactions with technologies—biomedical technologies in particular—might in part produce rather than simply reflect such differentiation.

Ontological Choreography

Reproduction has rich literal and metaphorical meanings that spill well beyond the biological and permeate the public sphere and intimate lives alike. It is ripe for sociological, philosophical, and historical analysis. In this book, I examine how so-called biological reproduction intertwines with the personal, political, and technological meanings of reproduction in the particularly telling site of assisted reproductive technologies. I examine how reproduction has changed dramatically over time in the sites dedicated to producing reproduction and how identities, social stratification, certain techniques, scientific knowledge, law, politics, and our experiences of bodies and reproductive and parental roles have been produced, reproduced, and challenged in these sites.

The term *ontological choreography* refers to the dynamic coordination of the technical, scientific, kinship, gender, emotional, legal, political, and financial aspects of ART clinics. What might appear to be an undifferentiated hybrid mess is actually a deftly balanced coming together of things that are generally considered parts of different ontological orders (part of nature, part of the self, part of society). These elements have to be coordinated in highly staged ways so as to get on with the task at hand: producing parents, children, and everything that is needed for their recognition as such. Thus, for example, at specific moments a body part and surgical instruments must stand in a specific relationship, at other times a legal decision can disambiguate kinship in countless subsequent procedures, and at other times a bureaucratic accounting form can protect the sanctity of the human embryo or allow certain embryos to be discarded. Although this kind of choreography between different kinds of things goes on to some extent in all spheres of human activity, it is especially striking in ART clinics. They are intensely technical and intensely personal and political. All involved are allowed no illusions that they are dealing simply with technical matters or simply with personal or political factors.

Attempting to elucidate some of the specific choreography that enables ontologically different kinds of things to come together has inevitably led me to explore the ontological separations between things (for example, the separations between being a parent and having had a biomedical procedure) and to examine the reductions of one kind of thing to the other (for example, the reduction of a patient's biological age to the levels of her follicle-stimulating hormone (FSH) on day 3 of her cycle). These kinds of things occur at specific times, do specific ontological work, and are highly instructive. Mistakes, exploitation, danger, and bad philosophy occur—not because ontologically different kinds of things are combined (hybridity is not inherently polluting), because connectivities and interdependencies go unrecognized (purity is not always ascetic or aesthetic excess), or because one thing has been taken for another (reductionism and objectification do not always deny integrity and agency) but because these ontological processes get the choreography wrong in various ways. If a surgical treatment doesn't in fact help a woman get pregnant, if a legal decision leaves a child without a loving home, if a determination of FSH levels subjects a woman to age discrimination, or if payment for treatment excludes many people from that treatment, then the ontological choreography fails to lead to ontological innovation. Instead, treatment objectifies the patient, the law harms the child, the age test enables ageism, and the payment structure divides those who count from those who do not count. On the other hand, when treatment leads to pregnancy, the law protects a child, the FSH age test predicts an efficient treatment protocol, and treatment is financed for those who need treatment, the ontological choreography leads to new kinds of reproduction and new ways of making parents. In short, it leads to ontological innovation in the area in question.

Ontological choreography coordinates two different "things" that are especially salient in ARTs—the grafting of parts and the calibrating of time. An important element of ARTs is the grafting of the properties and processes that make up a thing onto the properties and processes of another thing. Thus, when body parts and instruments are mixed up to make a woman pregnant, the properties and processes of the instruments are annexed to the body parts in a way that makes a pregnancy become possible. The instruments have their use and thus their being in intervening to make body parts part of being pregnant. Ontological choreography in this case involves the physical places and configurations in which the instruments and body parts touch and also involves the coordinating, grafting, and often expanding of the very properties and processes that make up the things. This applies to political properties and "identity" properties, as well. If, for example, a Latina woman wishes to become pregnant but has gone through premature menopause, she may choose an egg donor who is also Latina to help her get pregnant. Certain identity properties of being Latina move from the donor through the egg and into the recipient and also from the recipient to the egg to make the recipient appropriate.

Second, different kinds of things and their different sizes, scales, materialities, and properties also exist in different kinds of time. It is common to think of modern industrial societies as having coalesced time into the synchronized, and always in principle sychronizable, forward-running, finely and evenly calibrated time of clocks, and the heavenly bodies and imaginary shared spatiality to which they have reference. This ought perhaps to be especially true in technical fields, but in fact in the field of ARTs, where there is a lot of ontological choreography, there are also a large number of relevant kinds of time. Devices for calibrating and coordinating different kinds of time are ubiquitous. For example, there are menstrual cycles and treatment cycles, which are cyclical and repetitive. There is the regimented, linear, and repetitive bureaucratic time of the working day into which appointments must be fitted. There is biological age—the so-called biological clock—which is linear, unidirectional, and nonrepetitive. There is the time of first-person selfhood, which runs backward to cohere the psycho- and physicobiographic precursors to the crisis of infertility, makes the space of the present, and runs forward to narrate the life course. And there are the different temporal histories that different groups of patients bring to the meaning of treatment.

An assumption that we all have coordinated clocks, and working schedules specified in terms of common understandings of the working day run some of these kinds of times. Measures of biological age, or calendars for treatment run others. Countless devices—including diaries, friendship and kinship events, and calendrical obligations—structure and prompt the time keeping of selfhood. In ART treatment, all these are layered on top of each other. Prospective parents might have to explain why they have sought treatment, give a sexual and reproductive history, and express a desire for children in the future, so as to initiate some kinds of treatment. Women might need minimum FSH levels as an indicator of biological age before embarking on a cycle of in vitro fertilization. And they might have to schedule a day to begin taking drugs to stimulate the development of eggs for in vitro fertilization at a time that falls on the right day of a menstrual cycle as well as during their doctor's working hours. Likewise, patients might bring different experiences of historical and political time to their treatments, which, in turn, take up a part in that historical time. A white middle-class couple might view their treatment as part of a platform of federally protected reproductive rights that began with *Griswold v. Connecticut* (1965) and *Roe v. Wade* (1973). An African American couple might see the right to have children and to forge enduring bonds of kinship as helping to alter a history that has been marked by slavery's centuries-long denial of kinship. Exploring the ontological choreography of these technologies is a major aim of this book and is the focus of part II.

The Biomedical Mode of Reproduction

Artificial reproductive technologies represent one aspect of an increasing tendency for people to turn social problems into biomedical questions—the so-called biomedicalization of U.S. society.[3] ARTs can be used as a lens through which to study some of these profound changes in the economy, in science, and in laws that regulate kinship in contemporary U.S. society. In chapter 8, I make bold claims about the trends that are emerging from the very heart of technology and about the role that is played by technology in society. I compare and contrast this biomedical mode of reproduction with parallels found in the capitalist mode of production—with an aim to provoke ethical, intellectual, and political inquiry.

In the area of the growing biomedical economy, I focus on five areas of change that seem important. The first is technology's shift from production to reproduction as its fundamental mode. Associated with this shift in mode is a shift in the source of profit: where once labor produced things that make profit, and workers could be alienated from the profits of their labor, now bodies reproduce things that make profit, and bodies can be alienated from the profits of their reproduction. Equally, I argue that for biomedicine, capital has a constitutively promissory core, depending on such things as the reproduction of life, and the development of future cures, rather than capital being accumulated, as is more typical in capitalist enterprises. Capitalist production typically is maximized by improving efficiency and productivity. Biomedical reproduction, however, maximizes profit not by being efficient but by the success of a procedure or a process and its reproductivity. And finally, the hallmark capitalist problem of how to dispose of the bulky and toxic waste by-products of productive processes also changes. For the biomedical economy, the signature problem is whether reproductive by-products (organs from a cadaver or embryos) can be disposed of at all.

In the realm of relations between science and society, I point to two key trends. One model of the ideal relationship between science and society places moral responsibility to search for the truth in scientists, who are in turn policed by the norms of decency that they have internalized from a democratic society. This model posits a trickle-down path of knowledge to a general public that has a reasonable understanding of science and is instructed in basic scientific literacy. This model does not work well for the expanding realm of biomedicine, however. For one, even if the individual scientist acts in a morally responsible way (which is equated with being governed by a disinterested search for truth and by a belief that truth and the good coincide), too many issues are designated as "bioethical" in a reproductive economy; they have to do with who we are, who we want to save, and who we want to become. The scientist is not able to police issues such as whether we should do stem-cell research, and a model that gives moral responsibility to scientists thus gives way to a model that places the moral responsibility for science on the public. But the members of the polity are not an audience to biomedicine. Their bodies are the substrate of biomedicine, and their involvement is not that of a consuming public but one of private involvement. It is not sufficient to call for the public understanding of science, as there is already a private implication in science.

The biomedicalization of U.S. society is also changing our notions of identity and kinship. Assisted reproductive technologies have introduced a collaborative reproduction that involves a gamete donor, an embryo donor, or a surrogate, thereby lateralizing "descent." And the use of grandparents as surrogates and children as gamete donors has made possible the reversal of linear descent when determining kinship. The widespread availability of ART procedures for reproduction without heterosexual intercourse has also helped normalize single and homosexual parenting and kinship. The biomedical and ethical standards that warn against too much sameness in kinship (incest and homosexuality—what I call "too much homosex") and too much heterogeneity in kinship (adultery and miscegenation—"too much heterosex") are no longer intercourse-dependent. Instead, those biomedical and ethical standards warn against cloning (too much homogeneity) and ooplasm transfer or xenotransplantation (too much heterogeneity). Disambiguating and stabilizing these kinship patterns involve using technical definitions when the sociolegal definitions are underdefined, and vice versa. In a biomedical mode of reproduction, society can no longer maintain that natural kinds are essential (what a mother is, for example, is under debate) and that social kinds are socially constructed. Rather, all kinds are specified and differentiated by strategic naturalzation and socialization, depending on which part is underdetermined at a given time and place.

U.S. ART clinics routinely produce ex vivo human embryos, which means that they are implicated in one of the most divisive and yet most distinctive of all contemporary U.S. political issues—the abortion debate. The ex vivo human embryo can be seen as the iconic bearer of the changes from the capitalist mode of production to the biomedical mode of reproduction. I describe each of these parameters in detail using the embryo and the increasing intricacy of the means of its disposition as examples. The embryo in ART clinics and their environs has the potential to shift the biomedical economy out of a strictly capitalist mode of production, where the embryo is either sacred life or a waste by-product of production (and hence an intractable site of oppositional political and moral cathexis), to the biomedical mode of reproduction, where the embryo is the focus of extreme care and procedure—about donating, implanting, discarding, or using it for research. Far from being a reckless or brave new world, this mode of reproduction demands a new world of extreme care.

The Methods

Infertility clinics and related assisted reproductive technology sites are dynamic cultural sites where human reproduction is a highly desired, stubbornly difficult, technical production goal. Many accounts of the new reproductive and genetic technologies either stress their epochal novelty or lament their complicity with early twentieth-century eugenics. In the course of a decade and a half of treatment in, experience of, and research into reproductive technology clinics and related sites in the United States, I have found great novelty, particularly of the ontological kind, as well as the persistent political and national tropes from which novelty is crafted. The few clinics I have visited outside the United States and my reading of other scholarship on ARTs suggest that ARTs play out in different ways in other national settings but that this combination of novelty and sameness in not atypical. Change is not sudden, forward looking, or always irreversible but instead is iterative, multidirectional, patchy, frequently conservative, and often the result of unintended consequences.

The clinical parameters of ARTs are determined in all sorts of ways by regulatory, technical, economic, and demographic currents. Nonetheless, the physical clinical spaces and the activities that are orchestrated in and around them are relatively self-contained, which makes them ideal sites for my primary method—ethnography. The much-mocked $n = 1$ and $t =$ "the eternal present of the ethnographic gaze" facilitated my pursuit of some of these questions and hindered others. I employed a range of ethnographic methods, including participant observation and semistructured and formal interviews.[4] I supplemented ethnographic work with a wide range of primary and secondary literature research in post–World War II reproductive science, biomedical ethics, U.S. political theory, and feminist and other writings on ARTs themselves. I researched media accounts of breakthrough events, starting with the earliest attempts at in vitro fertilization in the 1970s, and I read widely on global reproductive politics. Initially, capturing ethnographically the change over time in ARTs in the United States was difficult, but after a while, in conjunction with other research methods, a pattern began to emerge as an unregulated best-interests-of-the-child model shifted to a more regulated reproductive-privacy model.

The fieldwork on which this book is based was carried out between 1988 and 2004. The work done in the late 1980s to 1990 predated my enrollment in a Ph.D. program, and some of it was carried out while I was going through infertility treatments in the United States. I attended an international meeting on reproductive technologies and read voraciously, especially in the scientific and early feminist literature, during this time. I developed a questionnaire to measure how well-informed patients were, reflecting the concerns of many at that time that success rates, side-effects of drugs, and reasons for protocol choice were undercommunicated, and underdocumented. My fieldwork from 1992 to 1996 formed the basis of my Ph.D. dissertation. I did two internships in infertility clinics. The first began with a month of daily full-time work at the clinic, which was followed by regular weekly and then monthly visits to the clinic. The second began with one to three days a week, depending on the procedures that were scheduled on a given day, and dropped off after the first month to weekly and then monthly visits.

At one clinic, I mainly shadowed staff and did clerical and data-entry jobs. At the other clinic, I carried out elementary jobs (such as preparing equipment for patient exams and helping with microscope work during surgery), traveled with the staff team, and regularly attended staff meetings. At the first clinic, I mostly shadowed the primary physician, while at the second I spent the majority of my time with the head embryologist, although at both sites I spent considerable time with physicians, embryologists, and nurses. At the first clinic, I attended more than fifty patient examinations and consultations and attended surgery one or two days a week. I helped compile research data in a pharmaceutical-company-financed surgical trial that the clinic was participating in and shadowed technicians who worked at the animal facilities that provided hamsters to the clinic. At the second clinic, I attended over thirty patient exams and many surgeries. I followed egg-donation, surrogacy, and male-infertility surgery cases and helped out in the lab, including preparing for a lab certification visit.[5] I watched training videos about micromanipulation techniques with lab staff and attended sessions where technicians were taught how to blow micropipettes and do intracytoplasmic sperm injection (ICSI). I compared and contrasted these two clinics in my Ph.D. dissertation.

In 1997 and 1998, I conducted a number of interviews with patients outside the clinical setting. I attended a small number of infertility support-group meetings and followed a couple through treatment from the perspective of their experiences at home. In this phase of research, I went to other ART sites to observe the purchasing and sharing of fertility drugs and to attend meetings with other RESOLVE members. During a later period of fieldwork in 1999 and 2000, which was completed while I was undergoing treatment myself, I relied on the tools of ethnographic research that I had developed during my Ph.D. work. In 2001 and 2002, I completed a series of interviews (which serve as the basis of chapter 7) with patients outside of the clinic setting about their political orientations to their own treatment choices. In 2003, I began to do more systematic Internet ethnography of support groups, patient chat rooms, clinical data, clinic advertising, and gamete bank data. The passage of the Fertility Clinic Success Rate and Certification Act of 1992 and the comprehensive data that have been collected under its authorization since 1996 have provided another major online research tool.

During fieldwork that was conducted for this book, I visited another eight clinics, some just once and some over a period of time. Two were in the United Kingdom, and six were in the United States (two in the Midwest, one on the East Coast, and three on the West Coast). I interviewed Robert Edwards, one of the laboratory fathers of Louise Brown, the world's first IVF baby, as well as one of the embryologists at Bourne Hall, and I toured the Bourne facility. In addition, I attended several professional medical and scientific meetings in the United States and one in Europe. During two professional meetings, I met with many other practitioners and attended panels on scientific, nursing, clinical, psychological, ethical, political, and legal issues in the field. I took ethnographic notes on the reactions to presentations, on behavior, and on the interest that was generated by new or controversial topics. And I spent a good deal of time in the exhibit halls, where fertility drugs, surgical equipment, and egg-, sperm-, and embryo-supply services were showcased and marketed to practitioners.

Like many who have studied technical practices, I have some background in science, and I have remained more or less within my own cultures for this work. Although an ethnographer who is working within a familiar culture is less likely to impose an internal but ultimately irrational symbolic or functionalist coherence on that culture than if he or she is faced with the exotic otherness of a far off culture, there is a temptation to confer truth values that are in line with preconceived categories. This is particularly true in the United States, where the average ethnographer is studying those of a higher social status (doctors and scientists) and where participant observation becomes vicarious upward mobility, with the comfort of role distance intact. The opposing temptation for the ethnographer who works within a familiar culture is to assume that he or she is somehow able to improve on an actor's accounts of what is going on—even to expose what is really going on. My sense after doing ethnographic work in clinics over a number of years is that practitioners and patients—because they are intimately connected with, critical of, informed about, and committed to the technologies—are themselves reflective and reflexive participants in the generation of accounts of what is "really going on" in the clinic. I thus tried to avoid slipping into an easy scientism on the one hand, and to steer clear of righteous polemic on the other.

Some biomedical theorists have emphasized the role that is played by experience, especially women's experience, in reproductive technologies, while other theorists have concentrated, in celebratory or denunciatory mode, on the science and practice of the technologies. Clinics combine lived lives with scientific and technical practice, making it hard for an ethnographer to ignore the centrality of the intersection between lives and technology. Viewed together, actors and technology ventriloquise and animate each other; patients are not voiced by the technologies, and technologies are not animated by patients, despite what might be implied by writings that grant causal privilege predominantly to technology or to experience. The ethnography at the heart of this book has more people in it than a typical lab study precisely because I am interested in the intersection between subjectivity and technology and because narratives of the self are important in infertility. The ethnography is not about the lives of infertile women and men or of clinicians per se, which would clearly extend into a myriad of sites and spaces of memory and action well beyond ARTs. I hope that it is clear, nonetheless, that I care greatly about patients and also about nonhumans and quasi-humans (including numbers, nitrogen tanks, and nuclei) and that for me charity (interpretative and empathic) is enabled through, because of, and in spite of nonhumans. But I care about people more, cyborg though they be. My greatest respect is reserved for the first-person biographies that, with poignancy and pertinacity, are broken, repaired, enabled, and constrained in the site and that always are being narrated, whether in the patient's medical record, on the operating table, or in a quiet moment after an exam.

Reading This Book

The book is divided into three parts. Part I, Disciplinary Stakes, provides an account of the fields of inquiry that have the strongest bearing on this project: one is science and technology studies, and the other is feminist scholarship on reproduction and assisted reproductive technologies. Both fields are core to my intellectual formation, and both are rapidly evolving and intellectually invigorating loci of work on science and society. Readers who have little interest in these quasi-disciplinary roots and the stakes of disciplinary location can begin the book with the empirical chapters of part II, Ontological Choreography, which is the ethnographic heart of the book. There I develop the theoretical concept of ontological choreography from four key perspectives: normalization, gender, kinship and ethnicity, and agency. Part III, Economies, explores the economies of reproductive technologies from the perspectives of their political origins and their future possibilities.

Science and technology studies and feminist scholarship on reproduction and assisted reproductive technologies have many sensibilities in common, and several scholars are members to a greater or lesser extent of both fields. Nonetheless, the intellectual motivations and trajectories of the fields cannot be explained by a single narrative. The field of science and technology studies is anchored in historical and epistemological questions. It tries to shed light on the special truth status and world-making ability of science and technology by investigating its canonical sites—the laboratory, scientific controversy, and the rise of modern science. The field is predominantly (and in many variants, prescriptively) morally neutral and in North America typically involves studying (relatively) high-status scientists and their (relatively, even if contested) high-status knowledge. Over time and in some variants, the field of science and technology studies has begun to pay attention to users, the field, other sectors of society (such as law and the economy), and the transnational dynamics of science and technology. Each of these has been incorporated through being theorized as being in some meaningful sense part of the content of science.

Feminist scholarship on reproduction and assisted reproductive technologies, on the other hand, is anchored in concern with (especially) gender-based inequality and discrimination. It views science and technology as a locus of discrimination through the masculinist epistemology of scientific knowledge and authority of scientific practitioners and through a history of making women's bodies the object of scientific knowledge and experiment in the name of conformity to patriarchal class and family norms. Its canonical sites are women, bodies, and the users and victims of science and technology. Feminist studies of reproduction are usually morally normative. They typically deal with (relatively) low-status downstream users, minor players, or underrecognized women scientists and can be highly critical of scientific knowledge, scientific institutions, and individual high-status (usually male) scientists. Over time, moral preemptiveness has given way to a greater sensitivity to the moral complexities of technoscientific practice, and practices of agency, resistance, and other dimensions of stratification have been added to gender as foci of concern. These changes have come about as part of feminist poststructuralist critiques of the overly structuralist analyses of gender.

Because of these differences between science and technology studies and feminist studies of reproduction, attempts to link them too closely, would misrepresent the two fields. In particular, the differences between "studying up" (science and technology studies) and "studying down" (feminist studies of reproduction) make the fields close to impossible to coalesce without losing the narrative thread of both and the grounds for comparison between them that I find to be especially useful in understanding the particular character and intellectual content of both.

In chapter 1, Science and Society: Some Varieties of Science and Technology Studies, I present some of the evolving motivations, controversies, and common explanatory aims of the field known as science and technology studies (STS or S&TS) of which I am a part. I situate the findings and arguments of this book within this framework. The chapter ends with a call for a reinvigoration of our understanding of the relations between science and society in this increasingly biomedical and biotechnological era. In chapter 2, Fertile Ground: Feminists Theorize Reproductive Technologies, I explore the wealth of feminist scholarship on and around reproductive technologies to show how my argument draws from and contributes to this multidisciplinary body of work. I emphasize the shift in the last decade to scholarship that displays neither strong technological optimism nor strong technophobia but which grapples with the complexity of actors, interests, and outcomes. I also point to the importance of feminist scholarship in both incorporating gender and difference and also theorizing body, emotions, and intimacy in relation to politics and economics.

Chapter 3, Techniques of Normalization: (Re)Producing the ART Clinic, discusses the means of normalizing these miraculous and technological procedures—their ubiquity, mundanity, and ontological heterogeneity. It looks at ways of negotiating access to these sites, interpreting statistics, and following the etiquette of technological sex. Chapter 4, Is Man to Father as Woman Is to Mother? Masculinity, Gender Performativity, and Social (Dis)Order, explores the threats to and performance of masculinity in clinics and contrasts threats to gender identity for men and women. It argues that reproduction, for all the cognatic descent that is supposed to be obvious in human reproduction, is asymmetric. Men are both chided for and given little opportunity to express kinship other than by a biological essentialism that reduces more readily to genetics than it does for women. I show the frequently conservative nature of gender relations in the clinic and the manner in which these patterns of gender and heterosexuality are drawn on to facilitate social and technical innovations.

Chapter 5, Strategic Naturalizing: Kinship, Race, and Ethnicity, explores the means by which kinship and procreative intent, though underdetermined by the procedures, are deftly and strategically naturalized by patients. Notions of race, ethnicity, culture, and extended family are used by patients in ways that are sometimes naturalized and sometimes social to disambiguate the reproductive facts of the matter. Chapter 6, Agency through Objectification: Subjectivity and Technology, shows that contrary to the common argument that objectification and technology dehumanize, participants in ARTs, pursue agency through their own objectification. Under specific circumstances ART patients' pursuit of agency is part of the temporal narrative dynamic of their selfhood.

Chapter 7, Sex, Drugs, and Money: The Public, Privacy, and the Monopoly of Desperation, addresses a unique intersection of three realms of privacy—private lives, scientific autonomy, and the private sector. The intersection of these privacies is a powerful way of coordinating citizens, science, and the state. "The monopoly of desperation," whereby monopolistic tendencies come from the demand—the so-called desperation of patients—rather than the supply side, makes the development of ARTs in this quasi-private space extremely entrepreneurial yet in many more regards than critics have given it credit, exemplary in the field of biomedicine. The implementation of the Fertility Clinic Success Rate and Certification Act of 1992 has resulted in a data-collection instrument that now regulates assisted reproductive technology clinics. This data-collection, analysis, and publication effort arose from pluralist pressures—from medical and pharmaceutical professional organizations in the interests of self-regulation, and research aimed at increasing and improving the specialty; from nationwide patient-support groups that support access to infertility treatments as part of reproductive privacy; from legal activists in family and property law (adoption, parentage, inheritance, and child support); and from government agencies that cooperate with those who advocate semiprivatized regulation and accountability.

After working in these clinics for a number of years, I began to believe that technologies of reproduction (such as in vitro fertilization, preimplantation genetic diagnosis, and cloning and stem-cell research) were destined to be the twenty-first century's moral equivalents to the twentieth century's nuclear technologies of destruction. ARTs are among the most important practices through which we are remaking individual and collective identities, and they are the sites in which we declare choreographed reactions of horror, revulsion, and hope, and make us ask, "Do these ends justify these means?" Massive and largely unaccountable state support for weapons systems are unlikely to decline, but the rise of molecular biology and its biotechnological spinoffs over the last half century poses a fundamental challenge to the relations between science and society in a liberal democracy. If bodies are at center stage in biomedicine and biotech, keeping a strict separation between matters of the citizenry and matters of epistemology is not possible. Likewise, it is not possible to conduct science in a disinterested manner and then apply it in the interests of state security. There is simply no longer any principled institutional or epistemological way to draw the line between pure and applied.

In the concluding chapter, chapter 8, The Sacred and Profane Human Embryo: A Biomedical Mode of (Re)Production? I speculate about some of the changes in relations between science and society that I think that these technologies portend. To do this, I look at the disposition of human embryos in ART clinics at the fulcrum of contemporary political perspectives, human and nonhuman, technical and sacred. As discussed above, the biomedical mode of reproduction (which I contrast to industrial capitalist modes of production) is already beginning to have its own characteristic systems of exchange and value, notions of the life course, epistemic norms, hegemonic political forms, notions of security, and hierarchies and definitions of commodities and personhood. As enthusiastic, critical, implicated actors in the sciences of life, it is incumbent on us to try to understand these characteristics and make explicit their implications. By ending the book with this polemical chapter, I hope to leave readers committed to following the worlds, objects, and subjects that they make possible and that the life sciences foreclose. Above all, I hope to leave readers newly committed to demanding more science not less, but ever more careful, ever more empirical, and ever more multivocal.

Where Do I Stand?

I frequently get asked three related questions that do not quite fit into any chapter in this book:

• What is your personal investment in assisted reproductive technologies?

• At a time when legal, political, and technological recognition of fetal personhood is threatening the right to safe and legal abortions, how are you able to work in a site that contributes to fetal personhood and still support a woman's right to choose to terminate a pregnancy?

• In the end, where do you stand on these technologies?

Behind the first question—"What is your personal investment in assisted reproductive technologies?"—lies a cluster of issues:

• Why did you pick this site? (Did your personal experiences draw you to it?)

• Have you used these technologies? (Can you be appropriately empathic or critical, depending on the point of view of the questioner?)

• Why have you written the book in this way and not another? (Do you emphasize theory at the expense of the voices and experiences of suffering women? Do you emphasize women's experiences at the expense of interventions in science and technology studies that more "masculinist" field sites enable one to make?)

Although I am reluctant to put myself in my academic writing and thereby lose some of my own and my family's privacy, a persistent minority of my small readership has asked for an accounting of my personal connection to the technologies. I do not subscribe to the identity-politics idea that you have to experience something before you can write about it or that if you experience something, especially discrimination or stigma, you have to write about it. A couple of generations of scholars, however, have raised compelling epistemological and moral objections to academic writing that appears to present an objective "view from nowhere," and in the classroom we are daily made aware of the arrogance of such projects, our own included. The voyeurism of this book's topic (reproduction) and method (ethnography) and the epistemology and ontology of the branches of the major theoretical fields that inform it (strains of cultural anthropology, feminist theory, and feminist technoscience) lead me to address briefly my position and commitments in this field site.

Why do some personal factors matter and others do not? From whose point of view are my interventions to be measured? Can everything be expressed as flowing from certain autobiographical precursors? My interests in philosophy, science, and transnational women's health long preceded my personal medical history. My father worked for the World Health Organization, and I grew up in New York, Bangkok, and Geneva. I was weaned on health-care as governance, and women's-empowerment. I took the international baccalaureate, studied philosophy, psychology, and physiology at Oxford as an undergraduate, and put off choosing between C. P. Snow's two cultures for as long as possible. Secondary infertility, however, was a life-course experience and project that provided me with a focused field site. By the 1990s, when I began my Ph.D. fieldwork in infertility clinics, wildlife parks, and zoos (to pursue my interests in many reproductions and pronatalisms), I had already spent a number of years learning about and being part of the rapidly developing infertility industry.

Being a feminist academic and a former patient probably affects every aspect of my writing about reproductive technologies. Evidence for my ambivalence about particular technologies is ample in this book. I do not pronounce any reproductive technology to be intrinsically bad for women or necessarily eugenic because choices present themselves differently under different circumstances. The complexities that animate technology as it operates in different times and places, with different reagents, and with different professional staff become intertwined with different political and personal stories. Ambivalence about reproductive technologies has grown into a respect for the people and the objects who participate in a complex mutual choreography that allows unexpected agencies and political configurations to emerge. This book is dedicated to trying to tease apart and simultaneously honor some of these complexities. I am steeped in the ways in which identity is on the line, treatment can be abusive, the epistemology, and actuality of failure, and the local and global inequities of access to and role in treatment. I am also a witness to the transubstantiation of technology and body into life—the Promethean choreography of ontologies that the media call "miracle babies." I don't want to lose or to presuppose these complexities.

The second question that I am commonly asked—"At a time when legal, political, and technological recognition of fetal personhood is threatening the right to safe and legal abortions, how are you able to work in a site that contributes to fetal personhood and still support a woman's right to choose to terminate a pregnancy?"—references the embryonic and fetal monitoring and visualization techniques that are central to treatment. It also picks up on the fact that assisted reproductive techniques allow early identification of pregnancy, and hence hurry the anticipatory socialization that goes with that (not to mention presupposes that all feminists ought all to be pro-choice). Because I spend so much time around those (myself included) for whom preciousness and desire has been predicated over future imaginings and current longings and resource expenditure, I am predisposed to ascribing pronatalist purpose and even quasi-personhood to fetuses. Reproductive technology clinics are as much about intentional reproduction as they are about pronatalism, however, and intentional reproduction is central to prochoice politics in the contemporary United States. Adoption still overwhelmingly operates on a "best-interests-of-the-child" model that allows the state to remove children from some kinds of parents and to assess the suitability of other would-be parents. ARTs, on the other hand, have been astonishingly successful in a very short period of time in reframing the model to one where the intention to parent predominates. This reproductive-choice model means that despite increasing the sites of fetal personhood, ARTs also increase the sites and kinds of reproduction that make up reproductive freedom.

In her *Women of Color and the Reproductive Rights Movement*, Jennifer Nelson writes:

> In response to arguments made by women of color that legal abortion was not synonymous with reproductive freedom, reproductive rights feminists came to maintain that the right to bear children was as important to reproductive freedom as the legal right to terminate a pregnancy.... Women of color challenged the white middle-class feminist movement to recognize that the abortion rights movement needed to encompass "bread and butter" issues such as health care for the poor, child care, and welfare rights in addition to anti-sterilization abuse efforts.[6]

This corrective could equally be applied to the debate about fetal personhood and reproductive freedom that I am addressing here. The grounds on which I favor legal, affordable, convenient, and safe access to contraception and abortion may not be the same grounds that are favored by those who typically support such access, but my political position is not different. For example, it is hard for me to enter either-or arguments about whether a fetus is a potential person. A pregnancy's potential is a function of many different things that are relevant under different circumstances. Among other things, the I-Thou components of potential personhood—the longings for and attributions of the other's selfhood—are especially relevant for many patients and may be much less salient or even absent under other conditions of conception, such as accident, rape, or incest.

It is also hard for me to agree with arguments about ideals of the autonomy and sanctity of a women's body. Women (especially in reproductive and domestic contexts) and men (especially at work and at war) sacrifice their bodies both intentionally and unintentionally, as instruments of selfhood, coercion, the state, capitalism, and any number of other individual and collective subjectivities and subjectifications. The rigors of infertility treatments are just one aspect of this. My dissent with certain strands of prochoice thinking, however, compels me to support abortion rights. Both prochoice and prolife courses of action involve choices and valuing some kinds of life. Making safe or affordable abortions illegal preempts a recognition of the specific circumstances of different pregnancies and enforces some and not other narratives of love, violence, and loss. The availability of such services does not guarantee their uptake (which would preempt other accounts of love, loss, and violence) but opens up the space for a greater recognition of our interdependence in both love and violence.

The third question that I am commonly asked—"In the end, where do you stand on these technologies?"—is related to the first two and is in some ways the most controversial, given feminist, academic, and lay criticisms of these technologies. I make the case in chapter 7 of this book that assisted reproductive technology in the United States at the beginning of the twenty-first century has become in some ways a model medical specialty. The major political and moral issue is not assisted reproductive technologies in particular but is access to medical care in general. Both implicitly and explicitly, I argue that these technologies began as anything but a model for other areas of practice (there were few clinics, which had astonishingly low success rates, imposed grueling treatment regimens, and excluded most would-be patients because they were unable to pay or were judged to be suitable as parents) but have become unusually accountable to various stakeholders and have been established as a site of activism within medicine.

After a decade and a half of involvement with assisted reproductive technologies in the United States, I feel that the most pressing issues are not technologies run amuck, risks to women's bodies, the commodification of children, or the specter of eugenics. Weaponry is a much better candidate for technology run amuck than are ARTs, over- and undereating remain much greater threats to American women's bodies, and the privatization of the childhoods of middle- and upper-class children as compared to the relatively public childhoods of the working classes are far more troubling sites of the commodification of children and of eugenics. Several more urgent issues regarding ARTs have received political and scientific attention, including the sociolegal vacuum that is created by collaborative reproduction, truth in advertising about success rates and risks, health risks to unborn children from reproductive and genetic technologies, the sharing of scientific information so that practice reflects the best possible data, and the ending of the exclusion of would-be patients on the grounds of sexual orientation or marital status.

These are indeed areas of significant concern, but they are exactly the kind of concerns that patient activism, scientific activism, legal activism, and professional self-regulation have proven reasonably successful at combating. The low number of states that have enacted legislation to provide insurance coverage for infertility treatment is a more recalcitrant problem with ARTs, but nonetheless that number has steadily increased so that fourteen states currently recognize infertility as a disease and provide some insurance coverage. In general, the states that have been slow to adopt any kind of coverage for ARTs have histories of being anti-abortion or of being pro-labor that mean that they also fail to cover other services that most Americans have come increasingly to view as worthy of medical insurance coverage. Thus, reproductive services are insufficiently covered in most states that oppose abortion, and states with strong labor unions that exclude infertility benefits also tend to exclude mental-health benefits. These patterns reflect views about what things are considered to be illnesses in different regions. On a nationwide scale, the inequities that stem from disparities in access to treatment and in quality of treatment have proven to be more resistant to change than any of the inequities that were specific to reproductive technologies, such as judgments about who is worthy to be a parent.

The most urgent national inequity, by far, is the lack of universal health care in the United States. Until the allocation of health care is based on need rather than income and employment status, it will not matter how many insurance statutes are on the books at the state and federal levels. People who are without regular paid employment, who are unable to work, who work at home or in other nonpaying jobs, or who work in jobs with no "benefits" (extraordinary euphemism) will not have access to assisted reproductive technologies unless they happen to be rich. Regardless of how successfully ART doctors and patients argue that infertility is a disease and that treatment is not elective or how successfully they move the discourse from "the best interests of the child" (who would be an appropriate parent) and to reproductive privacy, enfranchisement will extend only to those who qualify for insurance or who can pay out of pocket. Emergency care and other government-funded health care for the unemployed is resolutely antinatalist, presumably because being uninsured is associated with not deserving to be encouraged to reproduce (any more than they already have). Implicit judgments about the reproductive worthiness of whole segments of the population are much harder to change than explicit judgments regarding the right to use infertility services.

It is perhaps anticlimatic to conclude that the single biggest problem with reproductive technologies in the United States is not patriarchy, eugenics, or infringements on the sanctity of life but a lack of access to health care based on need rather than on employment status and its identity-based proxies. But attempts to improve the ART field have made substantial inroads against some kinds of inequity, while other kinds of inequity have remained recalcitrant to improvement. A rigorous, publicly accountable, and pluralistically framed data collection that is similar to the kind that now characterizes U.S. ARTs would be a huge boon for all medical specialties. It suggests, however, that the push for need-based health care must be carried out at the most generic political level, for it does not come out of the (considerable) activism and self-improvement capacity of the field itself. To its credit, the field of ART continues to lobby for universal health care, sponsor events, and designate days that are devoted to drawing attention to health care.

In the end, members of a biomedical society should have help in getting pregnant (with all the modalities of caution that patients and practitioners have been building into the field of assisted reproductive technologies for the last fifteen years) if they wish to be pregnant and they cannot get or stay pregnant without help.

I

Disciplinary Stakes

1

Science and Society: Some Varieties of Science and Technology Studies

Reproductive technologies can be situated in a broad set of inquiries that explore the links between science and society in general and the special role of genetic and reproductive science and technology in contemporary society in particular. This chapter develops an account of the field of science and technology studies (STS) within which to situate this kind of intellectual inquiry. STS exists in a wide array of different forms and national contexts and draws on and interacts with other disciplines. I have tried to make clear my debt to a long-running series of intellectual innovations and disputes, even as my material pushes me to particularity.

What motivates practitioners to work within science and technology studies? The single thing that most unites those within the field of STS is an interest in the deep interdependence of nature and society. This distinguishing characteristic of the field flies in the face of the conventional divisions in academe between the social and natural sciences and between the ways that scholars think about social and natural causes. The majority of practitioners of STS have spent a considerable time in the natural sciences, know a fair amount about their former fields, and have a great interest in how it functions in society. As I discuss below, some STS scholars investigate scientific and technical knowledge, which is thought of as certified knowledge about the natural world: Who has it? What are its warrants? How is it acquired? What does it do? How does it order the world and the people in it? Others focus less on scientific knowledge and more on ontology, technology, or science as practice. These latter scholars document how assemblages of people and things together support individual identities, shape society and politics, and even determine what counts as nature. Topics such as the rise of experimentation, the importance of visualization, and the achievement of standardization in science and technology recur as loci for exploring these lines of inquiry.

A common methodological orientation accompanies this interest in the interdependence of society and nature. Despite a good deal of reflexivity on the nature of data and its interpretation, there is a valorization of empirical data collection either by ethnography and participant observation, or by original contemporary or historical archival and document research. This empirical methodology is intimately tied to the theoretical insights just described. Synthetic, a priori, and purely interpretative methods, for example, are all viewed suspiciously if they are not bolstered by empirical work. Versions of empiricism and positivism, thought of as not requiring any interpretation (as advocated in some natural and social science methodologies), are viewed as equally suspect. All these methodologies—whether they are analytic, are interpretative, or try to get at "pure nature"—are thought to be making the same mistake. They are thought to assume a fundamental disjunction between the way the world really is (nature and reality) and what it means, how it appears, or how it is represented as being (appearance and discourse). STS denies that this distinction is clear or that it can be taken for granted, and many STS scholars view distinctions such as that between reality and interpretation as achievements themselves worthy of empirical study. STS thus opts for empirical methodologies that are nonetheless assumed to be interpretative. Because of the tight knot between methodological and theoretical concerns, STS often reads like empirical philosophy or an empirical case study that has been trotted out to make a theoretical point. This unusual combination of theoretical ambition and empirical predilection is a hallmark of the field. When the two are done poorly, they satisfy neither those whose primary allegiance is to the historical record nor those for whom rigor in argument is the primary criterion of excellence. When well executed, however, they add insights and innovations to the telling of history and provide exciting empirical complexity and constraint to normative argument.

Many STS practitioners have an interest in contingency and interpretive flexibility (showing that objects of scientific inquiry could have been captured, interrogated, and described otherwise than they currently are and therefore could have had different characteristics). They also tend to have an interest in the stability of scientific facts: facts tend to stick, to travel well through time and space, and to work. After all, it is the power and efficacy of science and technology that make it interesting. Unfortunately, the field has frequently been misunderstood as being interested only in contingency, an interest that itself frequently is labeled *social constructivist, postmodern,* or *relativist* in a derogatory way that means something like "reality denying." STS scholars *could* believe that scientific knowledge was socially constructed in the sense of "made up to suit purposes other than the pursuit of truth," but most likely they would believe this only of certain claims that are made by particular people at particular times. It is much more common for STS practitioners to insist that scientific knowledge is constructed in the sense of requiring work, instruments, institutions, conventions to discover and sustain it, just as a house is constructed using work, tools, materials, and plans. This second sense of *constructed* has little or nothing to do with an independently existing reality, just as believing a house was constructed does not mean the same as (or anything like) believing that it doesn't exist. What I have found is that the metaphor of construction leads people not to discount reality but to attribute reality and causal power to many ontologically different kinds of things and to many different kinds of agents. This could perhaps be called promiscuous realism but certainly not reality denying.[1]

Practitioners of STS are also frequently asked whether they are for or against science. Because this question, like the reality question, lurks beneath a good deal of misunderstanding, it is worth briefly addressing.[2] I think that most people working in STS would agree that they encounter scientists that they admire more than others, findings that seem more compelling than others, and uses to which the sciences are put that seem more enlightened than others. In what walk of life is this not the case? Some scholars are more inclined to be apologists for science or hagiographers of particular scientists, and others feel that their responsibility is to be strongly critical of the sciences. In the majority of its variants, however, STS is about understanding science and technology and not about criticizing individual scientists or undermining science as a whole. Indeed, STS sees science and technology as in some ways *the* story of the modern world and *the* story of modern political order, and so it certainly doesn't underestimate its importance or its creative genius.

A separate but related affective and methodological question is whether those who study science consider scientists' approval of their writings to be relevant. Some practitioners care deeply for the opinions of scientists, especially those whose field they have studied. This can be especially important for historians of recent science, for whom actors' (or actors' descendants') approval is a measure of having "got it right." Others think that just as political scientists do not typically look to politicians to evaluate their work or give them the right to study that part of our shared culture, neither should STS practitioners be beholden to those with an interest in the outcome of their findings. Science should be studied like any extensive and well-funded part of society—whether religion, education, agriculture, politics, sports, the courts, the media, the market, or science—because it is interesting, because it has an impact on our lives, because debate is essential to democracy, and because outside scrutiny (within limits) helps to keep institutions honest.

So far I have introduced the field of STS through three widely shared sensibilities, all of which I inherit and work with in this book. The first was an interest in the interdependence of science and society and of nature and culture, the second was a methodology that could be called empirical philosophical, and the third was an interest in both the contingency of scientific facts and the robustness and efficacy of those facts. These are all fundamental orientations for my research and this book. I now turn to examine the work, motivations, and influences of major practitioners in the field who have most influenced the approaches and analytics taken here. I divide this chapter's classification of the field into strains that emphasize the problem of scientific knowledge, the ontology of science, and the politics of science and technology. The differences between them are more often a matter of emphasis than an absolute difference of focus, but there are real differences. Each contributes to the inquiry in this book. The account here is biased to North American STS, but STS also flourishes in the United Kingdom, continental Europe, Australia, and India and is becoming more important in East Asia and elsewhere. In the interests of brevity, I have not summarized the case-study details and findings of individual pieces of scholarship but rather have referenced what appear from my perspective to be canonical works that are representative of the different approaches. Readers can turn to several anthologies and review essays for further citations and exemplars as interest dictates.[3]

The Problem of Scientific Knowledge

The parts of science and technology studies that raise the problem of scientific knowledge probe the intimate connections between knowledge production and how we classify, order, and evaluate people, things, and nature. These branches of STS have their disciplinary roots in a mixture of sociology, philosophy, and history. I begin with the sociology of scientific knowledge (SSK) and its roots, but I also consider ethnomethodology and social-worlds theory as applied to science, where it is normative order in general that is more the focus of analysis. I also refer to the social construction of technology, which has close ties to the sociology of scientific knowledge. I then consider STS as practiced in the history of science and finish the section on the problem of scientific knowledge with a brief consideration of STS and the philosophy of science.

American sociology of science has paradigmatically concerned itself with the relations between modern Western science and modern liberal democracies. Beginning with the work of Robert K. Merton, whose classic essays on the normative structure of science began to appear in the late 1930s, sociologists of science tried to show that polities that are stable liberal democracies are also societies that encourage and enable good science. Merton argued that this was due to the similar normative structure of questioning, transparency, and disinterestedness that both possess in their ideal forms—that free societies and good science go together. A stable social order guarantees good science and vice versa. The totalitarian anti-intellectualism that derailed Nazi science can, according to Merton, be avoided by promoting the moral attributes that ensure against abuses of both knowledge and power.[4] These moral attributes, Merton believed, are communitarianism (or an attitude that sharing and openness should be nurtured), universalism (scientific knowledge should have the same content regardless of who possesses it), disinterestedness (science should be pure and not instrumental), and organized skepticism (an attitude of questioning should be directed to a focused area of empirical investigation). Collectively, these norms (which he called KUDOS norms) function to determine the "normative structure of science." Scientists follow these norms and expect sanctions of various kinds if they deviate from them. Merton proposed that these norms are absorbed by scientists during their scientific training and function thereafter as "internal policemen" that keep scientists from excessive competitiveness and interestedness.[5] Following these norms protects the scientific community from abuses of and by science. Similarly, fostering the equivalent norms in society at large would protect people from political excesses and abuses such as those evident in the totalitarian regimes of Europe.

Critiques of the foundations of Mertonian sociology of science began to appear in the United Kingdom in the mid-1970s and spurred the development of what is called the sociology of scientific knowledge.

Scholars working more within the Mertonian paradigm of the sociology of science have concerned themselves with structural and institutional topics such as the funding of science, the professionalization of contemporary science, the way that peer review and citations work, and access to the field by class, race, and gender. Although most STS scholars within sociology have roots in criticisms of Mertonian sociology of science, a number of more Mertonian scholars are actively contributing to STS. These latter scholars are recognizable by their focus on the interactions between knowledge and the social structures of science they are analyzing. Examples of this work include studies of the interactions between scientific knowledge and the state and studies of ways of delineating science from nonscience so that institutions that are central to modern liberal democracies (such as policy and the law) can deploy it. This scholarship brings to STS a strong precedent of examining the politics of scientific knowledge.[6]

The sociology of scientific knowledge began to be developed in the late 1970s by such scholars as Steven Shapin, David Barnes, and David Bloor.[7] It examines, in the words of Steven Shapin's title, the social history of truth. SSK developed partly as a critique of Mertonian sociology of science. It shared Merton's idea that ways of organizing society and knowledge about the natural world are deeply intertwined with each other even in modern scientific societies. But it departed from Merton by arguing that society's mores are mirrored not by the normative structure of science but by the technical content of science. SSKers argued that scientific facts themselves depend on equipment and methodologies that have a history and that have to be agreed on by scientists. Likewise, they pointed out that practicing scientists differed about how to follow agreed on methodologies. Similarly, scientific facts seemed to depend on conventionally agreed on standards of dissemination, credibility, and proof. Thus, because facts seemed to include some things that were social, questions could be raised about both the technical content of science and the social order: How do we know when we have a fact? Who is a reliable spokesperson about the natural world, and when and why do we or should we believe him or her? What does it take to stabilize a scientific fact so that it can be understood and used in reliable and reproducible ways by many different people in different places and in differing conditions.

SSK built up its analytic tools by borrowing from an eclectic mix of disciplines. From the similarly named sociology of knowledge, it took its interest in the social and cultural conditions that are needed for the construction and maintenance of facts, knowledge, and belief. SSK extended this inquiry into the systematic study of actual episodes of the production of scientific knowledge, which was previously considered out of bounds for sociological explanations.[8] Extending the sociology of knowledge into scientific knowledge was ambitious on two fronts for it took on the two bastions of resistance to social explanation and insisted on "sociologizing" them. The first of these was scientific truth. Truth is supposed to be one and to be the underlying reality on which social variants are played out, but SSKers insisted on the social and constructed nature of scientific truth, credibility, and authority and denied truth a transcendent, ahistorical, and foundational character. The second source of sociological ambition was to replace the great disembodied individual scientific minds of conventional historical narratives of science with collective accounts of scientific practice. In the accounts of SSK, great scientific minds emerge as major products of scientific culture rather than as the prerequisites for scientific discovery. This problematizing rather than simple valorizing of charismatic individuals in science is a shared theme among many practitioners of STS and is one of several strands that lend the field an antielitist and demystifying politics. Most sociologists of scientific knowledge have backgrounds in the natural sciences rather than in the social sciences or humanities and become sociologists by the nature of their program of research rather than by their training. Sociology is thus not the parent discipline of SSK but an adopted discipline.

From anthropology, SSK borrowed and was part of the repatriation and decolonization of the idea that social and natural orders—even in advanced capitalist societies—are stabilized together.[9] SSKers used anthropology's methodology because anthropologists had worked out ways of examining different cultures' beliefs about nature without presupposing that those beliefs simply reflected the state of nature. This anthropological approach allowed SSKers to analyze scientific practice and discourse to see where, when, how facts about nature are established without presupposing that modern science simply reflects true nature. Taking this same attitude to examining one's own scientifically informed belief systems is much harder, however, just because we do take for granted all sorts of external warrants, such as causal efficaciousness. When scientific findings are proven wrong, we tend to give explanations for why a method didn't work (such as poor methodology, false starting assumptions, or faulty data analysis), and these explanations resemble anthropological explanations. SSK took on the challenge to give anthropological type explanations for true facts about nature within a scientific world view.

SSK also drew on ideas from the later writings of philosopher Ludwig Wittgenstein as interpreted by Peter Winch to suggest that formal logical explanations have to come to an end or hit "bedrock" somewhere.[10] There is an infinite regress of ways to question knowledge. For example, if I say that the double helix contains cytosine, you can ask how I know this. In response, I can point to laboratory techniques for isolating and labeling components of biological samples. You can then ask why I believe those answers. I can reply by citing additional theories that implicate the same techniques or by giving a history of the development of the techniques. You can then ask why I believe those elements of my story, and so on, ad infinitum. "Ways of going on" or "forms of life," including scientific practice, have boundaries that are shaped by shared and largely unquestioned assumptions, without which meaning and progress would be impossible because people would be compelled to question and doubt everything. SSKers took up what they understood as Wittgenstein's antifoundationalism—the possibility of questioning and doubting any piece of knowledge. This led them to abandon the search for a scientific methodology that would provide a foundational level for scientific knowledge that was beyond doubt. They also incorporated Wittgenstein's alternative to a logical or methodological foundation for knowledge—namely, his turn to forms of life.

This antifoundationalism was combined with arguments found in various theoretical and empirical works in the history and philosophy of science. Arguments had been made in these disciplines for indeterminacy or underdetermination, and thus interpretive flexibility, of facts about the natural world.[11] Indeterminacy was evident in many famous examples, such as the claim that things that are and are not heaps of sand cannot be decided definitively. If we pile many grains of sand together, we know we have a heap, and if we pile a few grains, we know that it would be incorrect to call them a heap, but a large indeterminate range exists in between. Underdetermination noted that the "same" natural phenomena could be perceived and understood in different ways and that nothing in nature makes the case for only one true classificatory scheme. For example, only ground that rises higher than 1,000 meters above sea level could be called a mountain, but peaks between 700 and 2,500 meters might also be called mountains, which would be a different way of classifying nature but would be no less true to nature. Nature underdetermines our choices of theories, then. Combining indeterminacy and underdetermination with the anthropological view of the interconnections of social and natural orders results in a methodological principle of relativism. Given that more than one true way of describing the world is possible, no one classificatory scheme, theory, or set of facts is uniquely true to nature. Therefore, analysts of science should use the same means to study both accepted scientific knowledge and beliefs in disproved or alternative claims to knowledge. This methodological principle became known as the principle of symmetry and was formalized in the Strong Programme in the Sociology of Science at the Science Studies Unit at the University of Edinburgh.[12] In the terms of the realism versus relativism debates, this methodological relativism examines scientific belief systems without assuming that they have a special status because they are true. Instead, practitioners examine how scientists constitute what is true and how they make truth work for them. This kind of relativism is thus agnostic about realism because it has nothing to say about whether a real world is systematically revealed and represented by our best scientific theories.

The sociology of scientific knowledge approaches emphasize the importance of tacit knowledge and scientific practice over the explicit or formal renderings of the scientific method that are often assumed by historians of ideas and philosophers of science. This turn—from studying science as the progressive accumulation of truths about the natural world that are divined by great minds to the study of science and truth in the making—unites many different strands of science and technology studies.[13] It also helped to extend STS to fields of science and technology that exist first and foremost as practice, such as biomedical technologies. SSKers shared with Merton the sociological perception that it was not primarily great minds that made great science,[14] but they found little evidence in their empirical studies of scientific practice to support the idea that Mertonian norms were responsible for good science. Indeed, for every norm for which they found evidence, such as openness, they found counter norms, such as secrecy. When they looked to the practice of science, they found that technical standards and shared bodies of knowledge make up scientists' ways of going on. This was much as the historian and philosopher of science Thomas Kuhn had characterized periods of "normal science."[15]

Given the SSK desire to investigate rather than presuppose what makes something true, SSKers concentrated on areas where the division between true and false and between good and bad science is not yet firm. Early Modern science is a favorite focus because it enables historians of an SSK bent to explore the rise of modern science as a system of the production of truth about the natural world. Likewise, the great scientific standardizations of the nineteenth century have been well studied.[16] Another common site, and for similar reasons, is the study of scientific controversy, where the truth is not yet broadly accepted as falling on one side or the other of the controversy. In social constructionist technology studies, which overlap to some extent with SSK, the predominant focus is on the contingency of the ways in which one or another technology design comes to be accepted.[17] Throughout all this work, the crucial themes are credibility (who has it and how one gets it), trust (what must be unquestioned for any system of truth to be sustained), and the authority of science (the ability of science to produce assent to its claims). In SSK, these theoretical concepts have been used in the place of appeals to transcendent truth to account for the success of modern experimental science.

Ethnomethodology is a branch of sociology that looks at the constitution of normative order. The normative order refers to the complex of technical, institutional, and moral prescriptions (norms) that govern how to behave as a member of a given group of people—as a family member, a faculty member, a student, a jury member, a voter. A norm has a double meaning that relates to the concept of normal. That is, it describes things as they usually are and as we feel they ought to be. For example, body-weight charts are average height to weight ratios that are derived from a healthy population (the original adult U.S. ratios were measured from young, fit, army personnel). These charts confirm whether a person's weight deviates from the norm and provide a target weight range that is considered appropriate for a particular height. For anyone who wishes or attempts to conform to this norm, these normal body-weight charts also direct and justify action. Likewise, failure to conform to widely accepted norms has consequences that depend on the sphere of action and the nature of the transgression, including having a poor body image (in the case of being overweight), failing to pass exams or be promoted (in professional and academic fields), being imprisoned (for violation of legal norms), or being disliked (for a violation of accepted ways of relating to other people). Norms pervade everyday life as well as specialist forums, and we are all experts at recognizing and following them. Norms describe things, classify them in an evaluative way, provide guides to action, and justify retribution for failure to conform.

Drawing on the writings of phenomenologists like Alfred Schutz, ethnomethodologists concern themselves with the phenomenological "life world" of lived experience and eschew *a priori* theoretical accounts of concepts like truth and rationality.[18] Classical ethnomethodological accounts looked at legal reasoning, the performance of gender, and scientific practice.[19] Ethnomethodologists were dissatisfied with structuralist sociological accounts of action, which seemed to reduce all meaningful behavior to being the product of various structural social properties, such as gender, class, race, and profession. These accounts seemed to render life as experienced by the person living it as merely epiphenomenal and to make people "judgmental dopes," as Garfinkel expressed it. Ethnomethodologists tried instead to explore structural sociological explanations about appropriate assumptions about the nature of the world—the world that was enacted in the here and now, that was produced and reproduced in lived experience, and that did not exist as a set of abstract rules of behavior and bodies of knowledge. Michael Lynch, the most prominent ethnomethodological scholar in STS, has shown that science relies on the development and sustaining of specific immanent (present in the situation at hand) ways of experiencing and representing the world.[20] The importance to reproductive technologies of multiple levels of locally sustained norms is the subject of chapter 3. The connection between the performance of gender and the maintenance of social order, a locus classicus for ethnomethodology, is taken up in chapter 4.

Social-worlds theory is another significant sociological approach to the study of science, and it has some affiliations with ethnomethodology. In social-worlds theory, analysts pick and delineate their site of study through following those people and things that have shared commitments with the group being studied. It draws on the tradition known as "symbolic interactionism" and has moved easily into studying scientific practice, inheriting an already well-established emphasis on institutions and professions.[21] In the scientific realm, a social world in this sense includes the people, places, objects, knowledge, complex of values, economics, and so on that are found to surround and create the piece of science of interest. Thus, studying reproductive medicine (as Adele Clarke, a social-worlds theorist, has done) requires an analysis of those who practice reproductive medicine, those they interact with, and the flow of people, instruments, capital, and theoretical and moral ideas that make up the evolving field. Larger conceptual connections that link different social worlds (arenas) are also analyzed. Social-worlds theory contributes a key component to my research—namely, the inseparability of moral, economic, and knowledge economies. As STS has moved beyond the lab and the physical sciences to field, distributed, human, life, and biomedical sciences, this multisited, mixed methodological approach has become more necessary.

Recent developments in the historiography (theory of writing history) of science have meant that there is considerable affinity between the wider community of STS and certain historians of science. Indeed, the history of science has provided the theoretical innovation for much of STS. For example, a number of the founding contributors to the sociology of scientific knowledge are historians of science.[22] The areas of the history of science that have received the most STS attention include early modern European science, nineteenth-century field sciences, the history of modern biology, the history of medicine, twentieth-century physics, and the history of technology.[23] STS has inherited from the history of science its widespread view that the rise of modern Western science is a phenomenon of special historical interest and that the development of experimental science reflected and in turn shaped the societies where it developed.[24]

Historians of science who are also practitioners of STS have been among those who have reacted against what is called "Whig history"—namely, the telling of history from some later point of view or later stage of development of society. This kind of historical narrative is accused of being "presentist" and of appealing to events, outcomes, concepts, and understandings that were simply not available at the time to the actors being studied. This criticism has been raised against all kinds of historical narrative but has a particular form when used against narratives in the history of science. For example, writing the history of genetics in a presentist way might involve giving an account of steps that led inexorably to modern-day genetics and might interpret earlier findings such as Mendelian inheritance as evidence of the presence of genes. By contrast to this presentism, a historicist approach to describing Mendel's work would use actors' categories and not the modern-day concept of the gene and would not assume that the importance of Mendel's work was whatever aspects of it we can now translate into genetic theory. Although a presentist account might be a good way to write some narratives about Mendel's science (say, high school science textbooks), it would be a bad way to write history because it would not reveal much about living in Mendel's time or thinking in Mendel's terms. An additional point connects closely to the methodological principle of symmetry described above (analysts of science should use the same means to study both accepted scientific knowledge and alternative scientific knowledge). If the historian of science is trying to explain the history of the acquisition of certified knowledge about the natural world, then interpreting the history of science in a presentist manner would render invisible the processes by which the science in question changed or became certified and accepted as true. It would also make the historical trajectory of science seem a lot more progressivist and deterministic than might in fact be the case by masking the contingencies that led up to the present divisions and theories of the sciences.

Some, most notably Steven Shapin in his elegant essay on the topic, have argued that historical accounts of science that predate STS can be sorted into "internalist" and "externalist" narratives.[25] Internalist historical narratives tell the history of ideas or intellectual history (sometimes called "talking-heads history" because of its interest in the findings of a few "giants" like Newton and Einstein, whose ideas are portrayed as having had the power to change the history—idealism). Externalist histories of science relate problem choices and interpretations of findings to events that were going on in social and political culture at the time and place of discovery. An externalist account of Newton's science would examine why he chose certain problems and certain resources in terms of the political context at the time. Materialist factors, such as the economy, rather than ideas, become the wellsprings of change.

Their embrace of STS has led some historians of science to reject both internalism and externalism. The turn against disembodied intellectual history has instead emphasized the things that internalist accounts seemed to leave out, such as the actual practice of science, the instrumental and other material culture of science, and the various literary cultures of science. Externalism has been eschewed for its suggestion that the relations between science and culture and between science and politics are relations of reflection with the science simply piggy-backing on politics. Inspired by Kuhn's attention to the technical norms governing the content of science, people in STS have wanted to tell the history of science in ways that show interconnections between science and culture, without one presupposing the other. This attempt to give a fuller and more accurate, if more heteroclite and iconoclastic, portrayal of the history of science has much in common with the attempts of others in STS to understand science as practice and culture.[26] Not all historians of science who consider their work to be "post-Kuhnian"—that is, who take scientific practice and scientific content seriously and are historicist rather than Whiggish in their historiography—are practitioners of STS. The characteristic analytic interest of practitioners of STS in knowledge, truth, and objectivity may be absent in historians of science who otherwise share many STS tendencies.[27]

The philosophy of science also contributes to STS inquiries into the problem of knowledge. Most practitioners of STS, by virtue of their interest in questions of knowledge, occupy the philosophical wing of their home disciplines. This is true of my work when considered among other feminist writings on reproductive and genetic technology. Likewise, as noted above for the sociology of scientific knowledge, philosophical work has been important in the foundations of other branches of STS for providing insights into such things as underdetermination, the theory ladenness of science, and antifoundationalism. The philosophy of science, one of the core disciplines of STS, shares with the rest of STS the view that science is a good place to look to raise and answer questions about knowledge, truth, and objectivity. Some philosophers of science are often more interested in questioning epistemology, while keeping metaphysics out of bounds intellectually. Thus naturalist philosophers of science might ask about things that pose tensions for a position that presupposes objective truth, such as how to account for progress in science: if a scientific fact is true, why are the scientific truths of one generation superseded by later discoveries? What does it mean to get closer to the truth, if truth is not relative? When properties were ascribed to phlogiston, what, if anything, was really known, and how does it relate to our current understanding of oxygen?

Some philosophers of science have resisted with ingenuity the idea of multiple or contingent ways of describing the world. The things that might point to a constructed and contingent (but not random or purely social) understanding of scientific knowledge are made commensurable with a unique and true way of describing the world. Other philosophers, including several feminist epistemologists, have embraced multiplicity. For example, after the wide dissemination of Kuhn's *The Structure of Scientific Revolutions,* the so-called Kuhn wars erupted among philosophers of science over Kuhn's thesis about periods of normal science punctuated by paradigm shifts. Some philosophers maintained that Kuhn's thesis supported the idea that world views are incommensurable from one paradigm to another while the philosophers who are associated with resisting multiple worlds attempted to patch together epistemological and metaphysical continuity between paradigms. An important dynamic of the division in STS between those who argue for multiplicity and those who argue for the unity of the world is that each side tends to think of itself as capturing reality better than the other. Those pushing multiplicity, sometimes called "constructionists," think that their accounts capture the complexities of the world and believe that they can identify adequate resources for constructing and maintaining the relevant normative notions (such as truth and objectivity). Other philosophers of science, sometimes called "realists," think that relativist tendencies (such as an openness to incommensurability and multiple worlds) disqualify the other side from a claim to being realist. These fundamental differences are augmented by lesser differences in style and orientation.[28]

Philosophers who do not presuppose the fundamental unity of science and who have been influential in the wider interdisciplinary STS community include Nelson Goodman, who argues for underdetermination (that there is always more than one way to account for empirical observations), and Ian Hacking, who has shown the constructed nature of naming terms in science.[29] Similarly, feminist philosophers of science (like Helen Longino, Alison Wylie, and Sandra Harding) and feminist philosophers of mind and language (like Elizabeth Lloyd, Patricia Hill Collins, and Lorraine Code) have argued that knowledge is social or that the subject position or the context of life and work matters to what is known and how.[30] Finally, a number of STS-type philosophers from a variety of backgrounds have begun to produce empirical and theoretical studies of formal devices and languages for organizing or representing knowledge, including studies of aspects of information technologies and studies of such devices as bureaucratic forms.[31]

The Ontology of Science

Ontologies are theories about being or reality, and scholars of science and technology studies who are interested in ontology probe the connections between science, technology, and the world. Someone who is interested in the ontology of physics might want to know what a quark "really is" and how the answer to that question connects with the techniques and theories through which quarks are isolated and identified. Someone who is interested in the life sciences might ask what a gene "really is." These questions are clearly related to but can be contrasted with the questions raised by scholars who are interested primarily in scientific knowledge. Such a scholar of physics might ask how we know that there are subatomic particles and why others believe theories that postulate quarks as entities.

Actor-network theory (ANT) is a branch of STS that is associated most strongly with the work of Bruno Latour, Michel Callon, and their colleagues at the Centre de Sociologie de l'Innovation (at the École des Mines de Paris) and John Law in the United Kingdom.[32] Practitioners of ANT are based in sociology departments and interdisciplinary departments. ANT is as much philosophy as it is sociology and borrows from (as well as contributes to) anthropology and the history of science. The work of Andrew Pickering on "heterogeneous assemblages" is closely related to actor-network theory and applies ideas about contingent and constructed ontology to the physical sciences.[33] "Situated knowledges"—an epistemological term that is used in feminist theory and elsewhere—originates with the feminist STS of Donna Haraway and combines questions about knowledge, ontology, and politics.

Actor-network theory portrays modern science as an unprecedented source of innovation for understanding the natural world and for connecting people and things of different sizes and scales over space and time to make new worlds. ANT differs from the "critical tradition" in dissociating an understanding of science and technology from the necessity of being antiscience (or as Bruno Latour calls it, antimodern). ANT is remarkable for its theoretical boldness. I described a principle of symmetry above that urges analysts to explain episodes of scientific discovery with the same anthropological approach that is used to explain discredited or superseded science. Like the sociology of scientific knowledge, actor-network theory emphasizes practice ("science in the making") rather than "science made." ANT adds a second dimension of symmetry to that proposed by SSK, however—that analysts ought to treat society in the same conceptual way as they treat nature. ANTers argued that SSKers tended to see (contested) truths about the natural world as reflecting (battles over) political status quo, without problematizing the latter.

This second principle of symmetry (treating society and nature the same conceptually) expresses the fundamental aspect of the approach that I take to reproductive technologies in this book—that networks of people and things together make both new knowledge and technologies, and the social categories of identity and society. Pure social or natural facts are, according to ANT, end products of processes of separation and purification rather than preexisting distinct explanatory repertoires. The process whereby facts and technologies become stabilized and generally taken as true has been labeled "black-boxing." Scientific experiments yield results that can be expressed as facts within the appropriate theory. When these facts withstand tests and confirmation attempts and are promulgated as truths in scientific journals or informal lore, they can be said to have become black-boxed.[34] The facts stand as a self-contained package and are valid without reference to the conditions under which they were produced. Truth is routinely precipitated from science in the making in this manner and so should be thought of more as product than as preexisting.

ANT points out that no scientist studies pure nature. Scientists and their equipment always intervene and interact with the objects of study. Scientists set up scenarios or develop observational and recording skills against which the objects of study (whether nematodes, baboons, or muons) manifest properties and behaviors. Similarly, humans are not the only agents in the world. Nonhumans also have agency, a situation that is examined in Latour's essays on the rainforest, on Pasteur's microbes, and on lactic acid, and in Michel Callon's famous essay on scallops. The attribution of agency to nonhumans is a distinguishing characteristic of the ontological branches of STS. Because nonhumans and humans act, swap properties, and are together the condition of each other's identities in these strands of STS, each uses metaphors of hybridity. Thus, ANT uses networks and actants, Donna Haraway and others use "cyborgs" and "tricksters," and Pickering refers to heterogeneous assemblages.

Haraway, whose ontological STS work is described below, enjoins the analyst to be knowing or at least alert to certain political, epistemological, and ontological tropes. In ANT, on the other hand, the analyst adopts the naïve perspective of science in the making. That is, the analyst doesn't know what will be done next, what things will be connected to what ideas, and how this will happen, and her task is to document specific cases without regard to preconceived ideas about which elements are natural and which are social. ANT and the other branches of ontological STS portray science and technology as processes where genuinely new assemblages and alignments of things and people are brought about.[35] In ANT, then, the hybridity of social and natural things is the condition of producing objective facts about the natural world.

ANT has introduced a number of explanatory concepts that now have broad circulation. Among these are centers of calculation, obligatory passage points, enrollment, delegation, and networks. Centers of calculation are places or institutions at which particular kinds of facts are collected and disseminated, such as the U.S. Department of Health and Human Service's Centers for Disease Control and Prevention (CDC) for infectious diseases or the Stanford Linear Accelerator Center (SLAC) and the Conseil Europeén pour la Recherche Nucléaire (CERN) for work in experimental physics. Certain kinds of facts have to pass through an obligatory passage point to become stable and standardized facts. Enrollment is the process whereby something or someone is allied to a particular view of the world and comes to act or speak as evidence of that position. ANTers grant agency to both humans and nonhumans, so Pasteur is seen as having enrolled microbial, political, and scientific allies. Delegation is the process whereby an assemblage or a link in a network holds in place particular social relations and a particular representation of part of the world. If I go to a South Pacific island and bring back specimens to Paris, I have connected a lot of people, places, and things together by displacing a tiny bit of one place to another place.

The networks of actor-network theory are much disputed: what are they made out of, and what work do they do? Pickering's heterogeneous assemblages avoid the problems with the network metaphor by essentially abandoning it. My understanding is that the networks of ANT are the connections that are made between the things, facts, and people that make up the scientific network that is being studied. The network metaphor gets it strength from its theoretical neutrality: it allows the analyst to consider each thing that goes into science in the making without ruling out certain kinds of things ahead of time. The network metaphor is excellent for highlighting changes in scale—from local to global (as when knowledge is produced when a bit of a South Pacific island is brought to Paris) and from weak to powerful (as when a physician at the CDC classifies and controls an agent of infectious disease). These explanatory freedoms have proven fruitful for STS because they seem to allow for better empirical accounts of what actually happens in science in the making.

Haraway's blockbuster history of the intertwining of gender, race, nation, and nature in the rise of the field of primatology, her "cyborg" metaphor, and her exploration of the world of biotech in which the patenting of oncomouse was possible all form part of her oeuvre in the ontological branch of feminist science studies. Many of her major philosophical themes can be traced through her deployment of the idea of "situated knowledges," which she introduced as a means of preserving the strengths of empiricism and radical constructionism while avoiding their weaknesses. The institutions, people, international standards, scientific journals, equipment, experimental setups, and funding that are the very stuff of science show that objectivity actually depends on partiality rather than universalism and transcendence and on embodiment rather than disembodied knowing. Partiality, embodiment, and "mobile positioning" share sensibilities with recent feminist epistemologies based on motion rather than position—such as Rosi Braidotti's nomadic subject, Angela Davis's blues epistemology, Annemarie Mol's ambulatory idiom, and Elspeth Probyn's metaphor of departure, despite the somewhat static sound of Haraway's word *situated.*[36]

The Politics of Science and Technology

For many STS folk, scientific truth is a resource for talking authoritatively about, forging definitions of, and drawing boundaries around nature, politics, and identity. Coproduction is the branch of science and technology studies that most directly embraces this endeavor, and it is associated with the work of Sheila Jasanoff and her colleagues and students. A lawyer and contributor to national and international science policy, Jasanoff has studied the interactions among science, politics, and the law in an international comparative framework.[37] Social and cultural anthropology, women's studies, feminist science studies, queer theory, critical race studies, postcolonial theory communications, cultural studies, and information technology studies have all also played a part in this aspect of STS. Most of these subdisciplines began contributing to STS later than SSK, ANT, and their variants, but they interact vigorously with them, while adding new areas of focus and new analytic resources.

Coproduction shares many sensibilities with the sociology of scientific knowledge and with actor-network theory. Its constructionist framework refuses to take truth for granted or to separate politics and knowledge, and requires an explanation for any stabilization of facts and standardization of scientific practice. While most of the studies in SSK and ANT focus on the lab or on relatively discreet episodes of science, Jasanoff has taken STS into the broader political realm by looking at different national styles of accommodating, promoting, and using science for governance. Characteristic topics of inquiry of others working in STS on the politics of science include the relations between science and the military, lay participation in the production of scientific knowledge and the public understanding of science, the interactions between indigenous and scientific knowledge, and the effect of global regimes on national science institutions.[38]

In this chapter so far, I have described some shared sensibilities of practitioners of STS and attempted to tease apart strands of the discipline. I do not think that STS is the one and only true way to understand the world, however. Different kinds of work are better at isolating different kinds of phenomena and answering different kinds of questions. As suggested above, though, STS is helpful in empirically investigating knowledge, ontology, technology, bureaucracy, identity, and politics under a single lens in a particular area of practice. To my mind, this is important. On the one hand, STS allows us to undertake historical, sociological, and anthropological investigations of our institutions of science, which brings this fascinating strand of modernity into line with our other important institutions. On the other hand, it promises that we as a culture can become more educated about science and what it does and means. A science policy that simply decides which projects to fund, what kinds of oversight should be used to police scientists, and which problems to throw science at does not do justice to the ways in which science and society coproduce one another. Questions about the links between science and democracy, about choice and privacy, about the state versus the individual, and about the international division of labor are growing more, not less pressing.

In practice, most scholars in most fields recognize the shortcomings of their theories of choice and stake out ways to mitigate those shortcomings while retaining the strengths. And this is true in practice even of people who strongly excoriate methodologies other than their own. Fields also grow by new generations of students who often are led by one or two charismatic older scholars of the field—through scholarship, hiring practices, agreement about what they dislike about what went before, and the approximate means of addressing it. Along with many of the second generation of STS folks, some of the founders, and to some extent feminist scholars, I am less interested in laboratory science and more interested in science as it moves between different sectors of life, including the intimate and the transnational. I am also more interested in bodies and emotions and less interested in biography than many in earlier generations. I am more interested in how science (re)produces differences and stratifications among people and less interested in how it produces assent. And I am more interested in science's circulatory systems or economies and less interested in specific laboratory or institutional cultures. For each of these pairs, however, the second inquiry would not be interesting without the work that has been done on the former as a way of setting up the inquiry in question.

For work in reproductive and genetic technologies, the interdependence of society and nature is in some sense obvious. It is certainly much more intuitive to claim that the life sciences depend on social as well as natural phenomena than it is to make similar claims for the physical sciences. There is nothing particularly bold in my embracing STS as a framework, therefore, except that the natural and the social are still difficult to talk about, even where the material seems to demand it. I argue that the obviousness of the interdependence of science and society in the age of reproductive and genetic technologies has important implications. While STS from the late 1970s through the 1990s served as a crucial corrective to ideas about the infallibility of science and the transcendence of scientific facts, STS must provide ways to navigate politically and empirically where nature is in flux and science is anything but infallible. This means it must show that there is a difference between good and bad science even where uncertainty is the rule. And it must show that proceeding with caution—or precaution—in science and technology does not mean adopting the relativist position that enables gainsayers of science (those against emissions control or cloning, for example) to claim that because the science is not certain, we should not proceed.

Science and Society

In his famous 1938 paper "Science and the Social Order," Robert K. Merton observed that "It is true that, *logically*, to establish the empirical genesis of beliefs and values is not to deny their validity, but this is often the psychological effect."[39] Merton was not taking part in a precursor of the recent so-called science wars between those who practice and those who study the practices of modern science. At a critical moment in the history of the twentieth century, Merton was talking about reasons why some people were hostile to science. Scientists, not those who study science, were being accused of attempting to discredit others' practice through their study of that practice. Some who study science today might concur that their scholarly attempts to understand the empirical genesis of truth, beliefs, and values in modern science—using "detached scrutiny" or "organized skepticism," as Merton variously called it—are being read by some scientists as an attack on the validity of science and the integrity of its practitioners. Merton was decrying the anti-intellectualism and generalized hostility exhibited by National Socialism toward the democratic variant of science, and he was trying systematically to uphold the values of objective science despite the political and cultural expediencies of Nazi science. Any community, including communities of natural scientists, experiences discomfort and suspicion if it is subjected to scrutiny from without. But the temptation to let this natural suspicion lead to a shutting down of inquiry, to anti-intellectualism, should be strongly avoided.

Some who study science, including journalists, do a poor job of representing the subtleties of research and its findings, and others do it better. But there are good reasons why science should seek and be subject to "outside" scrutiny and publicity.[40] The findings of science move and inform more people when they are widely disseminated, and keeping communication flowing between the sciences and their various publics makes sense when taxpayers contribute significant amounts to education and research and live governed by the truths and technologies that science produces. At least in the United States, the general public still cedes the sciences tremendous cultural authority, even if some groups are for historical reasons inclined to trust it less. Many people attend schools and even universities, but not many inhabit the places or understand the knowledge that is produced at the cutting edge of scientific research. Science is about the independent and disinterested (correctly understood) search for truth and the love of knowledge for its own sake. It would be compromised if scientists and their findings were forced to subordinate their research agendas to the whims of public opinion. Likewise, excessive openness to rational bureaucratic scrutiny is not always in the best interests of science and is an impoverished understanding of what accountability amounts to.[41]

So the task becomes to balance the needs of the autonomy of science while demanding from it the same minimum standards of openness that we demand from the other major knowledge-producing institutions. As many scientists themselves have advocated, this helps to detect fraud and to avoid the calamities of sciences past. We all should shun the anti-intellectualism of cutting off our fields of expertise from outside scrutiny.[42] If many different kinds of people claim a stake in the links between science and society, more connections will be made among science, the objects of scientific knowledge, and the rest of the world. If we are lucky, this will lead to a securer future for science simply because more people will take it into account in more situations.[43] I argue in chapters 7 and 8 that the way to secure this balance in the field of assisted reproductive technologies, given U.S. political culture, is to move away from a model based on the "public understanding of science" and toward one based on the "private implication in science." I hold out the hope that striking this balance appropriately for different times, places, and subcultures of science will lead to better science because theories will have to withstand testing and application in more circumstances and from more points of view.[44] This book, then, is a defense of the ideals of science and democracy, with the caveat that inquiry must be situated to elucidate the nature of the science in question and of its publics.

2

Fertile Ground: Feminists Theorize Reproductive Technologies

I turn next to the rich, multidisciplinary feminist literature on reproduction, infertility, and reproductive technologies. This literature and that of science and technology studies are the fields in which this book is situated and to which I aspire to contribute. While STS offers the methodological and theoretical orientations from which I primarily draw, feminist literature on reproduction, infertility, and reproductive technologies offers detailed studies of field sites and content. STS and feminist scholarship on reproduction diverge considerably historically and, as is discussed in the introduction, they differ in affect and goals.

Reproductive technologies, and the infertility for which they typically exist to alleviate, pose a paradoxical tension for feminists. On the one hand, even during periods of decreasing birth rates and even where some choose voluntary childlessness, involuntary childlessness is recognized as being one of the greatest forms of unhappiness and loss an adult woman might have to endure. Would-be fathers experience infertility as a source of deep sorrow, and as threatening to desired kinship roles and masculinity.[1] Nevertheless, the burden of involuntary childlessness is taken to be especially heavy for women, and prominent feminists have long called for it to be taken seriously as a feminist issue.[2] A sense of a special burden for women is often exacerbated by the assumption that infertility is the woman's "fault."[3] On the other hand, feminists are also interested in disrupting the gender-role expectations and essentialist connections between motherhood and women's identity that greatly intensify infertile women's suffering. Contemporary infertility and its treatment are conceptualized and structured around a strongly coupled, heterosexual, consumer-oriented, normative nuclear-family scenography. When successful, treatment enables women to reinscribe themselves into that logic.

Feminists are well placed to understand the special burden that involuntary childlessness places on women, but they are ambivalent about supporting women who seek infertility treatments because of the implicit support that this seems to lend to conventional gender roles and gendered stratification. Not surprisingly, the relationship between infertility-treatment advocates (including infertile patients, medical personnel, and some social science researchers) and feminists has been strongly marked by this tension.

In this chapter, I trace the evolution of views in the explosion of feminist work on infertility in the age of the new reproductive technologies.[4] I argue that the high-tech medicalization of reproduction combines (or has been presented as combining) the economic, technical, rhetorical, personal, legal, and political elements through which the phases and conflicts of recent feminism have been articulated. In other words, reproductive technologies have been performed as the perfect feminist text for the last two decades. From early feminist writings that denounce the infertility business through increasingly sensitive work on the experiences and consequences of infertility, feminist treatments of infertility have come to embrace both sides of the feminist tension. This close relationship between feminist theorizing and reproductive technologies might erode as assisted reproductive technology clinics become the front line for such procedures as sex selection and stem-cell research, where the tension has less purchase.

To advance this argument that infertility has been performed as the perfect feminist text for many years, I use a crude but handy chronological distinction between phase 1 (roughly 1984 to 1991) and phase 2 (roughly 1992 to 2001) of feminist writings on reproductive technologies and infertility. Phase 1 scholarship includes the initial galvanization of feminists around the topic of infertility and the new reproductive technologies. Phase 2 scholarship includes work that continued but was also critical of the work that was done in phase 1. I argue that these phases somewhat overlap with second- and third-wave feminism. Phase 1 writings about infertility in the age of reproductive technologies exemplify second-wave feminist concerns. Just as materialist and structural functionalist feminists vied with mainstream liberal feminists during the period of second-wave feminism, radical phase 1 feminist critiques of infertility and reproductive technologies challenged liberal accounts of the technologies. Phase 2 writings about infertility and reproductive technologies not only exemplified emerging third-wave (or poststructuralist) feminist concerns, such as the recasting of agency; they also represented one of the sites of feminist activism and scholarship that precipitated third wave feminism.

Phase 1 (about 1984 to 1991): Feminists Engage the New Reproductive Technologies

Shortly after the birth of Louise Brown, the world's first test-tube baby, in England in 1978, feminist theorists and activists seized on the new medicalization of infertility as being an issue that was of special concern to women. A number of factors contributed to the rise of feminist concerns about assisted reproductive technologies. While mainstream or liberal Western feminists tended then, as they still do, broadly to support the development of biomedical reproductive techniques, on the grounds that such technologies augment reproductive choice, the techniques were opposed by the more technophobic, antieugenicist, and antipatriarchal sentiments of radical feminism. By the pronatalist 1980s, the 1970s feminist utopic longing for a world where technological progress would free women from their reproductive biologies was all but dead among academic feminists and others, and reproductive technologies were interpreted as increasing, not decreasing, subservience to biological destiny. Despite the antipathy that many feminists felt toward the views of religious conservatives and antiabortion activists, the possibility that their opposition to reproductive technologies might support these groups' agendas did not deter many feminists from decrying ART procedures.[5]

Phase 1 Feminist Critiques of New Assisted Reproductive Technologies

Phase 1 feminist writings that were critical of reproductive technologies grew out of and in turn developed several themes that were core parts of second-wave feminist scholarship on science, medicine, childbirth, and reproductive rights. Beginning in the mid-1970s, a number of influential feminists wrote about the crisis for women that was posed by the excessive medicalization of reproduction in the West.[6] As they saw it, pregnancy and childbirth had become mechanized and pathologized by a patriarchal and increasingly interventionist medical establishment, such that, as Sheila Kitzinger expressed it, many women were led to believe that if they were not around, "the pregnancy could progress with more efficiency."[7] Their agenda was clear: a rejection of masculinist technologization and a reclaiming of natural childbirth by and for women.

As the 1980s progressed, writers revealed how frequently women, particularly working-class and minority women, were victimized by medical experiments, negligence, and disrespect at the hands of doctors throughout history.[8] The connections between eugenics, medicine, and the control of women's reproduction were being patched together by a new generation of feminist scholars and activists,[9] who also thematized the use of reproductive medicine to subject women to ever-greater surveillance.[10] Likewise, by the beginning of the 1980s, feminists made up a significant contingent of the various critical science and technology movements.[11] These various feminist sources of concern were brought to bear on infertility and its treatments in phase 1 writings about ARTs.

Contributors to the first wave of important feminist anthologies and monographs on the new reproductive technologies critically addressed these issues as they resonated with infertility medicine.[12] In vitro fertilization (which produced so-called test-tube babies) provoked the most commentary, but artificial insemination, or therapeutic donor insemination, was also addressed.[13] These authors warned about the technological momentum of the new technologies and the apparent lag in legal and social means for incorporating the technologies responsibly. They noted the technological imperative for women to resort to these expensive procedures, which was exerted by the mere existence of prenatal screening and infertility treatments, and their "never-enough" quality.[14] The charge was renewed for the case of reproductive technologies that patriarchal control of women's bodies was achieved through medicine. In a small number of notorious cases of infertility treatment, women took hormones to stimulate their ovaries to produce lots of eggs at one time and died or almost died from ovarian hyperstimulation syndrome.[15] These cases and the physically demanding and experimental nature of the new medical infertility procedures prompted alarms that women's bodies were becoming experimental sites.[16] The extremely low and often misleadingly presented success rates of infertility clinics were cited as evidence of the emptiness of the promise of technological salvation for involuntary childlessness.[17] As was often pointed out, treatments such as in vitro fertilization did not, even when successful, cure infertility; they alleviated the condition of involuntary childlessness.

The search for safe, affordable, appropriate, and available contraceptives continued to unite liberal and radical feminists. Phase 1 writings did not produce a feminist consensus or technological optimism around any of the new reproductive technologies, however. Genetic screening and prenatal testing reeked of the potential for eugenic abuse and seemed largely inappropriate technology in the absence of research to improve understanding of and infrastructure for disability.[18] In vitro fertilization was too expensive, dangerous, and ineffective to give many women the increased reproductive choice that might offset the additional patriarchal control of women's bodies that the new technologies enabled. The resolution passed at a 1985 conference held by the Feminist International Network of Resistance to Reproductive and Genetic Engineering (FINRRAGE) in Sweden called for "a different kind of science and technology that respects the dignity of womankind and all life on earth."[19] The predominant message was that it was "not too late to say 'no' to these technologies."[20]

There were, however, two sources of technological optimism for the new reproductive technologies among radical feminists in this period. For socialist feminists, the march of history included inevitable technological developments. Their optimism, if it can be called that, derived from the desire to direct the use of reproductive technologies toward improving women's lives by alleviating class differences:

> Reproductive technology now exists. It cannot be undiscovered by turning the clock back. Fears about the misuse of these techniques cannot be answered by banning them, but only by their being democratically controlled. They are either going to be controlled by us for our own benefit or they will be used by private businesses to make money out of us and threaten our well being. They cannot be independent of class society.[21]

With prescience, a number of radical feminists also espoused the possibility of subverting these technologies to break down compulsory heterosexuality and facilitate lesbian parenting, either through self-insemination or through the development of parthenogenesis.[22]

Phase 1 Feminist Positions on Infertility and Stratification

In addition to having roots in second-wave feminist arguments about science, medicine, and reproduction, phase 1 writings on reproductive technologies were interpolated with more overtly political feminist conversations about privacy, the family, and stratification. Stratification by gender, class, race, age, country of origin, and able-bodiedness all loomed large. When phase 1 writings had any global dimension, they tended to deal with contraceptive technologies for the fertile Other and conceptive techniques for infertile Westerners. For example,

FINRRAGE, an international group that first convened in 1985, articulated the transnational reproductive politics and feminist struggles that are illustrative of this period.[23] German sensitivity to the eugenic potential of genetic engineering was evident, as were South Asian critiques of widespread patterns of son preference and the use of prenatal testing for female feticide.[24] International concerns also addressed abuses of sterilization, testing and dumping of contraceptives, and access to family planning and abortion.[25] These international issues modulated the predominantly North American, Australian, and Western European feminists' associations of technology and medicine with patriarchal and capitalist expansion.[26]

Early phase 1 writings about infertility and reproductive technologies were notable for emphasizing concerns of structural stratification rather than for acknowledging or alleviating negative experiences of infertility. Toward the end of phase 1, feminist ethnography began to examine women's experiences of infertility for purposes other than criticizing the technologies.[27] In the early part of this period, however, some feminists deconstructed women's desire for children and for access to assisted reproductive technologies as the result of ideological duping.[28] Because of its overwhelming focus on systematic structural discrimination, feminist work in phase 1 exhibited a denunciatory and morally unambiguous tone. Many felt that feminists in this period were asking women who were dealing with infertility to give up their individual desire to bear children in the name of the general goals of feminism.

A number of more liberal feminists and other social analysts of the new reproductive technologies tried to answer these critiques with a more mainstream approach. A second round of anthologies and monographs came out that were not exactly in the genre of "self-help," but did self-consciously address and include active infertility patients. These works recommended judicious stewardship of the deployment of the technologies, better information about treatments and the alternatives of adoption or child-free living, better clinical standards and results, and proper discussion of and regulatory attention to discrepancies in access to treatment.[29] Collected volumes on the ethics of reproductive technologies began to appear in this period, too, suggesting that the mainstreaming of the critique and reform of infertility medicine was in full swing.[30] Similarly, books that summarized current and pending laws, legal precedents, and landmark cases gave order to the public's and media's fascinated speculations about the new technical and social situations that the technologies permitted.[31]

Beginning in the mid-1980s, patients, activists, drug company representatives, professional organizations, and governments in a number of countries forged criticism and reform into a progressive, commonsense, educational idiom. One reason for this was the inevitably public and contentious nature of the issues involved because of their intersection with the abortion debate. The market for the new reproductive technologies, atypical in medicine, was also crucial. Most patients were high-paying consumers who were not suffering life-threatening conditions but instead were pursuing life-defining goals. Given a patient population who had the buying power to go elsewhere and had equal or greater social status than their treating physicians, the cutting edge of the field developed more like a consumer-oriented business than a state-sponsored social service. Transparency and reform were market strategies.

Patient activism, coordinated through organizations like RESOLVE, waged legislative battles for standards and patient information in several countries. In the United Kingdom, the charismatic and mediaphilic Robert Edwards met the criticisms of the technologies head on and began what has now become a characteristic of the field of infertility medicine in the United Kingdom: preemptive popular books and journal articles by prominent physicians that address the legal and ethical issues sparked by the technologies.[32] Professional organizations like the American Society for Reproductive Medicine and hormonal drug companies like Serono began to colonize the information gap. Balanced, informative, women-friendly publications on every aspect of assisted conception—stress, donor eggs, electroejaculation, surrogacy, fibroids, adoption, and the ardors of sex on schedule—are now features of every infertility doctor's office. One result of this remarkable mainstreaming of criticism and reform was that to some extent the industry stole the ground out from under the feet of the radical critics.

This liberal, ameliorist approach seemed too complacent to many radical feminists. Had reproductive technologies simply begun a debate about whether ARTs benefitted women, theorists probably would not have advocated a simple prelapsarian rejection of the technologies. The strength of the fear and revulsion that the technologies evoked for some has to be understood in terms of the multiple systems of stratification that were potentially reinforced by reproductive technologies. Concerns were raised about gender, class, age, race, nationality, species stratification, and the newly growing opposition between the mother-to-be and fetus. Normalizing the technologies, as implicitly or explicitly recommended by the liberal and mainstream trends just surveyed, risked rendering each of these dimensions of the technologies invisible. To many, that removal of fundamentally political matters of distributive justice from the political realm to the medicotechnical realm was simply unacceptable.

The class dimensions of reproductive technologies remained a major concern that the liberal approach could only hope to address by "trickle-down" access. The parents of the first test-tube baby, Louise Brown, were from the working class, and some infertility patients can receive treatment on the National Health Service at a few facilities in the United Kingdom and in other countries with state-subsidized health-care systems. Options vary greatly by region, however, and waits for appointments are often long, which can be disastrous for older women.[33] Disparity in access is even more striking in countries (such as the United States) that lack universal health care. The cost of infertility procedures varies between countries and regions, over time, by procedure, and within procedure according to the characteristics of the individuals involved. Nonetheless, a cycle of in vitro fertilization, if paid for out of pocket, costs patients several thousand U.S. dollars (table 2.1) and is prohibitive for most people. Because wealthy people have much better access to the procedures than poor people, the technologies only selectively increase reproductive choice. (See table 2.2 for infertility coverage by state.)

Given that there are limited resources available for medical and other social services and that those resources are unevenly distributed, some feminists questioned the spending of resources—both personal and public—on treatments like IVF.[34] For example, adoption was advocated as an alternative that was more worthy than reproductive technologies to receive scarce resources because it spreads resources better throughout society.[35] The interconnections between race and class were noted by some feminists, and the argument was made that basic child and maternal health services for poor or minority women and children should be a greater funding priority than techniques of assisted reproduction.[36]

The other infertility practice that raised economic and cultural issues of class in an especially poignant way was conventional commercial surrogacy.[37] As Gena Corea sarcastically expressed it,

Table 2.1
Approximate Costs of Assisted Reproductive Technology Procedures and Means of Payment

---

Nationwide, infertility accounts for approximately 0.1% of the total health-care budget, with assisted reproductive technologies (ARTs) believed to account for no more than 0.03%. In Massachusetts, a state with comprehensive infertility benefits, the costs are approximately $2.25 out of an average monthly premium of $562.50.

$3 to $5 billion is spent a year in the United States on infertility treatments, of which 20 to 35% is covered by insurance.

The average cost of a cycle of in vitro fertilization (IVF) in the United States is approximately $8,000 to $9,000, including medication, hospital, and anesthesia fees. Micromanipulation (ICSI) typically adds another $1,750 to $2,000 to treatment costs. Cryopreservation (freezing embryos) typically adds $600 or more.

A couple can undergo two cycles of IVF for approximately the same cost as tubal surgery, and results are typically better.

The per pregnancy cost for IVF, not including additional perinatal costs associated with multiple gestation, is currently approximately $30,000 to 40,000 in the United States.

Some clinics and one national network, Advanced Reproductive Care, offer packages to make treatment more affordable. Advanced Reproductive Care involves over 250 reproductive endocrinologists in 37 states. An example of a package is the offer to pay $27,000 for three cycles of IVF. If the patient is not pregnant within a year, $20,000 of that is refunded. Typically, "shared-risk" or refund programs do not include or refund costs associated with medications and monitoring of ovulation stimulation.

Some pharmaceutical companies offer discounts on fertility drugs based on financial need.

Some employers sponsor elective health-care plans that have an allowance that can be spent on ARTs.

In 2000, the average payment to sperm "donors" was $60 to $75 a time. The Ethics Committee of the American Society for Reproductive Medicine recommends payment of from $3,360 to $5,000 for egg donation, based on the numbers of hours that are involved in being an egg donor compared with sperm donation. They caution that payments for eggs of over $10,000 might provoke women to discount their own risk. Nonetheless, advertisements are frequently posted offering more than this to desirable donors. They emphasize that payment for egg donation should be thought of as compensation for costs and inconvenience incurred and not "baby-selling" (Ethics Committee of the American Society for Reproductive Medicine 2000).

Table 2.1
(continued)

The following typical fee schedule for required psychological and legal services for surrogacy agreements, over and above treatment costs, and payments to surrogates where applicable, is adapted from a program specializing in surrogacy arrangements:

Psychological consultation	$250
Legal consultation	$250
Surrogate contracts	$1,250
Psychological support for a surrogate mother	$660–$2,200 per case
Establishment of parental rights before birth (court order)	$1,500–$2,000
Miscellaneous legal expenses to finalize parental relationships	$600–$800
Legal case management to secure attorney	$3,000

Prices do not include any expenses related to contested parentage or alleged breach of surrogacy contract.

The typical fee schedule for a donor egg includes the following:

Medications	$2,000
Hospital	$1,150
Anesthesia	$400
Procedure	$5,150
Egg-donor payment	$2,500

As has been widely publicized, egg donors that meet certain criteria can be paid in the tens of thousands of dollars.

Sources: RESOLVE Web site; ASRM Practice Committee of 1997; private clinic details from fieldwork.

Women's bodies are not only the *recipients* of so-called treatments for infertility like *in vitro* fertilization. In the institution of surrogate motherhood, our bodies actually *become* infertility treatments. Women are hired to be artificially inseminated, to gestate a child, and then to turn that child over to the sperm donor, thereby "treating" the infertility of the sperm donor's wife.[38]

Two infamous cases of surrogates—Mary Beth Whitehead, who fought an unsuccessful custody battle for Baby M, and Alejandra Munoz, who was tricked into coming to the United States from Mexico and being a surrogate for a Mexican American couple in the belief that she was having "embryo flushing"—bought the class issues to the

Table 2.2
Infertility Insurance Coverage by State

The 14 states listed below have insurance laws that mandate the coverage of or mandate the offer to cover some infertility services. All other states have no relevant laws. The Employment Retirement Income and Security Act of 1974 exempts companies that self-insure from state regulation.

State	Date	Mandate to Cover	Mandate to Offer	Includes IVF	Excludes IVF	IVF only
Arkansas	1987	X[a]				X
California	1989		X		X[b]	
Connecticut	1989		X	X		
Hawaii	1987/2003	X				X[c]
Illinois	1991/7	X		X[d]		
Maryland	1985/2000	X[e]				X
Massachusetts	1987	X		X		
Montana	1987	X[f]				
New Jersey	2001	X		X		
New York	1990/2002				X[g]	
Ohio	1991	X[h]				
Rhode Island	1989	X		X		
Texas	1987		X			X
West Virginia	1977/1995		X[h]			

Source: Adapted from a Fifty States Summary of Legislation Related to Insurance Coverage for Infertility Therapy," U.S. Department of Health and Human Services, Health Resources and Services Administration, Maternal and Child Health Bureau, 2004, ⟨http://www.ncsl.org/programs/health/50infert.htm⟩.

a. Includes a lifetime maximum benefit of not less than $15,000.

b. Excludes IVF but covers gamete intrafallopian transfer (GIFT).

c. Provides a one-time only benefit covering all outpatient expenses arising from IVF.

d. Limits first-time attempts to four oocyte retrievals. If a child is born, two complete oocyte retrievals for a second birth are covered. Businesses with 25 or fewer employees are exempt from having to provide the coverage specified by the law.

e. Businesses with 50 or fewer employees do not have to provide coverage specified by law.

f. Applies to HMOs only; other insurers specifically are exempt from having to provide the coverage.

g. Provides coverage for the "diagnosis and treatment of correctable medial conditions." Does not consider IVF a corrective treatment.

h. Applies to HMOs only. The Arkansas, Hawaii, Maryland, and Texas laws exclude coverage for IVF using donor egg or sperm. The New York law explicitly excludes several procedures, including IVF, sex change, and cloning.

fore. Paid surrogates are typically of considerably lower socioeconomic status than the commissioning couple, and the payment received for conceiving and gestating a child and the signed contract of intention to give up the child are usually sufficient to override the surrogate's claim to custody. Surrogates have to make themselves somehow exempt from a general trend toward the sentimentalization and decommodification of children that occurred in the West in the twentieth century.[39] Drawing on class-based stereotypes, advocates of commercial surrogacy presupposed that lower-class women had weaker sentimental ties to children, were hardier and could easily endure the rupture of those ties, were happy to give to others what came naturally to them (pregnancy and childbirth), or simply had greater need and so would be willing to withstand the commercialization of feeling.[40] While feminists generally concluded that surrogacy was not the same as slavery or baby selling, many lamented the increase in the commodification of reproduction that was facilitated by reproductive technologies.[41]

Feminists also feared that other bases of stratification would be exacerbated by encouraging new reproductive technologies. Age-based exploitation made the news as teenagers struggled for privacy and reproductive rights and as postmenopausal grandmothers were recruited as gestational surrogates and made the headlines.[42] Religion continued to be a bugbear in some feminists' sides, restricting many women from access to birth-preventing or birth-enabling technologies and increasing pressure on infertile women by condemning anything other than monogamous and child-bearing roles for women. A great deal of enthusiasm, as noted above, was generated for the subversive use of reproductive technologies to enable single and lesbian women to become mothers, but feminists noted that most clinics implicitly or explicitly restricted services to heterosexual, married couples.[43] A less common source of concern was the specieism of reproductive technologies. Nonhuman animals, including primates, were used for testing hormonal drugs, and the hamster test for sperm viability required superovulated mother hamsters to be sacrificed.[44] This was also the period during which feminist scholars began to notice that a ideological wedge was being driven between a pregnant woman and her fetus. Fueled by antiabortion politics and with the help of the technologies of visualization and prenatal screening that had been developed with the new infertility treatments, a hierarchy was being set up between pregnant woman and fetus. The trend was toward considering a fetus to be separate and distinct from the woman on whose body it was dependent. This was interpreted as an assault on the bodily integrity, right to choose, and privacy of women.[45]

The Transition to Phase 2 (about 1992 to 2000)

The break between phase 1 and phase 2 of feminist writings on reproductive technologies and infertility was not a clean one, and many major participants of the first phase continued to be among the most significant theorists of the second. Toward the end of phase 1, the feminist literature on infertility and reproductive technologies began to reflect and build on changes that had been occurring in feminist theory. Ways to move beyond the "just say no" politics of the early work and address the "paradox of infertility" for feminism were developed that didn't require radical feminists to drop their interests in systematic matters of stratification.[46] Feminist theory in the 1970s and 1980s tended to be concerned with large-scale structural functionalist explanations for gender stratification.[47] A few highly influential feminist texts came out that offered explanations that were based on the psychological internalization of learned gender-differentiated roles (which is discussed in chapter 4).[48] With a number of theorists explicitly building on these works, a whole genre of feminist writing valorized womanhood itself and often equated it with motherhood or expressed it using maternalist metaphors.[49] Womanhood—essentialized around caring, connected, authentic, antiviolent stereotypes of motherhood—was proposed as the surest spiritual, moral, and epistemological basis for feminist change in the world.[50] This writing that valorized women's experience, which came out of structuralist understandings of gender, paradoxically gave analysts of infertility medicine the means to argue for the return of agency to infertile women. The argument was that radical feminist critics had been mistaken to deny the authenticity of the maternal instinct. Whether it was socially conditioned or innate was irrelevant; infertile women's desire to have children was more important and more substantial than simply a patriarchal mandate to reproduce.[51] This step opened feminist writings on infertility and reproductive technologies to developments in feminist poststructuralism, feminist science and technology studies, and feminist anthropology, which together cemented the transition to phase 2.

Feminist poststructuralism and feminist STS galvanized the thinking around reproductive politics by the late 1980s. Poststructuralists were showing the interdependence of what I have here been calling stratification, structure, agency, and experience. While no one wrote about these concepts as if they were completely separate, many treated them for analytic purposes as being governed by different causal logics. By showing that bases for stratifying people (such as gender, race, class, and sexuality) are intertwined with (instead of being caused by or causing) the culture, identity, and experience of group members, poststructuralists liberated feminist thinking about the problem of infertility.[52] If poststructuralism blended material and ideal, structure and agency, in ways that were stimulating for understanding the experience of infertility, feminist STS made the technologies come alive by insisting that analysts pay attention to their technical and material specificities.[53]

Feminist anthropologists Marilyn Strathern, Sarah Franklin, and other of Strathern's colleagues and students simultaneously transformed both feminist scholarship on the new reproductive technologies and the anthropological study of kinship.[54] The very word *kinship* had come to sound primitive, and a biological understanding of descent had become thoroughly naturalized in mainstream thinking.[55] In infertility treatments, the dissociation of childbirth from biological motherhood and the polygyny involved in gestational surrogacy and egg donation sounded like exotic forms of kinship indeed, and yet they were coming out of the cutting edge of Western scientific and medical practice.[56] The new reproductive technologies and the court cases prompted by custody battles over resultant children made biological motherhood uncertain just as DNA testing was beginning to make the notoriously uncertain facts of paternity more certain.[57] Perhaps of even greater significance to the broader shape of feminist theory, these anthropologists embarked on an exploration of how biological narratives of the facts of life become fundamental, how and when they remained stably thus, and under what circumstances they can be denaturalized. They were joined in their endeavor—to revitalize kinship studies and to interrogate the multiple meanings of blood and biology in designating kin—by an increasingly sophisticated anthropology of families that did not conform to the linear cognatic model. They thematized gay parenting, families formed by public, private, and transnational adoption, ethnically based and state-sponsored initiatives to remove children from their biological parents, and circumstances under which shared blood did and did not confer kinship.[58]

The early 1990s saw a shift away from the moral certainty that marked phase 1 and toward a tone of moral ambivalence that became the hallmark of phase 2. This change in tone in part reflected the sheer amount of public debate that new developments were receiving and the activist "biosociality" that blurred boundaries between groups that had been considered in phase 1 to have distinct interests, such as doctors and patients.[59] Historians of infertility in the United States and Great Britain also did much to disrupt the tone and content of the phase 1 paradigm, showing the continuity of old and new reproductive technologies,[60] the existence among the poor as well as the rich of involuntary childlessness,[61] and the surprising degree of agency that has been exercised historically by infertile women patients.[62] Rayna Rapp's and Barbara Katz Rothman's work spearheaded the move away from easy condemnations and toward multiplicities of women's experiences of reproductive technologies.[63] Like Adele Clarke's recent work, Rapp's work also cautions against simplistic right and wrong readings of physicians and scientists.[64]

The shift to phase 2 sensibilities also reflected other changes in feminist scholarship on infertility. Among other things, anthropologists and interdisciplinary feminist scholars (coming from cultural studies, science and technology studies, and women's studies) were beginning to pay detailed attention to the lived worlds of infertility and reproductive medicine.[65] It turned out, for example, that doctors were and always had been a mixed bunch: at least some were interested at least some of the time in the same goals as feminists. Indeed, several were feminists, and several had endured personal struggles with infertility. The motives of both feminists and doctors—insofar as intentionality and internal states can be inferred or imputed to anyone—and the distribution of power appeared more diffuse, more mundane, more up for grabs, less conspiratorial, and more contingent than they had been portrayed as being in phase 1.

Science, Medicine, Technology, and Infertility in Phase 2

Many of the issues raised in phase 1 continued to concern writers in phase 2, but the tone had changed. A new generation of feminist scholars documented the pathologization and technologization of reproduction, but as much to examine the active role of technology in determining the semiotics of reproduction as to denounce the adverse effects of technology for women.[66] Meanwhile, in the United States, for example, physicians and regulatory bodies including the National Institutes of Health (NIH), the American Bar Association (ABA), the Society for Assisted Reproductive Technology (SART), and the American Society for Reproductive Medicine (ASRM), all collaborated on lab standards, the implementation of the Fertility Clinic Success Rate and Certification Act of 1992, and the production of the major ethics reports.[67] ASRM also supported RESOLVE in its arguments to the insurance companies and to legislators that infertility should be covered as they would cover a disease or illness that needed treatment and not be considered an elective condition like cosmetic surgery. The major Anglophone journals, *Fertility and Sterility* (ASRM's official journal) in the United States, and *Human Reproduction* (run out of Bourne Hall in the United Kingdom) continued to publish timely opinions on controversial topics, often written by leaders of the field.[68] Likewise, although there are huge discrepancies among clinics, most improved their understanding of the broader meanings of infertility for their patients and began to offer such things as psychological counseling and even debt counseling.[69] From the patients' point of view, success rates were improving from the 5 to 10 percent take-home baby rate per treatment cycle that was the best that could be expected in the mid-1980s to at least double that a decade later. And by the end of the 1990s, serious attention finally began to be paid to the possible connections between superovulation and ovarian cancer and to the health, financial, and emotional burdens that typically result from high-order multiple births.[70]

Phase 2 writings did not exactly exhibit a lightheaded technophilia as regards the medical, scientific, and technical aspects of infertility in the age of reproductive technologies. They didn't even express a liberal or socialist feminist faith in reproductive choice and technological progress. Nonetheless, they granted the technologies a much less monolithic, oppositional, and inhuman role and a much more mediating and active role than their predecessors had. They saw in the development of reproductive technologies the potential to articulate new ways of embodying reproduction, some of which would disrupt conventional families and gender stereotypes, and they refused to read ARTs as simply signing and sealing preexisting oppressive social orders. The phrase "every technology's a reproductive technology" captured the dual sense of the new possibilities that were inherent in technology and the belief that women and members of historically oppressed groups should not reject but entangle themselves with "the most powerful games in town"—science, technology, and medicine.[71] Among some there was a new faith that social change arose from social contingency and that the fragility of the reigning social order was greatest at times of great biological innovation. The new front line was inside, not outside, the laboratory and clinic.

Stratification and Difference: Gender, Race, Class, Sexuality, Nation, Religion

If the front line of the medicalization of infertility was now inside the clinic and the laboratory, what was happening to feminist theorizing about infertility, stratification, and difference? Given the faltering moral certainty that came to dominate in this period, injustice and inequality might be expected to have taken a back seat to the cultural and ontological arguments that were preoccupying theorists. To some extent, this was true. Some work exhibits the ethnographic version of neutrality. Other work emphasizes "horizontal" difference (things like race and sexuality) because they are easier to conceptualize in the cultural poststructuralist terms of phase 2 and disproportionately ignores "vertical" stratification (primarily social and economic class) because it smacks too much of structure and phase 1 moral certainty. But these apolitical and politically correct antidotes to the structural black-and-whiteness of phase 1 were destabilized by the same observations that precipitated them: the interdependence of biological and social systems of classification and the interdependence of experience and stratification. These observations—made ethnographically, archivally, and theoretically—showed that exploring the experience of infertility and reproductive technologies revealed as much about how society is stratified as it does about what it is like to be infertile "from the inside" because the two depend on each other.

Findings about gender in feminist works that were written in phase 2 focused on the mundanity of gendered orderings of the world and the extent to which we are instinctively highly adept at complying with those orderings. Infertility patients display exaggerated stereotypical gender attributes at appropriate times during treatment—perhaps to signal their fitness to become heterosexual nuclear parents and probably also to rescue gender and sexual identities that have been compromised by the lack of fertility.[72] Patients had to act out these roles emotionally, economically, and legally to have access to treatments, which, if successful, allow them to reassert their station in this normative social order.[73] Although much remains to be done, men's roles in infertility and reproductive technologies also finally began to be pried apart from an unexamined assumption of "hegemonic masculinity."[74] These changes in theorists' emphasis from systematic to mundane aspects of gender roles around infertility were echoed in feminist theory more generally. During the 1990s, motherhood (increasingly called "mothering," to make it a activity and not a state of being) was still important in feminist theory, but it received neither the valorizing, essentializing kind of attention nor the denigrating, rejecting kind that it had enjoyed in feminist theory in the 1970s and 1980s. Instead, feminist scholarship was curious about ways that women and men work with and against mothering stereotypes, are liberated and oppressed by them, and change them as they go.[75] Writing on patriarchy and infertility no longer ignored experience but found the expression of the one through the other.[76] More conventional gender-equality battles that arose from developments in reproductive technologies (such as whether postmenopausal women should be allowed to have children at ages at which men still frequently became fathers) received much more coverage in mainstream venues than in feminist theory.

The discussions of both class and race in phase 2 writings about infertility and reproductive technologies focused around prenatal screening and surrogacy. Rayna Rapp's work insisted that class, race, religion, and individual life experiences were essential to the meanings and uses of prenatal screening and disability. Her empirical research left no doubt that race and class determine not just access to and uses of technology, but its meaning too. Wholesale condemnations or promotions of the technologies would thus be misguided; political action needed to be embedded within a greater level of context specificity. Writings about conventional surrogacy continued to raise issues of class in a poignant manner. Helena Ragoné's monograph on the experiences of conventional surrogates confirmed the observations of phase 1 writers about the class dimensions of commercial surrogacy, but it also paid attention to the nonmonetary motivations that surrogates had for participating and the pleasures and frustrations of the class mobility associated with being pregnant for a couple with a higher socioeconomic status.[77]

The signature legal case for phase 2 feminists who were interested in stratification as evidenced in assisted reproductive technologies was a gestational surrogacy case: the 1993 *Johnson v. Calvert* case, which affirmed a lower court ruling that a gestational surrogacy contract was legal and enforceable and held that the woman who intended to bring about a birth and raise the child as her own is the natural mother under California law.[78] Feminist scholars took up the case to illustrate the way in which race is used as a marker of hierarchies of fit motherhood.[79] This work critically interpolated race into the question of whose body can be hired for reproductive labor, and it brought out the connections that were exemplified in the *Johnson* case between racial classification, alienated reproductive labor, perceived phenotypic resemblance, and natural descent.[80] As in feminist theory more generally, then, phase 2 work was increasingly sophisticated on matters of "difference," but to its own detriment it lost some of its critical hold on socioeconomic status.

Feminist theorists in phase 2 continued to be concerned about the growing conflict between mother and fetus in industrialized countries, evidenced by increased social, legal, and medical salience and rights being accorded to fetuses.[81] New and theoretically important attention was being paid to the question of privacy in this period, as general feminist theory merged with feminist scholarship on infertility and the new reproductive technologies. These works plotted the relations between technologies that permit knowledge of infertility and pregnancy and the ways in which that knowledge inevitably changes the aspects of a women's body and experience that are private and protected as such and the aspects that are subject to public legal or medical intervention. Lauren Berlant argued that fetal personhood had become vital to contemporary U.S. ideas of citizenship and that this public fetus had paradoxically privatized and rendered intimate and apolitical the prevalent form of citizenship.[82]

In one way or another, many phase 2 feminists were questioning what is sometimes taken to be fundamental to feminist reproductive politics in the West—namely, the desirability of making reproduction a private matter. Some were noting that the 1980s shift in reproductive responsibility toward women was not simply a result of a feminist victory in establishing the principle of a woman's right to control her own body.[83] They argued that privatized and medicalized reproductive services were still tied into broader state policies of selective pro- and antinatalism. For example, those in the middle classes are encouraged or at least not discouraged in their efforts to have children, whereas those without work have the opposite fate.[84] Other scholars documented the fact that state concerns about infertility and the medicalization of childlessness were at the heart of twentieth-century politics, nationally and internationally.[85] Non-Western feminists writing during phase 2 also emphasized "the relation between the order of the family and the order of the state."[86] It could even be argued that state concerns about infertility flow from the idea of population, which is the basis of political power in the modern nation state.[87] Phase 2 feminists began to show that modern nation states still take an interest in reproduction, despite their framing of infertility in the quasi-private realms of medicine, reproductive choice, and the market.

In some ways, then, the culminating achievement of phase 2 feminist scholarship on infertility and reproductive technologies was to bring work on infertility into a framework of the transnational politics of reproduction.[88] The extent to which reproductive labor, despite occurring within the private realm of the family and home, is nonetheless part of the broader systems of meaning and stratification that make up a given society had become increasingly apparent.[89] A number of studies specifically dealing with infertility in non-Western countries showed that these dynamics also held for infertility. Thus, for example, in patriarchal societies, infertility deprives women of their access to the public goods of economic protection and status because they are unable to produce children.[90] Others showed the ways in which state, religious, and ethnonationalist aspirations and anxieties play themselves out on the bodies of the infertile.[91] The advent of so-called procreative tourism brought to light the need for transnational analysis of infertility within Western liberal democracies.[92] In 1995, Faye Ginsburg and Rayna Rapp edited a volume that was a pioneering integration of essays on transnational reproductive politics based around the notion of "stratified reproduction."[93] This work—together with compelling new scholarship on the historic and continuing connections between U.S. domestic stratification, racism, reproduction, and kinship—opened the way to linking concerns about reproductive technologies to political culture and theory.[94]

The framing of infertility in a transnational politics of reproduction that was achieved by the end of phase 2 bought together recalcitrant feminist worries about stratification, the experience of infertility, and social theory in a global context. This marks a great improvement in our understanding of infertility and reproductive technologies. It also lessens the antipathy between feminists and advocates for the infertile because it sees the experiences of infertility and stratification that are produced by the politics of reproduction as being inextricably connected and two sides of the same feminist coin. Feminist theorizing about infertility does not just reflect concerns that are being pioneered in other areas of feminist thought and action. The cumulative feminist work on infertility and on the new reproductive technologies has become one of the most fertile places for the production of new feminist theory.

Feminist Studies of the New Reproductive Technologies?

As reproductive technology clinics offer more and more genetic services, the field is masculinizing.[95] Preimplantation diagnosis, stem-cell research, and cloning are all biomedical fields that in their human applications depend on reproductive technologies, the services of reproductive technology clinics, and the bodies of reproductive women. Nonetheless, they are all dominated by that branch of normalizing decision-making discourse known as biomedical ethics. Biomedical ethics performs a series of important translating and sorting functions, enabling medical practitioners to cushion themselves from unwitting atrocity and variously inclined and informed people to line themselves up with particular positions.[96] It remains to be seen whether the understanding that has been developed in the two phases of feminist scholarship can continue to be brought to bear on these new developments without being sidelined. Work on the relations between biotechnology and culture suggests that there is plenty to build from, but the prognosis is still uncertain.[97] This book is intended as a contribution to that goal.

II

Ontological Choreography

3

Techniques of Normalization: (Re)producing the ART Clinic

In this chapter, I introduce the first of the four pieces—normalization, gender, naturalization and socialization, and agency—of ontological choreography that I develop in part II. The approach that I take in this and the next three chapters, the empirical heart of this book, is primarily ethnographic.[1] The ethnographic gaze is limited in that it focuses on one site at one time (the $n = 1$ and $t =$ "the eternal/primordial present of the ethnographic gaze" problem that is mentioned in the introduction). The ethnographic method thus tends to be relatively blind to things that take place over time or that involve reflections of or relations to other systems of meaning and stratification. Elements that are important to the experience of infertility in the United States—such as the life course, religious and political convictions, health insurance, and the drug market—were only available ethnographically to me insofar as they were explicitly invoked or directly influenced action or material within the clinic. Ethnography has much in common with other natural-history pursuits that rely on empirical perceptual information, although it lacks the equivalent of field guides and other taxonomic tools for establishing interobserver reliability. Just as field scientists become knowledgeable about a field, ethnographers see more in their immediate surroundings over a period of time and develop an ability to see patterns of which the ethnographic site is a part.[2] In part III I turn my attention and to some extent my methodology to some key aspects of assisted reproductive technologies in the United States that are central to the story of ARTs.

This chapter discusses how the clinic restricts access to treatment, how women's body parts stand for the couple in treatment, and how the clinic's statistical culture pressures women to agree to open-ended numbers of treatment cycles. This chapter also documents struggles in the presentation of self and couplehood, including repair and maintenance of spoiled or threatened identity, and complicity with epistemic and technical things during treatment. I thus aim to complicate the feminist conversation on access, objectification, and technological imperatives in contemporary reproductive medicine that was begun in chapter 2. I argue that the stated aims of infertility medicine —diagnosis and treatment for the involuntarily childless—are not achieved simply as results of the application of medical knowledge about the human reproductive system. Rather, the diagnostic and treatment options and knowledge about human reproduction are mutually dependent. Various frames of reference and modes of practice enable infertility to be understood and diagnosis and treatment to be performed. There are many strategies of demarcating the objects of study and treatment and several codes of conduct toward them. Many of these strategies can be understood as processes of normalization, including bureaucratic routinization.

Chapter 3 argues that infertility medicine is not a straightforward matter of applying medical knowledge about human reproduction. First, it illuminates the socionatural world of the (re)production of reproduction in infertility medicine as a part of contemporary culture[3] by highlighting the coordination of moral, technical, and scientific discourses and techniques that give the site meaning and effect. Second, the chapter uses the related strategies of normalization and routinization to illustrate the interconnections between social lives, material reality, and expertise that must be wrought to settle the existence and character of the infertility clinics.

My definitions of *normalization* and *routinization* are exemplified in the course of the narrative, but I start with some preliminary comments about the explanatory work that is being done by these notions. Normalization includes the means by which "new data" (new patients, new scientific knowledge, new staff members, new instruments, new administrative constraints) are incorporated into preexisting procedures and objects of the clinic. It also includes the ways in which the grid of what is already there is produced, recognized, reproduced, and changed over time. This use of the concept of normalization incorporates both "normal" and "normative" and is beholden to Foucault, Goffman, and their extension in the science studies and feminist literature, as discussed in chapter 1.[4]

Naturalization, which is explored in greater depth in conjunction with socialization in chapter 5, is an important part of normalization. As briefly discussed for different literatures in chapters 1 and 2, it refers to the rendering of states of affairs and facts in a scientific or biological idiom and to the means by which aspects of the site are rendered unproblematic or self-evident in the sense of seeming "natural." This meaning of naturalization encompasses what I describe in chapter 1 as "bedrock"—ways of going on that do not seem to be driven by underlying causal or rational logics and that apply as much to the practice of science and formal reasoning as they do to tacit forms of socialization. It also encompasses the ways in which scientific, biological, or "natural" idioms normalize and control the physically or socially deviant, pathological, and dangerous. Examining naturalization invites an analysis of the role that is played by specific configurations of bedrock in establishing the moral, epistemic, and technical taken-for-granteds that are essential to the practice of infertility medicine.

The routinization that is examined here is the skilled local knowledge that is exercised by practitioners and patients in conjunction with medical technologies. The acquisition of and differentials between kinds of skills and expertise are important, as is the use of routinization as a means of normalizing. The quasi-technical use to which I am putting these notions deprives them of their full richness but allows me to link these conceptually separable notions. What is normal is often stabilized by what is natural in this site. For example, a mother-and-father family is normative for the clinics because it is assumed to be the natural state of affairs, so clinics do not need to invoke the social convention of marriage in selecting their patient couples; it is sufficient that patients present as stable heterosexual couples. What is normal or normative also helps fix what is naturalized. For example, compliance with treatment protocols is normative, and failure to suppress even those psychological side-effects (like stress) that are seen as integral to the Western experience of infertility is frequently naturalized as a side-effect of taking hormones. Likewise, a routinized skill (such as reading a scan screen to determine numbers of follicles) is often used to determine both what is natural and what is normal and enables the diagnostic separation of the normal from the abnormal. What skills and expertise are routinized depends in turn on what is normal. Values for diagnostic tests, possible explanations for responses to drugs, and so on are routinely recognized and recorded from the range of possible (normal) values. The processes of normalization, naturalization, and routinization that are characteristic of infertility clinics conspire to produce a real world that is replete with moral, social, technical, and intellectual textures.

Access

Among the practices of normalization that are particularly salient to the infertility clinic are the filtering mechanisms that operate from within the infertility unit that restrict who can be infertility patients and how they can gain access to the site. These mechanisms are responsible for populating the socionatural world in question. Feminist writings on reproductive technologies (described in chapter 2) have documented how the politics of sex, race, able-bodiedness, and class and the related politics of being an experimental subject contribute to the development and use of assisted reproductive technologies. Infertility clinics all over the United States, despite their demographic variations, tend to treat white, middle- or upper-income couples, which is a characteristic of so-called boutique medicine. Pharmacies supply infertility drugs that often come from Serono's Mexican factory, where the predominantly female workers are poorly paid, have access to substandard reproductive-health facilities, have few amenities and benefits for themselves and their children, and are targeted for public population-control campaigns. Despite the limitations (addressed above) of ethnography in assessing or even seeing these patterns, many mechanics of exclusion, stratification, and objectification are mundane and easily seen in the fabric of the functioning clinic. The ethnographer can aim to show how "external" political effects get created out of, contested by, and sustained in everyday local practices.

As an ethnographer, I myself had to undergo socionaturalization into the offices, operating rooms, and laboratories of the clinics where I worked. The largely implicit and informal training that I underwent to fit into these settings was itself revealing about the filtering processes of normalization. Getting access to and settling into the first clinic where I worked was stranger and thus more arduous to me than later access. I obtained permission from my home university to conduct human-subjects research, permission from the clinic to observe, and a letter from the hospital's Human Investigation Committee that stated that my research interests were innocuous enough that it would waive a formal committee deliberation. After making my way through layers of secretaries and gatekeepers at the hospital and clinic, I had a preliminary meeting with the director of the clinic, who was enthusiastic about the research.

On the first day of my fieldwork, Dr. T hastily greeted me, and his nurse gave me a white coat. I was informed that my title would be "a visiting scientist," and that is how I was introduced to patients. The white coat and descriptive title normalized my presence. It assimilated an unclassified outsider into the things that are normal in clinics—doctors and scientists in white coats. The coat and title also naturalized my presence during physical examinations and consultations to that of a neutral researcher who was there to observe rather than there "in person," so that my presence did not constitute a breach of privacy.[5] Furthermore, wearing a white coat socionaturalized me up: it conferred on me not just fittingness but some of the high mobility that is associated with the status and expertise of physicians and research scientists in the United States.[6] Before I entered consultation and examination rooms, one of the doctors first obtained verbal permission from the patient for me to be present. Observing surgery usually required prior written permission from the patient. Permission was granted on nearly every occasion. On the one occasion when I was denied entry, the patient said that she did not want to be watched by junior doctors in training. Her parallel was with the experience of a retinue of physicians at a teaching hospital, even though she had been told the nature of my research and my human-subjects description.

Observing surgery presented additional hurdles. Wearing the white coat and being in the company of a physician allowed me to enter the operating room area, where I was directed to the women's changing room to change into scrubs. Little or no explicit instruction was given on what to do there. Operating rooms are governed by strict rules of entry, sterility, and appropriate behavior, and the rituals of belonging in the operating room are far deeper than those related to the rest of the clinic. My first attempt at appropriate comportment was marred by several blunders. On my first day observing surgery, I negotiated the stacks of clean linen, hats, and shoe covers, observed the difference between the clean and dirty entrances to the changing room, and walked out of the changing room toward the laparoscopy operating room. The head scrub nurse tied my mask for me and found Perspex glasses to protect my eyes from the laser surgery. On the second day, I was unaided and covered my mouth with the mask while leaving my nose free. I entered the room before surgery began and was reprimanded by the head nurse. As she decided whether to instigate formal procedures of decontamination, the IVF nurse reminded her that Dr. T (the physician who was operating that day) never covered his nose in the operating room and that I might have copied him. They concluded that if his nose was sufficiently clean then mine probably was too. I was allowed to proceed but cautioned that only surgeons could take liberties with sterility procedures.

During my second extensive ethnographic internship, I committed new sterility-related blunders. The second clinic was more entrepreneurial than the first clinic. This was reflected in the relations between the hospital administration and the clinic, in the physicians' relations to research and professional bodies, and in the patients, flexible payment strategies, and treatment plans that the clinic developed. The second clinic was decidedly less "boutique" than the first and treated a wider group of patients for a broader group of conditions. This ethos was also evident in my role in the clinic. I was frequently assigned treatment-related tasks in both the clinic and the operating room. In addition, the second clinic performed male infertility surgery that I had not observed at the first clinic, and so several of its procedures were new to me. A role that I sometimes had was to examine semen to see if it contained any live, motile sperm. This was critical for vasectomy-reversal patients as well as for patients who congenitally lacked the vas deferens. The presence of live sperm suggested the likelihood of a good outcome for reconstructive surgery and gave the operating team a chance to take a sample during the operation for freezing and possible subsequent fertilization. I would sit at a microscope on a bench inside of the operating room and be handed a slide containing semen. If I saw any live sperm moving in a forward direction, I would call out. On one occasion, the surgeon called me over to the patient's exposed vas deferens to look through the operating microscope at the vasectomy site. Although I was allowed to touch my microscope on the lab bench, I hadn't realized that the operating microscope was kept sterile and touched a part of the operating microscope with my nonsterile surgery jacket. I was chided, and the nurse showed me how to hold my jacket away with one hand while leaning over and looking through the microscope.

Scholars of medicine, science, and technology have debated the necessity and nature of sterility procedures during surgery.[7] In my time observing surgery, there was debate among practitioners too. Surgeons and nurses broadly agreed that things must be kept as clean as possible to avoid infection, but they also felt that total sterility was unachievable and unnecessary and that procedures taken could vary from hospital to hospital. In one OR, female staff had to wear nylons underneath their scrubs to prevent so-called peritoneal fallout, and on one occasion this topic came up while we were in the OR staff waiting room between procedures. One of the nurses said that she thought that it was based on an old-fashioned sexist stereotype of women as dirty and polluting because they have more orifices for things to fall out of than men. Amid a good deal of laughter, someone else questioned the actual likelihood that men were cleaner or shed less from this area of the body than women. On another occasion, a physician compared the concentration of germs under the operating lamp and inside air conditioning ducts unfavorably to the germs around the patient's and practitioners' mouths and noses. The concern taken over the latter but not the former was discussed as irrational, and the doctor who left his nose uncovered was evoked as an example of the contingency of the measures. Asking whether sterility procedures were either good sense about sepsis or ritual enactments of hierarchy among staff and depersonalization of the patient would have been the wrong question in this setting. The practitioners themselves did not cling to a ready-made distinction between good sense about sepsis and ritual functions and occasionally took time to discuss that distinction.

My induction into fieldwork required me to be scientific in the clinic and sterile in the operating room, which neutralized what my presence potentially threatened in each site. A patient in an operating room who is under general anesthesia with an open peritoneal cavity or scrotum has vulnerabilities of a very different kind from someone in a clinic who is awake and describing his or her sex life and life-course desires or lying down with her feet in stirrups. The attributes of "scientific" and "sterile" defined "neutral" for these sites and codified the acceptable ethnographer. What mechanisms decided who was in the clinic as a patient, though? What were the "good," "right," or "natural" attributes for a patient?[8] I found that good patients by and large conformed to normalizing concepts that were common in the wider society and modulated in particular ways specific to the clinic. Norms of heterosexuality, ability to pay, appropriate comportment, and compliance were invoked and enforced in the clinic specific ways. This translation of societal norms into the everyday practice of the clinic in turn reinforced and changed (through applying them in these new contexts) the various expressions of these norms in wider society. These norms both placed restrictions on and contributed to the fashioning of the patient.

I found both de jure and de facto restrictions on access to the clinics focused around normative heterosexuality and civic comportment. Infertility clinics in the United States have variously codified guidelines that determine who may be treated, unlike their European counterparts, which have tended to standardize this and many other issues surrounding ART.[9] Some sperm banks in the United States have been developed with the explicit intent of enabling lesbian and single women to become biological parents, and clinics are increasingly offering IVF to lesbian couples so that both mothers can be biologically connected to the child. But most clinics operate with an implicit, albeit variously interpreted and in some cases residual from earlier practice, heterosexual coupled norm.[10] I was present on several occasions when practitioners discussed patients either informally or at staff meetings if there was any doubt about the appropriateness of a patient.[11] The default frame of discussion was usually the perceived needs of a future possible child, who would depend on the patients' ability to provide a stable family environment. As I describe below, the ability to provide a stable family environment was frequently discussed in terms that had to do with class, via references to patients' assumed wealth, stylish dress, sophisticated demeanor, or professional rank. There was broad agreement that anyone seeking treatment should be heterosexually partnered although not necessarily married. On one occasion, there was great speculation that a woman who was coming to the clinic for treatment was not in a stable partnership or even in any partnership with a man. After discussion, the staff decided that she comported herself in such a way that a "don't ask, don't tell" policy was appropriate. Everyone agreed to proceed as if she were heterosexually coupled and to ask no questions. In any case, she would eventually need either to produce a man for the purposes of obtaining fresh sperm for fertilization or to get sperm from a sperm bank that would have to cooperate with the clinic. After the staff meeting, one technician who was ambivalent about whether the "best interests of the child" had anything to do with the orientation of a child's parents expressed pleasure that the woman in question might successfully be "passing" the entry requirements. This elision of class, normative heterosexuality, and good parent is not surprising to feminists, and the desire of the physicians to attract (high-paying) patients inclined them to accept one of these qualities as the sign for the others.

The requirement of stable heterosexuality was invoked in implicit ways as well. While I was working at a clinic that did only fresh sperm treatments except where male infertility was diagnosed, women who identified themselves as single or lesbian called and requested information about the sperm-donation program. These callers were referred to a commercial sperm bank. The frozen donor sperm in this clinic was positioned within a matrix of diagnosis and treatment that implicitly restricted it to the couple or woman who needed ARTs because of physiological infertility. The frozen sperm was viewed as a stand-in for the component in successfully achieving pregnancy that is assigned to the husband's or male partner's sperm. If the husband's sperm was diagnosed as the reason that the couple couldn't get pregnant and the problem was considered too serious for treatment, the use of donor sperm was suggested. Private sperm banks like the one at this clinic are typically neither licensed nor set up for the sperm to be mobilized outside of patient couple context.

Most patients have done some contact work with the clinic where they will be treated before the first appointment. Their ability to pay is assessed during this initial contact period and provides another related means by which the appropriateness of the patient is judged. At one clinic, prices for typical treatment trajectories were quoted over the phone prior to scheduling for an appointment. Nonanglophone callers were routed through hospital translation services, adding an additional step to successful access. If the prospective patient gave evidence of being able to afford the treatment or of having adequate insurance coverage, an appointment was made for an initial consultation with one of the doctors. Scheduling an appointment then initiated a material trail that told nurses and physicians when to expect the patient, when to retrieve or prepare a file for the patient, and so on. Those not covered for the treatment and not prepared verbally to attest that they could personally cover the costs did not proceed to the scheduling phase. Patients who had already been seen but who had reneged on a payment did not get subsequent appointments.

These mechanisms ensured the reproduction of the typical white, middle- or upper-income patient couple referred to earlier, especially during the early period of growth of ART in the United States when even fewer states than today offered insurance coverage (see table 2.2). Diversity, class, and age vary from clinic to clinic and reflect the ambient demography of the clinic to some extent, as I discovered by working in clinics in different parts of town. The patient population also depends greatly on the level of insurance offered in a given state. Nonetheless, the correlation between race and class maintained an imbalance in the racial distribution of infertility patients, even though Americans and people living in America of all classes and races experience infertility. As well as being disproportionately white and middle-

or higher-income, patients are older than the average primiparous age, even for their socioeconomic class. The high age of patients again reflects an elision between class, ability to pay, and biological need for treatment: older people tend to have more resources, and those more able to afford treatment are also those more likely to postpone childbirth, thereby increasing their chances of infertility. Access to infertility waiting rooms is not this prohibitive in all countries and (as I recount below) has changed somewhat in this country as well.[12]

The norm of "the ability to pay" references those people who are taken by the wider society to be valuable enough to warrant the resources required by the pursuit of infertility treatments. In the early to mid-1990s, many physicians described to me an element of their job satisfaction in terms of the kind of patients they were able to work with, and one reassured me that most people who cannot afford to pay already have several children. Among other things, these comments invoke the stereotype that the population divides into those who have too many children and those who don't have enough, those who should use contraception and those who can use infertility treatments for conception. These kinds of remarks decreased considerably over time, however, and by the turn of the century I no longer heard these kinds of comments. Nonphysician team members, however, continued to express criticisms of the exclusivity of infertility treatment, echoing norms that reflected the ideology of science more than the upward mobility of an elite subspecialty. As one lab technician expressed it:

> [Availability] should be assessed by need only.... I think it's very unfair that only the wealthy can have this sort of procedure.... It's an ongoing concern of mine.... I don't feel the knowledge we have should be given out only to those who can pay for it. I think it should be if we have the knowledge, it should be given to everyone universally.

For some infertility physicians, then, having an elite clientele was an active and valued element of their working culture, whereas for some lab technicians (who typically have a science Ph.D. or M.S. and not an M.D.) it was a moral issue and contravened a responsibility of the possessors of knowledge to surrender that knowledge without prejudice. Legal, market, and patient-led pressures throughout the 1990s contributed somewhat to the decline of this elite ethos and spurred efforts to broaden the patient base. As more and more clinics opened up in the United States, the supply of infertility procedures in most parts of the country came to outstrip demand.

Normative judgments about appropriate patients shifted correspondingly over the period of my fieldwork. Clinics in wealthy catchment areas changed less, but in less uniformly affluent areas the preference for elite patients gave way to some extent to a tolerance of savvy patients. Patients who were well versed and realistic about their chances of success (and so who would help statistics, which in turn attracts patients), as well as patients who were adept at lobbying for and getting insurance coverage were warmly welcomed by the mid- to late 1990s. In states that were covered by new and reasonably generous insurance benefits, I noticed a significant loss of interest in the appropriateness of patients as a concept, whether socioeconomic, relational, or psychological. This normalized the appropriate patient in a new sense by making assisted reproduction more akin to other medical procedures.

By the late 1990s, I heard fewer references to the appropriateness of couples based on their assumed fitness or the high-income homes they would provide future children and more references to the right, need, or right time to reproduce and the corresponding privacy that such a decision merited. There was a subtle if partial shift from children's rights to adults' reproductive rights and a corresponding shift of ethical frame from social interventionism to privacy. Of course, both these framings for reproduction are commonplace and have been in a dialectic relation in U.S. culture and politics of the last quarter century, with foster care and adoption falling largely unchallenged on the social interventionist side; contraception, sexual activity, and child number and spacing usually being perceived as being private matters; and abortion being hotly debated in between. In the field of assisted reproduction, this shift from children's rights to adults' privacy rights has occurred unevenly. Surrogacy and its associated custody disputes are often dealt with in a social interventionist manner, perhaps because of the class dynamics that are involved in commercial surrogacy and the legal adoption requirements for surrogacy procedures in most states. IVF and artificial insemination are treated as matters of reproductive privacy. Pressure from adoption circles toward an acceptance of open gamete and embryo donations is already beginning to be felt, however (see table 7.1), and stockpiled frozen embryos as well as preimplantation diagnosis, cloning, and stem-cell research initiated with embryos left over from infertility clinics put all IVF procedures in limbo between the private and the public.

A paradoxical but important concomitant of these changes around the norms that govern the appropriate patient has been an increase in the use of biological norms. In the early years of assisted reproductive technologies, a self-diagnosis of infertility was sufficient biological qualification. A woman patient's age was not particularly "biological," even though its correlation with rapidly declining success rates had quickly become clear, and age cutoffs were up to the discretion of individual clinics. As implicit and explicit socioeconomic norms became less important, however, a series of biological norms arrived in their stead. The need to attract more patients combined with successive improvements in success rates, and increasingly arcane third-party reproductive options opened up the hope of biological childbirth to many more people. But the push for clinic accountability disinclined clinics to accept too many patients who had low likelihoods of success. Accountability toward patients dictated against treating patients for whom the procedures were unlikely to succeed. And pregnancy statistics that would attract other patients to the clinic would also be considerably worse if large numbers of such patients were treated. Similarly, as cycles and not patient courses of treatment became the new standard of accountability, the more diagnostics that could be done prior to embarking on a cycle of treatment, the better. These related demands of accountability led clinicians to begin to try to diagnose ahead of time the likelihood of success of particular procedures. By the end of the 1990s, several factors had been isolated that were somewhat predictive of the success of in vitro fertilization and turned into tests that could hone a patient's treatment as well as aid in counseling against certain kinds of treatment.

Patients now frequently take a menstrual cycle not just to diagnose the source of the infertility, such as blocked fallopian tubes or a low sperm count (as happened in the 1980s and early 1990s), but also to carry out predictive tests on women's bodies that determine whether they are candidates for treatment and exactly what kind of treatment would be most likely to succeed, if any.[13] These predictive criteria are used in treatment-related decisions about whether a patient should use her own eggs or use donor eggs but don't themselves count as cycles that must be entered in the Society for Assisted Reproductive Technology (SART) registry.[14] These measures are used in conjunction with but become proxies for a woman's age, and their results are strongly predictive of "ovarian response" (the numbers of eggs a woman will produce in a cycle when she has been given drugs to produce several eggs for IVF), pregnancy, miscarriage, and genetic anomaly rates after treatment.[15] Ovarian response dramatically worsens with age, associating these measures in practitioners' and patients' minds with the eclipse of a woman's reproductive potential and with hidden harbingers of menopause. Because of this, women have linked access to clinics and the possibility of using their own as opposed to donor eggs to an assault on their biological age. Many women report that they feel an additional insult to their already compromised gender identity when they receive this indicator of biological age.

As is discussed in chapter 1, a key tenet of the field of science and technology studies is that science and technologies should be understood in ways that attend to materiality, nature, truth, society, and social conventions. While this analytical "symmetry" is the goal, the changes that have occurred around reproductive technologies over the last twenty years have been asymmetrical. Social conventions, biological parameters, the reproductive body, and the ability to pay have always been and continue to be critical to treatment and to our understanding of the field, but the balance among the different elements has changed. Standardization and partial democratization of the procedures have led a primary hierarchy based on socioeconomic criteria to be replaced by a primary hierarchy based on biological criteria. Both hierarchies elide the social and the natural, but as evidenced by the shift in "woman's age" (a key boundary object in both periods) from a social, chronological measure with biological correlates to a measure of biological responses to a series of tests, symmetry can be markedly asymmetrical.

Once a patient has access to the clinic, patient comportment becomes an important dynamic. At all points, the members of a couple undergoing treatment must behave appropriately or risk forfeiting their status as patients. When the members of the patient couple do not stay in control, remain civil, and manifest their stability as a couple, their public persona as an appropriate patient couple breaks down. I observed behavior that compromised civility, ranging from mild transgressions (such as the failure of male partners to show up) to serious ones (where the patients argued with the doctor or argued with and insulted each other). In extreme cases, one partner insulted the other's motivation for or need for treatment by attacking his or her sexuality, genitalia, gender identity, or potential parenting skills. In these circumstances, which occur infrequently (most couples conform to the cooperative, civil participant mode), physicians can be loath to recommend treatment, and I have seen meetings end with the physician suggesting that the couple think about their options and call a secretary about whether they want to proceed. Occasionally, a couple can reconstitute the patient/physician relation and rescue a consultation that is going badly by emphasizing their underlying goodwill toward each other and their enthusiasm for treatment. During one consultation, the wife commented that she had often wondered if the size discrepancy of her husband's testicles lay behind their inability to get pregnant. The husband became embarrassed, and they briefly exchanged openly hostile remarks, but they repaired the breach and redeemed themselves by explicitly explaining their imperfect public facade as a reflection of anxiety about the treatment options and the draining effect of being unsuccessful thus far.

The norm of civility seemed to be doing two kinds of work. One staff member expressed the first as follows:

> There's no point everyone going through all this, not to mention the cost, if there's not going to be a functional home for the baby at the end of it. If they're doing it to save the marriage, that's not the reason to get into this.

Civility here was standing in proxy for the baby-centered heterosexual nuclear family mentioned above that is normative for the society in which the clinics are positioned.

The second reason for the emphasis on civility had to do with the management of patient stress and "psychological factors" and the implicit gendering of these notions. Infertility clinics expect infertility treatments to be stressful, and almost all clinics have in-house or associated psychologists to counsel patients (at an additional cost).[16] Infertility is classed by practitioners as a major life crisis, especially for women. The desire to have children is taken to be "natural," in both the biological and religious senses, and the treatments are acknowledged to be inherently stressful because of the time and financial commitments required and the success rates. The hormones that regulate the female cycle and that are administered in large additional exogenous doses during many forms of infertility treatment are also psychophysical surrogates for stress in contemporary Western culture.[17] Naturalized stress (stress that is thought to be naturally provoked by high female hormone levels) and the inherently stressful nature of infertility treatments meant that stress was overdetermined, gendered, and generally in need of containment and management if it was not to get in the way of practice as usual. In some places, the expression of stress was appropriate and even seemly (such as in consultations with the psychologist or nurse coordinator), and in other places, evincing high levels of stress indicated unsuitability for and a lack of appreciation of treatment (such as in most consultations with physicians). The psychological staff (whether on staff or working privately) in turn significantly normalizes the stress and emotion by framing it as "grief over loss" of expected reproductive potential and everything that that entails in this culture over the life course. While the idea that every American psychological problem is one of "loss" is something of a cliché in a country with such excess, its recognizability makes it a helpful externalization and normalization of hard-to-control emotions.[18] The informational booklets that are produced ubiquitously in ART medicine (by professional organizations, support groups, and drug companies) offer advice about handling the "emotional roller coaster," dealing with friends and family, reacting when others get pregnant, and coping during holiday seasons.[19]

Rendering the Body, Playing the Numbers Game

Appropriate comportment simultaneously guaranteed a patient's or couple's fitness to receive treatment and maintained the flow of expertise and deference from the site. As well as being normalized in terms of access and comportment, patient couples were normalized to fit diagnostic and treatment protocols. Couples had to be itemized in terms of specific physical sites at which practitioners could intervene with medical equipment and procedures. Patient compliance in this regard thus consisted in behaving in accordance with the reduction to (mostly female) body parts that this required, in accepting the epistemic logic, and in choreographing the material rerouting on which the procedures were based. This meant that treatment had a number of paradoxical effects: "couples" became, disproportionately, the female partner; treatment was single-track and yet open-ended so that stopping was almost impossible; and "natural cycles" were ceded to disciplined cycles almost regardless of the diagnosis for the infertility.

Epidemiological statistics estimate that the male partner is solely implicated in 10 to 30 percent of infertility cases worldwide, the woman is solely implicated in 30 to 40 percent of cases, and both partners are implicated in 15 to 30 percent of cases. Male infertility surgery is performed for vasectomy reversal, retrograde ejaculation, congenital absence of the vas deferens, semen collection in spinal-injury victims, varicoceles, and other conditions.[20] Problems with sperm are managed outside the man's body with micromanipulation techniques like ICSI. Regardless of whether the man has been a patient or simply has to ejaculate at the right time, any resultant embryos still need to implant in the woman's uterus, thereby obliging her to become a patient. Women also take most of the drugs and undergo most of the ultrasounds, hysterosalpingograms, surgeries, and other invasive procedures. The success of clinics is wholly dependent on the establishment of in utero pregnancies and the reinstatement of a 'normal' pregnancy trajectory. This makes the female reproductive organs a necessary focus of the treatment. A treatment works if and only if a woman gets pregnant: success is operationalized not in terms of the child that the couple desires or in terms of the particular problem to which the infertility is attributed, whether in the man or woman, but in terms of the normal functioning of the woman as pregnant.

As discussed in chapter 2, the rigors of repeated invasive techniques, hormonal hyperstimulation on women, and the associated culture of perseverance have been much criticized in infertility medicine. Patients feel that if the treatments exist, they have to try everything. Why is it so hard to give up? The lack of alternative operationalizations and an epistemic culture based on statistics were both elements that contributed to the culture of perseverance in clinics. One of the physicians dashed in one day and told me that if I ran, I could catch up with a patient and possibly interview her. It would interest me greatly, he thought, because she had just decided to stop treatment even though she was not pregnant. "Is that so rare?" I asked. "Put it this way: you may not see another case while you are with us." I did run, I did catch her up, and she did agree to be interviewed. And she *was* the only person I saw during my time at that clinic who gave up treatment without a recommendation from the physician, without severe financial pressure, and without being pregnant:

> The way we understand it, if God wants us to have a child, we will have one. There are many opportunities in our church for making a positive difference in children's lives even if you don't have your own. One way or another, God will put children in our lives.

The unusual commodity that this woman appeared to possess was a network for interpreting her infertility outside of the medical one. She credited her Mormon religion. At this point in her treatment, it was as powerful as the medical operationalization.

Clinic psychologists give couples advice on pursuing alternatives, such as life without children and adoption. Nonetheless, the structuring of treatment is open-ended in the sense that another combination of hormones, another cycle of artificial insemination, another go at IVF always can be embarked on, age and finances permitting. Guidelines recommend a number of times that various approaches should be tried, and physicians systematically relay them to patients. The success rates that are associated with each procedure are also relayed. But physicians do not always refuse to treat patients who want to supersede those guidelines, and in practice a rationale for not counting one or another cycle could be found, especially before the establishment of the SART registry.

Because all success projections are based on statistics, cumulative probabilities play a large part in superseding the guidelines. Independent probabilities on a given cycle may be low (which remains true for each given cycle), but if one in three or one in five cycles are successful, physicians and patients interpret those probabilities into a rule of thumb that justifies trying three or five times. Projections based on statistics license doing the same thing again in the face of failure. This is a distinctive epistemic element of treatment cultures. In practice, this epistemic standard is extremely difficult to live with because it disregards mechanism. If the very same thing works only one in five times, then there is no positive answer to the question of the cause and effect of the infertility. This epistemic difficulty is managed on a daily basis so that it doesn't interfere with practice.

When a new cycle is begun, its parameters are fine-tuned. Sometimes this fine-tuning is in response to specific things like poor oocyte maturation or past hormonal responses to the drugs, and sometimes it is done "just to be doing something different," as one physician expressed it. It implies that there was some reason why the last cycle didn't work that can be taken care of this time around. Nonetheless, for the cumulative-probabilities argument to work (after all, that was why another cycle was started), the two cycles must still be relevantly the same. The combination of cumulative probabilities plus fine-tuning means that the treatment plan is infinite. Cycles can always be ruled crucially not the same and so not counted; statistics will always say that this might be the time that the same procedure will work. Accepting the epistemic rationale on which infertility medicine is predicated involves accepting this open-endedness.

Natural Cycles?

Threats posed by ARTs to patient agency and autonomy are common topics of discussion in many fora (see chapter 6). An initiative to increase patient autonomy and reduce costs in infertility medicine by decreasing the extent of the disciplining of the woman's body in in vitro fertilization—"natural cycle" in vitro fertilization—has nonetheless failed to become integrated into most practices, ceding instead to the predominant disciplined superovulatory cycles. A "natural cycle" involves undergoing in vitro fertilization, except that the woman doesn't take the hormonal drugs that are designed to make several eggs mature at once. Instead, her hormone levels and "natural" single dominant follicle development (the typical monthly ovulation) are monitored with blood tests and vaginal ultrasound. The advantages to the patient are that costs are lower per cycle and that the ovaries are not hyperstimulated but continue functioning as usual. When the follicle reaches the optimum size as measured by its appearance on the ultrasound screen and blood hormone levels, human chorionic gonadotropin (hCG) is given to bring on final maturation and to have the patient ready for egg retrieval about thirty-five hours later, just as in "stimulated" cycles. The patient is taken to surgery, and ultrasound aspiration is performed to collect the egg. This is done under local anesthesia or sometimes just with oral pain killers. It can even be performed in the office, reducing costs still further. Thereafter, the egg is cultured with the partner's sperm in the lab, and if fertilization occurs, the embryo is loaded into a catheter and transferred back into the woman's uterus two days later.

At a conference in 1990, one of the British pioneers of natural cycles[21] told me that he was confident that natural-cycle IVF would become the protocol of choice for some diagnoses. Not only did it give women more biological and personal autonomy, he thought, but it also had better success rates and fewer side effects. The director of the program that I was studying concurred with all of this and said that he also assumed that natural cycles would sweep the country. Likewise, patients initially greeted the news with approval.

The British doctor explained that the rationale for superovulating women (apart from the fact that it is done in the rest of animal embryology, such as with mice) was that success rates seemed to rise arithmetically with the number of embryos that were replaced in the woman's uterus. Thus, the argument went, if one embryo was replaced, you had a certain chance of success, and if two were replaced, a greater chance, and so on, up to around four embryos, where the success rates flattened out. He maintained, however, that this was an artifact of the way the data were collected: the data were assembled based on women who had been superovulated. This specialist's claim was that a superovulated woman who produced only one egg that fertilized was adversely affected by the hormones in the drugs, would not be able to sustain a pregnancy, and therefore would not get pregnant. In fact, he said, the low rate of implantation when only one embryo was transferred related not to the advantages of transferring several embryos but to the disadvantages of superovulatory drugs on implantation in women who responded poorly to these drugs. He said that his data, collected on women who had had natural cycles, showed a success rate comparable to the best results achieved for IVF with superovulation.

Despite these claims for the procedure and the doctor's predictions, superovulation is still the protocol of choice, the arithmetic graph is still sometimes drawn to explain the chances of pregnancy to patients about to undergo IVF, and the plateau at three or four embryos is still used to limit numbers of embryos transferred and to determine how many embryos to freeze. Every few years, a new "natural-cycle" protocol is promoted, touting the same advantages of cheaper price, lower intervention, and greater "naturalness." Several clinics now are offering natural cycles where immature eggs are harvested, relying on improving techniques for maturing oocytes outside the body, which has been a technical hurdle until very recently. Immature oocytes are available from the simple excision of ovarian tissue and do not need exogenous hormonal stimulation because they are not matured in vivo. If indeed these natural cycles have so many advantages, why do so few patients undergo natural cycles? The short answer is that physicians rarely push it and treat superovulation as the default protocol. A longer answer includes the experiences of the staff and of patients who undergo natural cycles, the increasing standardization and efficacy of the new generation of recombinant drugs, and the huge financial and infrastructural involvement of the drug companies in every aspect of the field.

In the natural cycles that I observed or read the records for, a significant number of cycles were canceled because the patient spontaneously ovulated, the single egg was lost at retrieval, the egg was not properly mature, or the egg didn't fertilize. The strain of having all one's eggs in one basket (almost literally) negated the feeling that patients had of increased agency. With superovulation, there was some slack; embryos could be frozen for a later cycle, and if some eggs were not mature or did not fertilize, there would still be some that were fine. Relying on one's own body also increased the feeling of failure if something went wrong. For the staff, monitoring and retrieval are more nerve-wracking than usual in natural cycles because they might lose or damage the single egg. As a lab technician told me: "No one likes going through natural cycles; they're too stressful.... they're too hit or miss." It is possible that the sense of being hit or miss arises only by comparison with the superovulatory protocols where things go wrong but not usually at the point where eggs are retrieved and fertilized. Certainly, this comparative failure was always pointed out and seemed to alarm patients and thus possibly unnecessarily (that is, not as a direct result of comparing overall success rates) to discourage them from undergoing natural cycles.

In addition, the recombinant versions of the superovulatory drugs that have been on the market since the late 1990s (see chapter 7) are more reliably available and more reliable in terms of their biological activity, so that stimulation has become more controllable and more successful. This has increased the standardization of the stimulation portion of IVF and makes the relative haphazardness of natural cycles even more apparent. These drugs have also been lucrative for the big drug companies, Serono and Organon, that produce the main variants of the recombinant hormones. Given that these companies underwrite the annual meetings of the professional societies and provide information as well as these vital drugs, the financial and infrastructural embeddedness of these companies (and by extension, their products) is substantial. If every clinic moved from superovulated cycles to one or another form of natural cycles, these companies would lose a tremendous amount of current income, and the field itself would lose a major underwriter. While possible, this would clearly take a substantial reorganization of the epistemology, technique, and funding of the specialty. In addition, for natural cycles to be successfully incorporated, they would have had to generate a "livable-with" dynamic at the level of practice for both patients and practitioners. This would require their thorough integration into the evaluative structure of the clinic, not a simple grafting onto existing protocols. The protocols using immature eggs may have greater success at the experiential level because they fit more closely to the superovulatory model, given that several oocytes are harvested at once. But again, this would not be a trivial undertaking.

Beautiful Philosophical Objects[22]

In this section, I talk about variously instrumentalized and embodied procedures that are philosophically or epistemically normative. They all preserve information about spatial distribution or quantity and so allow the objects of study to be transported and mobilized as results, diagnoses, or treatment indicators, with only the most elementary of mathematical or quasi-mathematical transformations. The procedures are normative because of the epistemic work they do: semen, ovaries, uterus, and hormones relay nothing useful about infertility. Blood, urine, or semen in the office might simply be a mess, not a source of knowledge. Skills and collecting and analyzing devices render quasi-mathematical properties that are not the same as the objects of study but are faithful to the object of study and preserve this fidelity through the necessary transformations. The procedures do epistemic work because they can be transported to and within the office and are subject to transformations: hence, they can yield knowledge.

The aim of ART clinics is to help a woman get pregnant within a certain range of procedures offered. To help a woman get pregnant often involves diagnosis if the reason for seeking treatment is a failure to get pregnant, the patient's body is assimilated to a familiar topography, and then the body's relevant physical or functional deviations from normal are located. This assimilation is achieved through a number of procedures—the ultrasound, the hysterosalpingogram, and postcoital tests—that produce an isolated image that fits into representational parameters of normality. Semen analysis and blood hormone tests involve counting aspects of what is rendered visible by the production of a mathematized image. The image-producing residues are thrown away, not turned into traces. In the pelvic exam and during diagnostic surgery, the female body is turned into an object of study in situ, with the parameters of normality embodied in the physician's skilled looking and manipulation of instruments.[23] Treatment operations are not always discrete from diagnosis (for example, laparoscopies are typically diagnostic and simultaneously occasions for treatment), but the logic of the treatment step is to bypass, override, subvert, or simulate to compensate for the digression from normality.

New patients are asked how long they have been trying to get pregnant, whether they have seen an infertility specialist before, what their age, height, and weight are, how often they have sexual intercourse, what their alcohol consumption and smoking habits are, and what occupations they have. The woman is asked if her periods are regular, and the man is asked if he has any potency problems or visible abnormalities of his genitalia. Their answers are entered on a diagnostic form, which becomes the central inscription around which the treating physician structures a plan of action, arrives at a diagnosis, and initiates treatment. The pelvic exam (described in chapter 6) is also a routine part of diagnosis. In this procedure, the physician is interested in establishing the gross anatomical normality of the woman's body. Nothing is (usually) extracted from the body, no images are derived from traces of the pelvic area, and no grid for counting, aligning, or extrapolating is imposed on any organs to decide whether something is awry. Nonetheless, judgments of normality and abnormality are routinely made, fibroids are distinguished from uterine walls, and non-ovulatory cysts are distinguished from ovulating ovaries. Diagnostically unimportant differences such as the angling of the uterus or the length of the vagina are ignored for diagnostic purposes, even if they are noted for subsequent procedures like embryo transfer. The physician is aided by asking questions such as "Does this hurt?" In deciding whether the results of a pelvic exam are normal, the physician has to decide which physical parameters constitute sameness with some norm and which constitute difference. The physician might hold an idealized topographic representation in his or her head that is derived from an anatomy textbook's depiction of a woman's pelvic area.[24] This would then be a standard against which to make judgments about relevant sameness versus individual difference.

It is more likely that such topographic knowledge is embodied in the physician's skilled "recognizing things as thus and so" rather than as an ideal mental type. First, the graphic properties of a picture on a page do not necessarily coincide with those of a pelvic area. What topographic transformation could possibly yield an accurate prescription for all and only those pelvic areas that conformed to the image? And how would such a transformation from representation to gestural recognition be implemented? Second, an account that posited a mental representation against which judgments about sameness were made would make very impoverished use of the physician's resources. Physicians have built an enormous body of practical knowledge by doing the same thing time and again. They have a gestural repertoire of recognition. The graphic parameters of the textbook diagram provide a recognitional grid—a set of answers to the question "What am I looking for?" in the form of graphic relations and linguistic labels. But these do not function separately from the ways to look for and find those structures. The physician's talk during the exam consists of phrases like "Here's the right ovary; the cyst seems to have gone," suggesting an itinerary through the area. Much as someone in the street is first seen as *that person* rather than a prototypical person with *that person* properties, so the physician seems to recognize the pelvic organs as specific. In making the woman's body into an object of study and treatment, then, not all procedures involve explicitly producing or working from representations with graphic properties. In some procedures, the physician embodies, through his or her skilled gestures and manipulation of instruments, the normalizing work required to render what is seen or felt as something that is, in the very seeing or feeling of it, normal or abnormal. The quasi-mathematical properties of sameness and difference in this respect are merged with the patient's body through the process of skilled looking and feeling.[25]

Women patients usually undergo several ultrasounds in any given treatment cycle. Ultrasound images are rudimentary but diagnostically powerful. The image on the computer screen preserves many spatial characteristics of the phenomena that it is transmitting, so that it reconstitutes a seeing event. In monitoring superovulation, the absence of nonovulatory cysts, and the relative size and number of follicles, it produces the information that is required to proceed in treatment. In hysterosalpingograms, radioactive dye is squirted into the uterus to see if the dye spills freely out of the fallopian tubes. The images indicate whether the tubes are open and therefore could permit an egg to enter and normal fertilization to occur. This procedure involves the production of an image that provides diagnostic evidence. Unlike ultrasound procedures, viewing of the hysterosalpingogram image is delayed until the film is developed. Its particular interest as a diagnostic image is that it captures the results of a dynamic experiment: the dye is what shows up on the plate, so all the cavities that the dye reaches are illuminated. If the dye is constrained in a shape that looks more or less like a textbook diagram of a uterus and tubes, then the tubes are probably occluded; if, however, the dye flares out into the pelvic cavity, apparently flowing freely, then the tubes are determined to be open, and dye evidently passed easily through them.[26] An event that occurred over time is captured in the image and relays information about the functioning of the tubes. In the operating room, a laparoscope is inserted into a woman's peritoneum through two small incisions, which avoids major surgery and enables the surgeon to visualize the peritoneum of women infertility patients. The monitors that display the pelvic region during laparoscopic surgery, called "slave monitors," give a high-resolution real-time image for the surgeons to use to guide their gestures. The laparoscope allows "as-if-unmediated" looking without the hazards that are associated with major surgery. Each one of these procedures operates with different kinds of data, visual conventions, and sometimes tactile conventions and presents and preserves different kinds of information with different kinds of robustness.

The Variety, Gendering, and Culture of Skill

Possessing the routinized skills by which the epistemic norms of seeing, feeling, and knowing (like being able to read ultrasound scans or hysterosalpingograms or to carry out pelvic exams or surgery) are enacted is a large part of what it means to be a practitioner in this site. Infertility medicine is like other arenas of expert technical culture in that marking and differentiating skills and expertise (and the social, hierarchical roles that go with those notions) are an intrinsic part of the culture. I discuss just one example of this pervasive phenomenon. A consideration of who can do the looking, embody the treatment skills, and assume responsibility during laparoscopic surgery illustrates well the divisions of labor that normalize the practice and accountability of infertility medicine. At one hospital, I watched laparoscopies in a dedicated endoscopy operating room over a period of several weeks. At almost all of the procedures I watched, the same nurse was one of the people who worked the laparoscopic instruments. Toward the end of my internship at that clinic, I interviewed her between two operations. In this interview, I was interested in how the nurse would describe what she did relative to what the surgeon did and how she would describe her own skills and knowledge. I asked whether she felt she could do all the things that the surgeon could do. She replied:

> I feel the roles are very distinct. Sometimes I feel I could do some of the stuff—do the assisting and that kind of thing. I wouldn't be able to do the surgery per se. I mean, it looks like I could, but I wouldn't want to take that responsibility

because I don't have the training, and they know exactly what they're looking at. Even though I can say, "Oh yes, that's an ovary, and that's this, and that's that," I'm not real, real sure.

She limits her ability here first in terms of professional roles of the surgeon versus the nurse, which she says are distinct. In this hospital, gender was an almost perfect predictor of role: the nurses were women, and the surgeons were men. Feminists have pointed out that where gender predicts the differentiation of professional roles, the sexual division of labor usually works by underrewarding and underremunerating what women do and by simultaneously justifying the pay differentials in terms of the properties required to undertake the job. Those properties get at least part of their relative value from the strength to which they are naturalized to "women's roles," like caring.[27] After marking the primary distinction between nurses and surgeons, the nurse modifies this role-based account of skill by saying she could maybe do the "assisting" done by a second surgeon in more complicated cases. Not just being a surgeon but taking the primary surgeon's responsibility and having his or her training are issues that she isolates as critical. She makes a distinction between what she probably could do and what the real parameters of skill are. She holds instruments, operates the camera, occasionally makes cuts, uses the coagulator, cauterizes, and lazes just like the surgeon, but only on the surgeon's say so. The skill is not confined to the gestures, then, but includes knowing and deciding which ones to do when and bearing the responsibility that implies.

The nurse distinguishes between *being able to name* things (such as an ovary) and *knowing* (or at least from being "real, real sure") what they are. This is a familiar theory of knowledge in which perception and recognitional capacities are eminently correctable and not guaranteed as certain knowledge. The surgeons know "exactly what they're looking at." She does not offer what she thinks is needed in addition to her recognitional capacities to have certain knowledge—perhaps a combination of experience with like cases, a knowledge of and responsibility for this patient's history, a particular interactive role with the recognized organs during surgery, and so on. But that role is taken to coincide with this crucial epistemic distinction. If role is gendered, then so is access in this setting to certain knowledge.

Despite the restrictions that the nurse places on her expertise, at other points in the interview she describes episodes in which she made active decisions during surgery. This is what has been described as the difference between nominal and real divisions of labor.[28] Her "job description" institutionally and as she apparently understands it does not include surgical decision making. Nonetheless, at times, there is some role overlap between herself and the surgeon. This lends flexibility to the whole system, providing a guard against the breakdown of one part of the system. It also illustrates a paradoxical aspect of technical practice: the staff is required to be flexible to keep intact the strong partitions between the roles that structure infertility medicine. It was clear in this interview and in other occasions of talking to staff that role overlaps and skill and responsibility mobility were occasions in which they took great pride. But role overlap itself might be a mechanism that ultimately reinforces the categories by which role differentiation is maintained.[29]

Statistics as Normative Epistemology

Statistics are a major product of infertility clinics, and they regulate and calibrate the kinds of treatment procedures described above. The whole "epistemic culture" of the unit is based on the production of statistics,[30] meaning that the effectiveness of a procedure and its indications as a therapy are expressed statistically rather than, say, as a matter of experiment or a matter of fact. Fertility clinics are increasingly in competition for patients, and reputations—which are built primarily on successfully initiating pregnancies—draw patients. A patient who calls an infertility unit often asks what the center's success rates are, and clinic-specific success rates can be looked up online and are also published annually.

The implementation in 1995 of the Fertility Clinic Success Rate and Certification Act of 1992, also known as the Wyden law after its sponsor, Representative Ron Wyden (D-Oregon) (this law is described in detail in chapter 7, and its provisions are summarized in table 7.2), imposed a high degree of standardization on clinic reporting obligations. In the three-year period between the act's passage in 1992 and its implementation, however, members of the American Society for Reproductive Medicine (ASRM) and the Society for Assisted Reproductive Technology (SART) and their clinics voluntarily compiled statistics but without (the illusion of) any transparency and standardization between clinics. During this period, the activities that comprise a clinical statistical epistemic culture were raw and open and the subject of constant debate and suspicion.

The new reporting and certification obligations standardized many of the things that bothered clinicians during the interim voluntary reporting period but introduced their own areas of nontransparency. For example, it is now nearly impossible for potential patients to obtain per-patient statistics because the unit of accountability is now expressed as live births per cycle of treatment and not per patient. As indicated above in the section on tests for biological age, the goal now is to improve per-cycle success rates, which are driven downward when a clinic uses a response to treatment as itself diagnostic of how to proceed in the event of failure. This has led to a rise in the development and implementation of diagnostic procedures that can be done prior to starting a cycle of treatment to determine a patient's likelihood to succeed at all or to succeed under one protocol versus another. A number of these tests measure ovarian function (and genetic tests are increasingly being used to judge egg quality and thus likelihood of miscarriage in the event of a pregnancy), so an unintended effect has been to estimate a woman's biological age by her ovarian response, which causes many patients to feel anger and despair. Because ovarian function is biological and is more intrinsically treatment related than the earlier vagaries of statistics collection, though, these pretreatment diagnostic procedures are unlikely to spark the outcry and support that the Wyden law mobilized. Indeed, popular books and articles have pointed to women's rapidly declining biological fertility and poor success with ART as evidence of the risks of postponing childbearing.

During the interim voluntary reporting period, clinics invested a lot of energy into compiling statistics and generating success rates. Competitors' rates were viewed with interest and sometimes with suspicion, and ways in which reported rates might not be comparable were much discussed.[31] All of these phenomena provided further motivation to standardize and make public clinic success rates. A lab technician who was responsible for the compilation of statistics at one clinic described the open-endedness and anxieties that the statistics generated:

> [Centers] are so competitive.... They will say "O.K., well, I'm not going to include this patient because she has this or that. Throw her out." Or do a study where they're just looking at a certain group of patients and report the statistics on that. And centers that are very large will divide their patient populations into all different groups: "We're doing a research study on patients with male factor, so we've taken all those male-factor patients out. So when we report our statistics, we only report this group over here." ... The problem is that there is so much dishonesty in reporting things so that now there are actually going to be programs set up just for that fact.

Statistics are generated, recorded, and manipulated in several places, and a fairly large proportion of the lab technicians' time is spent on them. In the lab, records of all procedures are kept in binders and are organized by date, diagnosis, procedure, and age and identity of the patient. These are hand written and form the evidential base for statistical claims and for regulatory checks by outsiders. Prior to national standardization, statistical policies were developed and modified as part of the self-definition and modification of the clinic as a whole. On these occasions, what to count as a homogeneous group of patients, which cycles to count in the analysis, and whether to treat patients who are likely to adversely affect overall statistical results were discussed. Weekly team meetings with the director were occasions when the parameters of statistical analysis became negotiable. Pregnancies were listed, and failures were discussed to try to identify the reason that the procedure did not work. Sometimes a possible reason for failure suggested to those present that the unsuccessful cycle should not be included in the statistics.[32]

In this period, clinics exercised their own judgment about which treatment cycles to report and which could be left out. Although this is no longer the case, at the time it was considered unproblematic to leave out of the statistics certain treatment cycles, such as those in which the patient dropped out of treatment before finishing a cycle for nontreatment-related reasons. But other cases raised general points about the clinic's policy. For example, at one meeting, a patient over forty with multiple sclerosis had not gotten pregnant. The staff felt that she should not be counted because most clinics would not treat a forty-year-old patient with MS, which made comparisons between clinics unfair. They discussed whether they ought to restrict treatment to patients under a certain age and with specific diagnoses that are associated with good success. This approach was rejected as being out of line with their clinical (as opposed to research) orientation. An element of the discussion was whether it was ethical to the patient to treat them when success rates for that age group were low. A consensus was reached that any patient had a right to treatment as long as the success rates were realistically portrayed at the outset. Nonetheless, it was decided to separate patients of forty and over in compiling the statistics so that their decision of who to treat would not be adversely affected by their desire to maximize success rates. This decision reconciled the two things that defined the pragmatic goals of this clinic—to treat as many patients as possible and to present as high success rates as possible. The predominant ethic was clinical priority based on patient choice.

Above I discussed some aspects of living with a statistical measure of success from the point of view of the gap between statistics and mechanism when applied to an individual patient. A more radical problem arises for a clinic when the statistical rates themselves seem not to be holding up for a given procedure. In this case, a double epistemological crisis is experienced: the practitioners no longer have the justificatory standard for initiating treatment and continuing in the face of failure. In addition, because statistical justificatory standards say only that *x* works in *y* percent of the cases and not why *x* works when it works, there is no way of knowing what has gone wrong and so what to correct. This double crisis is further compounded by the nature of statistical reasoning itself: practitioners are aware that the statistics alone predict that dry runs should occur, but in the short term they have no recourse in the theory of statistics for knowing when local lapses are just statistical dry runs and when they indicate that a procedure has in fact been relevantly changed in a detrimental way. There are statistical methods for saying how likely a dry run of a given magnitude and duration might be, given sufficient cases, but most private infertility clinics are small and cannot quickly amass that number of cases. They must respond to local failures to maintain success rates before they have the relevant data to make this judgment.

While I was working at one clinic, such a crisis was being experienced for the in vitro fertilization results. The clinic had reported its best-ever success rates for IVF in the previous year; the rates were well above the national average. In the first five months of the current year, however, their success rates for IVF had dropped to half the number for the comparable period of the previous year (their results for other treatments were as good or better than the previous year's results). The numbers of patients who had to be told that they were not pregnant and the sense of backsliding were taking a toll on the staff, and by the time I arrived, much effort was being directed into deciding what was going wrong. This is how one lab technician described the crisis that this represented:

> Our IVF program: we had a wonderful pregnancy rate last year. This year our pregnancy rate has dropped drastically. And then the question was, "What is different? What has changed?"... Calling around other centers, talking to them, everybody goes through what they call the "dry spell." We had never hit one before, and because we are a small program, it hit us much harder. Other

centers will do fifty cases in a month, and they can tell. If they do a series of twenty patients and they get no pregnancies, they go, "O.K., what's going on here?" Well, it takes us six months to do twenty [IVF] patients, so we have a very long period of time. And then when we change, we kept trying to change things to see if that was it. Well, then, you would have to wait months and months to see if that change had an effect.

Despite this epistemological uncertainty, local action was taken. These are some of the things that were done to identify the cause of the drop in success rates and some of the explanations that were considered. The lab technicians looked in the lab:

We went through everything. We tried looking at every single chemical we used, every single plastic ware, anything the embryos come in contact with, looking at the gas we used in the incubator, looking at temperatures of everything, checking to see if there could be anything that is different. A few things I found is that one of the tubes that we use, the company changed the way they labeled them. So I contacted the company and discussed with them the changes, got them to give me tubes that were unlabeled and tested those to see if—. They said everything else was exactly the same, except the very last step where they put the label on the tubes was a different procedure. So I had them send me the tubes prior to them labeling them, to test it to see if that was the problem. That didn't seem to be the problem. I started calling other centers and asking them, you know, "What could be the problem?"

One of the physicians found something awry in the operating room while she was doing an egg retrieval for IVF:

One of the things we did find was that there was a problem down in the OR where the heating blocks that we use—for some reason, the temperature had gotten changed on us. Something had happened to it. And so now we monitor that on a very close basis, checking the temperature the entire procedure to make sure that it doesn't go too hot. And our pregnancy rate has gotten better since we found that.

The nurse coordinator, who keeps track of the treatment protocols and individual characteristics of each patient, thought it might have been partly due to differences in the individuals who had gone through IVF—a slew of older patients for whom you would expect worse results. She wasn't convinced, though:

I could go through them and say, "Oh, this patient had this problem, this one had this one, and this one had this one," but there were a few patients in there that were good patients that should have gotten pregnant and didn't get pregnant. So we felt that maybe it appeared to be more than just coincidence.

After I had been at the clinic a couple of weeks, the June and July figures were processed and were much better than the ones for earlier in the year. The director was ready for closure and settled on the theory that it was the malfunctioning heating blocks that had been the problem. He informed me both that the June figures had "saved this half year's statistics"[33] and that the few pregnancies that had occurred before the heating block was discovered "for one reason or another did not involve putting the test tubes in the heater." Both the physicians seemed sure that this was at the root of the problem, expressed anger about it, and appeared confident that with the new checks in place things had been taken care of. The lab technicians appeared to be somewhat more ambivalent about this single-cause theory, and one of them said that she would not regain confidence until the numbers themselves bore out the theory that the problem was resolved:

> We're hoping that was the problem. I'm still not 100 percent convinced that was the only problem. I'm hoping that was it and that we've corrected it, but I can't guarantee at this point, we haven't done enough cases for me to regain confidence.

Despite the fact that the success of procedures generates the statistics, the statistics justify the carrying out of infertility medicine—the choice of a particular treatment for a particular patient. Without the statistical justificatory system working, all the other repertoires of normalization and making things relevantly the same cannot be calibrated and validated.

Privacy

A central organizational problem for infertility medicine is dealing with the sexuality of the object of study (the patient's body and specifically the reproductive organs) when timing and locale require that procedures of diagnosis and treatment be done clinically and relatively publicly. If this boundary between the private and the public is not negotiated well, the routinization of these procedures as part of infertility medicine breaks down. This arises for all medical practice, and the rituals for establishing this boundary in the examination room have been described as an "etiquette of touching."[34]

Circumstances for blurring the distinctions between public and private are so pervasive in infertility clinics that there is a well-worked-out choreography of privacy. Some things cannot be rendered innocuous by imposing a clinical, scientific semiotics on gestures that might otherwise be read as private or sexual. For these, spaces that are usually considered public can be transformed into private spaces. We are familiar with the casual granting of privacy where it is not strictly demarcated—store dressing rooms, public lavatories, library space—but in this clinical setting, etiquette covers spaces that are only sometimes private, and the behaviors for marking the transformation are well developed and strictly adhered to.

The multiple lives of the examination room at one of the clinics make a good example. For a regular pelvic exam, the examination room door is closed while the patient gets undressed, making the room a temporarily private place. The privacy is marked by the closed door and a light that is turned on to indicate that the room is in use. The privacy is broken after a lapse of time by knocking. No answer is required for the knock; it simply registers the physician's imminent entrance. This requirement of privacy stands in contrast to a requirement of accountability that must be met after the physician knocks. The physician announces her- or himself with something like "Can I come in?" as he or she opens the door to the examination room. When I was observing, the patient would be asked at this point if she minded having me in the room. Before the pelvic exam can begin, the physician must have a female, professional witness. At some clinics I was considered an adequate witness, but at others I was not because my impartiality was not institutionally accountable. Typically, a nurse simultaneously helps the physician and witnesses the propriety of the exam.

For artificial inseminations and in vitro fertilization, the male partner's semen must be produced. Collection is not always done on the premises, but the exigencies of timing often necessitate on-site collection. An examination room is used for masturbation by male patients, and the protocol that is followed is similar to that used for women patients who are undressing, except that there are mechanisms for enhancing the spatial and experiential privacy of the room. At one typical clinic, the examination room farthest away from the nurses' station was assigned for male patients to collect. It was out of the line of sight of any offices.[35] A magazine rack held magazines for female patients. Pornographic magazines were in the room for semen collection, but they were in a drawer rather than out on open shelves. For collection, the "in-use" light is put on, and no one can break the seal of privacy except the patient himself, which he does by exiting. He is instructed to lock the door from the inside to avoid error.

When the collection is done, the man exits and hands the sterile and semitransparent container with the semen to one of the lab technicians. The technicians fill out forms or use the microscope or heating equipment in a room across the hall. When they hear the door open, they move to the door of the room to make sure that the man can hand his container straight to somebody who will deal with it technically. The transmission from private and sexual to an appropriate clinical object is thus smoothly ensured. Sometimes the man makes a wry comment about the process, and sometimes smiles are exchanged. Patients sometimes joke and commiserate with each other about the process once the man returns to the waiting room (one such incident is described in chapter 4).

Timing

In an infertility clinic, time is organized around the working day and progresses linearly, like most other bureaucratic office places. Calendars and other scheduling devices are calibrated to mark incremental passages of time. For the patients, visits to the clinic must be fitted into working days and coordinated with their own working schedules. Treatment, however, has to fit into the very different time scale that is demarcated by the woman's menstrual cycle. To be treatable, a woman must be cycling normally, but she has to cycle measurably and on time; the time scales must be reconciled. For women patients, the conflicts between these time scales are salient aspects of their experience of infertility medicine. Each new cycle takes patients back to zero in their quest to get pregnant, despite the cumulative treatment process; it also resets people's hopes, which frequently are dashed again in a month. Treatment is successful when the cycling is subverted and the progressive, developmental time scale of a pregnancy is embarked on.

Further complicating the experience of time for many women patients is the so-called biological clock. As women grow older, the statistical likelihood that they will become pregnant goes down. Thus, each cycle moves patients no further along the treatment trajectory, but the start of each cycle means a diminished chance for the subsequent cycle. This biological clock pressure is felt mainly by women who are told that they are at the outside of the normal chronological age range for the procedures to be successful (which is anytime from the late thirties on, after which point the statistical drop-off rate becomes significant) and by women who have had a negative or ambiguous ovarian response test result.

The treatment cycle in infertility medicine is also at odds with the cumulative structure of payment. You pay for all services received, even though many are performed only once per payment. In vitro fertilization, donor insemination, insemination with husband's sperm, and fertility drugs all work in the cycle that they are given in or not at all. There are no holdover effects.[36] This puts pressure on staff and patients alike and is part of the rationale in infertility medicine for performing surgery, which lasts, if successful, for a much longer time despite its poor success rates in establishing pregnancy.

These various time scales—bureaucratic, cyclical, and biological—pose several logistical challenges. Appointments have to be made in working hours, but they also have to be made on the right day of the cycle. This external disciplining of the couple is achieved with the help of some simple devices such as the hand-held device that makes the translation from the day of a patient's cycle to the appropriate date on the linear, shared schedule. Calibrating is done from the reliable visible sign of the first day of a period, which is called day 1, even though it is the shedding phase from the last cycle. Where more precision timing is necessary, standardized protocols are used. Thus, if the eggs are retrieved in an IVF case at a certain hour, the semen sample must arrive two hours later to be spun down and added to the culture dish. In superovulated cycles, hCG is given toward the end to bring the final maturing process into line with the working hours and to allow the surgeon to schedule insemination or egg retrieval an exact number of hours later. This frequently means that patients have to have injections in the middle of the night and that men have to masturbate on demand at bizarre times of the early morning. It also means much communication between the operating room, laboratory, office, and patients. Heavy use is made of the phones and electronic links to achieve this. Finally, the cycle is not completely domesticated; midcycle often falls on a weekend. At most clinics, staff members assume that they can be on call on any day of any weekend.

The Aesthetics of the Lab

I turn finally to what I came to think of as the aesthetics of the clinic laboratory for processing gametes and embryos.[37] Standards for dealing with the human embryos that were being grown and maneuvered in the lab marked out work with embryos as different from any other work done in the lab, including that done with gametes. The standards also seemed to enable the lab technicians to get on with their work with the embryos in a matter-of-fact way, without having to take time to philosophize about the human potentiality of the embryos and without being overwhelmed by the responsibility of the preciousness of this potential to the couple from whose gametes the embryo derived.

Egg retrieval for IVF is frequently carried out under low levels of lighting. This is supposed to protect the eggs from potentially harmful exposure to light. All work done in the lab when embryos or potential embryos are out is done is semidarkness. There is apparently no conclusive evidence that light harms the eggs or embryos, but the care is felt to be appropriate and to "make sense," as a lab technician expressed it. The care that is taken to reproduce what might be thought of as the maternal environment contrasts starkly with the spinning, washing, and freezing of sperm and the freezing of surplus embryos. Sperm is very robust and is well known to be capable of fertilizing eggs after these indignities. Freezing embryos is trickier, but freezing puts the life potential on hold so that practitioners are not responsible for preserving the chain of potentiality while embryos are frozen. As soon as an embryo is unfrozen, however, the same care is once again employed.

At one clinic where the lab was not adjacent to the operating room, the eggs recovered from a woman in the operating room during a cycle of IVF were transported to the lab in an isolette that was appropriated from the intensive-care baby unit (figure 3.2). The isolette was originally used to solve the problem of transporting the eggs at the right temperature and with minimum disturbance across the distance between the operating room and the lab. The chief practical assets of the isolette were that is was mobile and was an internal workspace where the eggs could be counted, pipetted, mixed with semen, and transported back to the office for embryo transfer. The serendipitous use of a newborn baby isolette added to the aura of extreme maternal solicitude toward the embryos at this clinic, and lab technicians referred to their job of tending to and examining embryos as "feeding and diapering the babies." Sarah Franklin has pointed out that the provision of a maternal environment in the lab for nurturing the embryos is the opposite of the entry of science (instruments, techniques, and knowledges) into the woman's body in the earlier parts of the treatment cycle.[38] This crossover of the properties of techniques between the woman's body (and woman as mother) to the lab creates the matrix in which reproduction can be conceived of and produced medicotechnically.

During egg retrieval, the eggs are sorted by level of maturity as soon as they are retrieved. During the subsequent fertilization and development of embryos in the lab, developmental details are added to the patient's record sheet, such as whether the egg fertilized, whether it fertilized normally and started to grow normally (specifically, whether more than one sperm fertilized the egg and whether cell division was progressing normally), and what stage of development the resulting embryo has reached at given times of the next forty-eight hours before transfer. The evenness of the cells, the similarity of cell size, and the presence of "blebs" (irregularities) are also recorded. The embryos are given a grade to reflect how "pretty" they are. Round, evenly developing embryos are good, and uneven, "misshapen" embryos with so-called blebs are not good. The "not good" embryos were routinely referred to collectively as "crud" at one clinic. Segregating the embryos in this way turns out to have almost as much to do with managing the disposal or freezing of unused embryos as with the likelihood that a given embryo will lead to a successful pregnancy if transferred back to the woman's uterus. When I asked about this means of distinguishing embryos, I was told that IVF doctors are divided about whether "good" embryos lead to more pregnancies than "bad" embryos if transferred immediately and that most embryos whether good or bad will not implant and that many bad ones do. Good embryos apparently survive the thawing process better than bad embryos, however, so only excess good ones are frozen.

Very often more embryos were produced than could safely be transferred in a single cycle given the risk of multiple pregnancy. Some embryos were considered not worth the price of being frozen and were neither transferred nor frozen. A criterion was used to justify treating some embryos with the esteem that was due to their life potential and value to the patients while disposing of the ones for which there was no treatment use, without contravening any moral or legal restrictions. Only "bad" embryos were disposed of while I was at the unit, and the idea of disposing of a "good" embryo was greeted with moral opprobrium. Both the provision of a caring environment for embryo growth and the morphological criteria for saying "good" (life sustaining) enabled the lab technicians to deal clinically with potential life and to accommodate wastage.

Hybrid Culturing

The nurturing ways in which embryos are cultured in the lab and their legally and ethically coded preciousness while they are on a trajectory with life potential can be juxtaposed with the ways that "waste" embryos are treated and with the technical procedures of sperm preparation or embryo freezing. This mixture of reverence and technicality that tracks the dynamic boundaries of life (what is waste, what is on hold, and what is alive) is a dynamic that is replicated in other aspects of infertility treatment. For example, a reverence toward women as child bearers is witnessed in the ethos of the clinics, but this is juxtaposed against the rigorous procedures that they must endure to become mothers. The hybridity of the version of contemporary parenting that is offered by these clinics results from the enactments of these juxtapositions in which the technical, social, and moral are constantly conjoined and then separated and then conjoined again. The normalizing processes described in this chapter show how reproduction is reproduced in this site in a manner that grows from versions of life, parenthood, and fertility that are already present in the wider culture. The reproduction of reproduction is not achieved through the imposition of an independently determined set of techniques on preexisting social categories. The social categories and techniques develop together and thereby change what life, parenting, and fertility mean in cultures with infertility clinics. The workings of expert sociotechnical hybrid sites as revealed by ethnographic data promise to reveal culture in the mixing and the making.

4

Is Man to Father as Woman Is to Mother? Masculinity, Gender Performativity, and Social (Dis)Order

In the most famous passage of her classic book *Gender Trouble,* originally published in 1990 and reissued a decade later, Judith Butler made the following argument that placed feminist poststructuralism and feminist science and technology studies in intimate conversation:

> And what is "sex" anyway? Is it natural, anatomical, chromosomal, or hormonal, and how is a feminist critic to assess the scientific discourses which purport to establish such "facts" for us? . . . If the immutable character of sex is contested, perhaps this construct called "sex" is as culturally constructed as gender; indeed, perhaps it was always already gender, with the consequence that the distinction between sex and gender turns out to be no distinction at all.[1]

Feminist biologist and STS scholar Anne Fausto-Sterling has brought this radical constructionism and poststructuralism about biological sex and gender to bear on thinking about masculinity as it takes meaning in biomedicine by treating scientific texts as one example of instructions of "how to build a man."[2]

In her statement of what is meant by *gender performativity,* Judith Butler draws on ideas of drag, performance, and parodic relation to the real to explain her alternative to essential, natural, stable gender identities. She argues:

> Gender ought not to be construed as a stable identity or locus of agency from which various acts follow; rather, gender is an identity tenuously constituted in time, instituted in an exterior space through a *stylized repetition of acts.* . . . This formulation moves the conception of gender off the ground of a substantial model of identity to one that requires a conception of gender as a constituted *social temporality.* . . . The possibilities of gender transformation are to be found precisely in the arbitrary relation between such acts, in the possibility of a failure to repeat, a de-formity, or a parodic repetition that exposes the phantasmic effect of an abiding identity as a politically tenuous construction.[3]

This chapter explores gender and particularly masculinity in assisted reproductive technology clinics. Like Butler, I look at instances of "social temporality" and the "stylized repetition of acts" that make up gender performance and reality. With Fausto-Sterling, I look at situations where the biomedical ontology of gender, with biology as foundational to and hence prescriptive of socially scripted gender roles, is underdetermined. I explore the aspects of the performativity of gender that help repair or supplant the ontology of kinship and bodies and choreograph kinship, economics, and status hierarchies. In this chapter, gender is cast as being as performative in these predominantly heteronormative biomedical sites—as it is for the homo-, inter-, and transgender sexualities with which Butler and Fausto-Sterling are primarily concerned. I suggest that the performance of gender and especially masculinity in ART clinics is captured well in the above quote from Butler—"the possibility of a failure to repeat, a de-formity, or a parodic repetition that exposes the phantasmic effect of an abiding identity." Patients going through treatment need to pass, socially and biologically, as the gender to which they have already been assigned, and male patients must do so in a way that supports an elision from sexuality to reproduction to parenthood. They do this by performing masculinity under conditions where the elision from sexuality to reproduction and parenthood is compromised (there is infertility) and sometimes where biological masculinity itself is threatened (there might be a male-factor diagnosis for the infertility). In the case of ARTs, the ontology of gender is also destabilized by the possibility of failure to perform hegemonic gender by those who precisely in this site must try harder than ever to perform and norm gender.

Gender, both biologically and socially understood and enacted, is a fundamental principle of categorization in ARTs. Furthermore, the deficit and stigma of infertility are often experienced substantially as gender deficits and gender stigma, even when expressed in ways that are also part of ethnoracial and other kinds of identity.[4] In other words, ways of understanding gendered life-course and social roles are compromised by infertility. Since social gender roles are fairly prescriptive in fertility clinics and since biological sex is naturalized in ARTs, these sites perhaps should provide a counterexample to the idea of gender as ontologically and experientially performative. Quite to the contrary, however, in the deadly earnest world of ARTs, parodic performances of masculinity—an exaggerated calling on of highly scripted kinship roles and stereotypes of biological and paternal masculinity—are often used by men and women patients as a way of repairing spoiled biomedical and social sex and gender identity and kinship. These stereotypical gender performances or callings into being (gender scripts often are attributed to husbands and partners rather than acted out by them) are also used as ways to earn access to treatment in their own and practitioners' imagined eyes. "Passing"—in terms of having an intact or reparable gender identity—can be important in specific ways and at specific times in treatment. For example, following a spinal injury, evidence of underlying fertility and virility such as a high sperm count despite impotence is used to repair stigma and can also make it permissible to seek to reproduce despite damaged potential to reproduce. Parodic performances of hypergender-appropriate behavior are also sometimes used to script and navigate treatment itself. Acting in a way that induces a partner or others to comment on fitness to be a parent, for example, displays a worthiness to receive treatment and can (as is described in chapter 3) sometimes make the difference between getting an appointment and not.

It is close to dogma for those of us who study gender from within the rich field of feminist and queer theories of gender that structuralist and poststructuralist accounts of gender are incompatible or at least grossly in tension with one another. Here, in trying to show the appropriateness of poststructuralist accounts of gender performativity to the clinic site, I show how structuralist understandings of gender are used to make sense of the scripts out of which deformed masculinity is repaired. Gender identities that are disruptive of heteronormativity show the constructedness of gender, but identities that seek to achieve authentic heteronormativity by parody, enforcement, or substitution of one kind or another show the constructedness of gender just as strongly. In other words, hegemonic gender roles and norms are not structural in the sense of ontologically fixed, although they do show a lot about biological and social gender structuration, stratification, and differentiation at a given time and place, as structuralism promised they would. If the ontological commitments of structuralism are loosened and the choreography of ontology in the site are more carefully pieced together, structuralism and poststructuralism have a lot to offer one another. The supposed determinism of structuralism and voluntarism of poststructuralism turn out to be as phastasmic as predetermined or entirely voluntary enactments of gender are themselves.

Masculinity in ARTs and ART Scholarship

To assess gender performativity and passing in preassigned gender categories and their relations to structuring scripts of paternity and masculinity, I explore the norms for, performance of, and representations of masculinity in these sites. In the context of reproductive technologies in Western democracies, masculinity has been something of a remainder category, partly because this area of scholarship was originally dominated by feminist and woman-centered work, which in turn has played a substantial role in the unfolding culture of these technologies.[5] The scholarship on reproductive technologies has also been influenced by the fact that reproductive technologies claim women as their primary patients, even when there is male-factor infertility.[6]

Understanding exactly how men and masculinity are implicated bodily, biologically, and otherwise in ARTs tends to elude the ethnographic gaze because of this treatment asymmetry: men just aren't around as much because they aren't required for most of the procedures. Thanks in large part to the fascinating work of Marcia Inhorn and Susan Kahn, masculinity is more prominent in the growing body of work on reproductive technologies in countries where patriarchy and patrilineal kinship are more official norms.[7] But perhaps because a cognatic biogenetic model of kinship (two parents, both of whom contribute equally to the child's genetics) dominates in U.S. biomedicine and individualist egalitarian norms surface in the biomedical encounter, scholars have assumed that sexism needs to be unearthed and shown to be a present and important dynamic. As is argued in chapters 2 and 3 of this book, reproductive technologies aimed at alleviating infertility have on the whole been harder on women than on men and have tended to reproduce certain gender (and other) inequalities. In the feminist literature on reproductive technologies, then, examining masculinity has mostly been a poorly theoretically integrated afterthought to the examination of women's experiences and the examination of gender from the crucial perspective of the perpetuation of sexism, heterosexism, classism, and racism of ARTs.

Despite these interrelated reasons for the sidelining of masculinity in scholarship on reproductive technologies, Gay Becker has shown that it is possible to focus on the poignancy and struggles of masculinity and men in ARTs without losing sight of the patterns of sexism that they in some cases reinscribe.[8] Men are half of the heterosexual couple that is the normative though not descriptive (or even prescriptive, now that the majority of U.S. clinics treat single, gay, and lesbian couples) template of the reproductive union on which these practices are predicated. Like women, men suffer from the stigma of infertility and experience strong desires to be parents, although to a greater extent than women they appear to vary in how much they are involved in and struggle with their infertility and treatment. Men with male-factor infertility have been found to experience their infertility as more stigmatizing than women do or than men do in an infertile couple without a male-factor diagnosis.[9]

Men's identification with their fertility has been analyzed as being part of patriarchy and part of a documented cultural obsession with genetics.[10] I have found, as I describe below, that this identification is also a reaction both to the elision between potency and genetic reproduction for men and a response to the reductive nature of ART clinics for most men. The flip side of the overimplication of women as patients is that men are often reduced to an ejaculatory role, even with a male-factor diagnosis for infertility. This reductive aspect of treatment for men means that genetic fatherhood is both what is compromised and what will occur if the ART procedure succeeds. It is grammatically appropriate that men would be genetically essentialist when there is no gestational role available to them and when the only way to overcome their faulty contribution to the couple's reproduction is through the genetic route or through displacement by a donor. That men tend to support their infertile wives when there is no male factor suggests that men are not simply genetically essentialist. If they were, then they would be expected to leave or otherwise be unsupportive of their partners, which I found not to be true.[11] Over time, it is possible that male-factor stigma will decrease, as the condition becomes more public and more medicalized, which both significantly reduce stigma.[12] I also found that individual men, much as individual women, tend to feel less stigma over time as they decide to accept biological accounts of their infertility and talk to fellow sufferers, friends, and family members. Some men observe that in the United States men are discouraged from talking about their personal lives, which contributes to a perpetuation of feelings of shame.

Masculinizing Assisted Reproductive Technologies

Masculinity is a salient aspect of the division of labor among medical staff at many assisted reproductive technology clinics. Physicians who specialize in the field of reproductive endocrinology and infertility medicine are often male, and the field has not been feminized in the United States to the same extent as obstetrics. This means that many clinics, despite having a largely female nursing and technical staff, are headed by male physicians. One physician in whose clinic I worked referred to the rest of us—staff, patients, and technicians—as his "harem." Both male and female physicians with whom I have talked tend to think that gender stereotypes that are associated with medicine play significant roles in the choice of medical specialization. Among the reasons that were cited for this relative lack of feminization of the field were the cutting-edge technological and entrepreneurial aspects of the specialty and its capacity to attract male physicians who were interested in "breakthrough medicine." The example in chapter 5 of the physician who wanted Kay and Rachel to have twins from two different mothers at the same time illustrates this pattern. Whatever the causes, a predominance of male physicians and female staff and an emphasis on female embodiment in infertility treatment have resulted in sites that are gender differentiated and stratified. Male patients' masculinity is expressed as another component of this agonistic texture.

Another important part of the increased salience of masculinity in ARTs was the explosion of research into and progress in male-factor diagnosis and treatment of infertility in the 1990s. Electroejaculation (adapted from domestic livestock fertilization practices) for spinal-injury victims began to be practiced, as described below.[13] But most accounts of male-factor ARTs credit the intracytoplasmic sperm injection (ICSI) technique with having revolutionized male-factor infertility treatments. In ICSI, a single sperm is manually sucked up into a micropipette and placed in a mature (metaphase II) oocyte for in vitro fertilization. One precursor to the ICSI technique was called partial zona drilling (PZD) and involved using a chemical reagent or sharp instrument to open a hole in the zona pellucida or outside layer of an egg so that sperm could more easily enter and fertilize the egg. PZD seemed to help in some cases of low sperm count or poor sperm motility but not in severe cases. Subzonal drilling involved placing sperm inside the gap in the zona but again did not help in severe male-factor cases and tended to result in polyspermia, or the fertilization of the egg by more than one sperm. The young reproductive biologist Gianpiero Palermo (now at the Cornell Medical Hospital in New York) and the team headed by Andre Van Steirteghem at Brussels Free University (Vrije Universiteit Brussel) pioneered ICSI in the early 1990s. The announcement of the first successful birth in 1992 of a baby using this method[14] helped ICSI to spread around the world with a remarkable speed aided by an ethos among lab technicians (many of whom are prominent reproductive biologists in their own right) of training each other through informal and formal personnel exchange and skill sharing. I was urged in the early 1990s by Robert Edwards, one of Louise Brown's "lab fathers" (an expression that is revealing about masculinity and ARTs), to travel to Brussels, spend time with the team, and learn their techniques.

The most resistant cases of male-factor infertility are those where there is apparent azoospermia, or a complete absence of spermatozoa in the ejaculate. In obstructive azoospermia, there are no spermatozoa in the ejaculate because the route from testes to epididymus is congenitally missing or blocked at some point by vasectomy or trauma. In nonobstructive azoospermia, there is an absence of germ cells altogether, or a "maturation arrest" where germ cells fail to make it through meiosis. Some of the nonobstructive cases are the results of chemotherapy, cryptorchidism (undescended testes), or mumps, and some are idiopathic, or congenital. ICSI alone could not help couples conceive with the male partner's sperm if there was no sperm in the ejaculate. Testicular sperm extraction (TESE), in which physicians surgically extract spermatozoa or spermatids directly from the testicles for use in IVF, combined with ICSI (TESE-ICSI) made it possible by the mid-1990s to establish some pregnancies where there is no sperm in the ejaculate.

A debate raged briefly in the late 1990s about whether round spermatids (that is, undeveloped spermatozoa) could fertilize an egg with TESE-ICSI, in part because the histologic and microscopic comparative studies of sperm-maturation arrest and incomplete spermatogenesis had not been assessed and standardized. Technicians could not always discriminate among different cells and classified spermatids and spermatozoa differently one from the other.[15] It was eventually resolved that at least some elongated, postmeiosis spermatids had to be present for fertilization and that ICSI could not help cases of complete maturation arrest or complete absence of germ cells. Most cases, however, became potentially treatable, as even the most severe male-factor cases still involved "minute foci of spermatogenesis."[16] It has been my impression that most people have been genuinely surprised by the success of ICSI. Van Steirteghem and team members wrote a supplement to *Human Reproduction* in 1996 describing the development and methods involved in ICSI,[17] and a 1998 "Male Infertility Update" summed up the scope and significance of ICSI to the field of ARTs for male factor infertility:

Since the birth of the first ICSI child in January 1992 . . . this assisted fertilization procedure is now widely applied worldwide to alleviate severe male factor infertility. . . . It can be used (i) for patients with quantitative sperm anomalies, since only a single live (motile) spermatozoon is needed to micro-inject each fetilizable metaphase II oocyte, and (ii) for patients with qualitative semen anomalies, since ICSI bypasses most steps in the oocyte-sperm interaction required for fertilization in natural conception or conventional IVF. ICSI can also be used (i) with epididymal spermatozoa obtained by mircrosurgical or percutaneous epididymal sperm aspiration for most patients with obstructive azoospermia and (ii) with testicular spermatozoa retrieved from a testicular biopsy by open excisional surgical procedure or fine needle aspiration in the patients with obstructive azoospermia and some patients with non-obstructive azoospermia.[18]

Debates are ongoing about the long-term health of ICSI babies and particularly about whether male children will inherit male infertility. Again, results have been remarkably good thus far, with rates for congenital anomalies the same as rates found in conventional conception and in conventional IVF, despite some ambiguous evidence that there might be more de novo sex chromosomal aberrations of paternal origin.[19] In a more clinical development and mirroring the explosion in the late 1980s and early 1990s of using IVF for factors other than tubal disease, ICSI has begun to be used in cases other than the severe male-factor cases for which it was originally developed.

Since the end of the 1990s, I have seen ICSI used for a range of male factors but also as a default fertilization method in IVF. This has become especially common where women are classified as having had poor ovarian response, often judged as five or fewer eggs retrieved after superovulation. It sometimes appears that the reason for this is to get practice at doing ICSI and to maximize the chances of having at least one egg fertilize. There appears to be no evidence that in the absence of severe male factor there is any advantage to doing ICSI, regardless of whether the woman responded well or poorly to ovarian stimulation. If her eggs are compromised anyway, ICSI might increase fertilization but would be unlikely to increase implantation and live birth rates. I have also heard physicians express anxiety that ICSI might damage eggs and decrease the resultant quality of embryos, especially if done by a less than expert technician. The excitement around ICSI has currently led to its overapplication, but it has also thereby put techniques developed for male body parts firmly into ART practice, including routine IVF in many centers. This has somewhat lessened the asymmetry between men and women whereby women's body parts and bodies were previously the almost exclusive logic of and focus for the medical gaze in ARTs.

Ethnography of Masculinity

To probe the nature of the performativity (the ontological choreography) of masculinity, it is helpful to start with existing ethnographic work on masculinity. This comes from feminist anthropology and from a growing body of work on the ethnography of manhood. Taken together, these somewhat offset what Matthew Gutmann has referred to as "the awkward avoidance of feminist theory on the part of many anthropologists who study manhood."[20] Similarly, I hope to mitigate somewhat the tendency on the part of many feminist scholars to assume—whether explicitly or through a lack of attention to the topic—that male subjectivity is privileged, pregiven, and immune to destabilization in a way that is not true of female subjectivity. Typically, ethnographies of masculinity have focused either on displays of manliness and machismo or on men's roles in relation to women and children.[21] One of the points of interest of this site is that it is always about both (threatened) gender and sexual identity and (threatened) paternity for male patients, and so both kinds of concerns arise simultaneously.

The narrow range of class perspectives that are represented in the patient population (most, but not all, patients are from middle- and upper-income groups) and the heterosexual dynamic of almost every aspect of treatment mean that the sample is somewhat narrow and less than representative of contemporary American masculinity.[22] While this means that my arguments are limited in the extent to which they can be generalized, the patient perspectives that I'm addressing correspond fairly closely to the subject position that supposedly benefits from a taken-for-granted privilege.[23] Similarly, these repertoires of masculinity, even though they sometimes are in tension with one another, reveal insights about reproduction itself and how it plays out in the workings of masculinity in reproductive technology clinics. The two different repertoires of gender identity and paternity also allow a classic ethnographic and feminist question to be asked about differences between men's and women's experiences of the technologies. These characteristics in turn suggest enduring differences between male and female subjectivity in the societies of which these technologies are a part.

The ethnographic vignettes that I discuss fall fairly clearly into one or the other of two repertoires—one displaying virility and one displaying the potential to be a good father. The claiming or attribution of masculinity occurs commonly in infertility clinics as elsewhere and tends to come with a heightened biological reductionism (when virility is the trope) and with social reductionism (when evidence of the ability to be a good father is at stake). Masculinity is performed in front of women as well as men and is sometimes contained in the representations of men by women, but is not defined by its relationship to women or femininity. Sometimes virility is talked about in an explicitly parodic, joking manner, not because men and women in these sites are more crass or more sexist than others elsewhere but because of the functionality of this strategy. In ART clinics, parodic gender performativity can literally make kinship.[24]

Virility

On one occasion when I was in an exam room assisting an infertility doctor, a man and woman dressed in his and hers Polo shirts and loafers were talking to the doctor about their course of treatment. From preparing their charts, I knew that the husband was about fifteen years older than his wife, who was in her mid-thirties. It was his second marriage and her first, and neither had children. While we were in the exam room, one of the lab technicians came in with the sperm-analysis results. The man's sperm count was in the "very high" range. The male doctor scanned the results and smiled and told the husband that he had a "very good numbers" and a high sperm count. The man was visibly delighted and grinned and nudged his wife. One of his comments was, "Not bad for an old guy, eh?" The doctor joined in for a moment, conferring an indulgent smile on the husband, and then resumed the pelvic exam of the wife.

The husband soon asked again what a normal sperm count was and reconfirmed that his count was considerably higher than normal. The doctor cooperated, giving a normal range but saying that as long as there were sufficient minimal numbers for fertilization, absolute counts did not matter much in in vitro fertilization. The wife seemed a little embarrassed by her husband's interjections but seemed pleased, and it occurred to me that she might be anxious about the implication that the infertility problem must lie with her. When the exam was over, the husband again mentioned that his count was high, still grinning a lot. The doctor gave a last smile and left the room. This patient must have been enormously relieved that his sperm was fine, especially at his age, and that his high sperm count made it unlikely that there was a male factor in the couple's infertility. He was expressing relief and repairing the threat to his masculinity. Taken out of context, however, boasting about a high sperm count sounds like parodic and over-the-top "locker-room" talk.

Representations or marking of virility by others than the male patients themselves also happened on occasion. Once I was sitting in the waiting room of an infertility clinic where several patients were waiting for appointments of various kinds. The doctors were running late at this clinic, and a kind of anonymous camaraderie was growing. Two of the women began talking to each other about their respective cases. After a few minutes, one of the nurses came into the waiting area, handed a sterile specimen cup to one of the waiting men, and beckoned for him to follow her. He got up and disappeared into the main part of the clinic. When he returned to the room a few minutes later, minus the specimen cup, the two women patients who had been talking to one another (his wife was not among those present) clapped. One of the applauders, on her third cycle of treatment at the clinic and the person who had initiated the conversation with the other women, explained to him that they had been impressed by how quickly he had produced his specimen. The man looked both embarrassed and pleased, and several others in the room who had until then appeared not to be listening smiled at him before returning to their magazines or glancing at their watches. Those paying attention—not usually a polite thing to do in that setting but made permissible by the unusual camaraderie—understood that the man probably had been summoned to masturbate so that the clinic could inseminate his partner's eggs with his sperm. The weary fellow patients were recognizing the oddness and difficulty of performing on demand and with an audience and praising him for his speed in ejaculating. What might have been ribald sexual humor in another circumstance was meant as and received as humorous, gentle encouragement and praise for his virility and cooperativeness.

Conversations about men and their ejaculatory role in infertility treatment are also common when women who are undergoing infertility treatments socialize with each other in the absence of their male partners. These meetings can be purely social gatherings or the formal meetings organized under official support groups like RESOLVE. They might occur to share fertility drugs or unused hyperdermics, to administer intramuscular hormone shots to another group member, or to visit a group member on bed rest following surgery or embryo transfer. I accompanied a woman who was going to visit one of her support-group friends because a shipment of her drugs was late arriving from the United Kingdom and the friend had some leftover Gonal-F from her own successful in vitro fertilization. After the handing over of the drugs, the topic of conversation turned to men in infertility treatment. The two women discussed their own husbands and the partners of several of their mutual friends. Eventually, they turned to the nature and quality of the pornographic materials that were provided at different clinics to aid masturbation. They commented, with much mirth, on how "gynecological" and unarousing they found pornographic videos and magazines and on the different tastes of the men in their circle in these matters. They were happy to share an enormous amount about the techniques that each man (or couple) used to obtain his semen. The conversation ended with the hostess happily proclaiming, "Hey, whatever they get off on!" There seemed to be no anxiety about their partners' arousal by the representations in these materials, despite this being a fear that women occasionally lightheartedly express before beginning treatment. There was also little or no political critique of the materials themselves, although I knew that at least one of the women was antipornography. The setting seemed to normalize stereotypical heterosexual male fantasy and indeed mandate it. The women had internalized this aspect of treatment, just as they did the other norms and bodily discipline of the site. Displaying ease with the world of commercial male fantasy was a sign of knowledge of and compliance with the treatment culture and regime.

In the last two cases described, men were primarily choreographed as ejaculatory extensions to their partners, and by parodying masculinity (through humor regarding orgasm and the consumption of pornography), participants facilitated treatment and normalized public exposures of sexuality. The display and rescue of virility are entirely different when there is a diagnosis of male infertility. In these cases, the management of stigma seems to be most important. Women patients often mentioned that they found childlessness more stigmatizing than the physiological failure to conceive, although one woman pointed out that because the sexually transmitted diseases that can lead to female infertility connote promiscuity in North America, infertility can be independently stigmatizing. Women also often reported a poignant desire, unrelated to stigma, to experience pregnancy and a preference for a child related to them genetically. For men, there seemed to be little evidence in the clinical setting of stigma when the cause of infertility was associated with their partners, and men often said that they were (and were often reported as being) supportive and active in efforts to diffuse stigma. This included such actions as enlarging circles of disclosure over the long term (telling more people about the infertility) or protecting their own and their partner's privacy. Just as for women patients, non- or less stigmatizing aspects of infertility were still key focal points of loss for men. Several men expressed disappointment at their diminished chances of genetic fatherhood together with their partners and "womb envy" that they might not ever have the opportunity to experience their partner's pregnancy.[25] Men who were diagnosed with male factors influencing the infertility, however, often experienced intense stigma. In the clinical setting, as far as I could judge from attempts at repair, this stigma came from the association with impotence, feminization, or lack of sexual prowess as well as from the childlessness because of an elision between sexual potency and fatherhood. In the two examples I give below, the patients in question strive to repair their threatened masculinity through macho displays of alternative measures of virility. In one of the examples, the patient also engaged in public boundary work to separate himself and others like him off from male-factor infertility patients with different or more severe etiologies.

I saw two different ways of handling the threats to masculinity posed by questionable virility. One involved what I came to think of as "exemption for reasons of hypermasculinity," and the other involved a search for underlying or residual biological masculinity. One subgroup of male-factor infertility patients contains those with paralysis due to spinal-cord injury. Such injuries happen most frequently to young men, and more than 10,000 such injuries are estimated to occur each year in the United States.[26] This is a sizable population. Sperm quality declines after injury, and impotence and ejaculatory dysfunction accompany paralysis. Only about 5 percent of spinal-cord injury victims are able to achieve pregnancy with their partners without help. In the last decade and a half, clinics have begun offering electroejaculation to men with spinal-cord injuries, originally combined with intrauterine insemination and after the development of ICSI in combination with IVF. Poor sperm quality—thought to be a result of a combination of lack of ejaculation, scrotal temperature increases, and hormonal and neurological disruptions that result from the injury—is a persistent problem. Young men who sustain spinal-cord injuries are now commonly asked to produce semen samples for freezing for potential subsequent use after the initial period of injury trauma is over but before semen quality has substantially declined.

Specimens can be obtained after some kinds of injury by using penile vibrators, but the more recently developed transrectal probe often works better for patients, especially if they have no hip-flexion reflex left (the reflex checked for by scratching the sole of the foot). One hospital staff member, who was married to a man with quadriplegia, told me that her husband used electroejaculation with a transrectal probe, which resulted in the birth of a child. They were back in the clinic for a second attempt. The nurse was enthusiastic about the technique and persuaded her husband not only to undergo it again, but also to become a spokesperson for the procedure. The press took up the story, emphasizing the man's former career as an athlete and the sports-injury cause for his condition. The clinic staff considered him to be an exemplary spokesperson for the procedure because of his unassailable masculinity (conferred by the nature and circumstances of his injury and his identity as a former sports star). In this case, a loss of virility was explained as being the result of previous hypermasculinity.

A second example occurred at a public patient recruitment and information seminar. These seminars, often elaborate affairs held at expensive hotels, are regular features of U.S. infertility medicine. Wealthy and populous states like California, where this one took place, are served by a surplus of highly competitive clinics, which use these events primarily to attract new patients but also to educate existing patients. From the patients' point of view, these events are a site of anticipatory socialization where they will be initiated into the choices, language, and practices of infertility medicine. Most of the patients at these seminars have already undergone preliminary work-ups, may have a diagnosis in hand, but have not yet undergone specialized assisted conception techniques. Others are still at the diagnostic phase, and some have recently moved to a new town or shifted allegiance from a different clinic and so are shopping around for a new place. I went to several such seminars hosted by the infertility clinics at which I worked, including a one-day seminar that was attended by over a hundred couples. Because male infertility was featured on this occasion, many of the couples who were present had male-factor diagnoses. The clinic showcased its new expertise in male-infertility surgery and in ICSI.

Patient seminars are remarkable for their choreographed coupling. Virtually all attendees sit in pairs, including the few gay or lesbian couples, with their bodies often turned toward each other and their hands touching. At this seminar, I followed the group that went to the sessions on male infertility. At the male-factor sessions, all the couples were sitting in heterosexual pairs, and the couple huddling was noticeable. The physician who was responsible for introducing the sessions and running the question-and-answer period began with an introduction to the varieties of male-factor infertility and the major treatment options. To my astonishment, the earliest slides in the physician's presentation showed a construction crane raised high and the same crane lowered and no longer working. The audience laughed a little but not much. Using these two visuals to suggest the nature of male infertility through an analogy with the inability to sustain an erection seemed both insensitive and curiously inaccurate because few male-factor diagnoses involved erectile disfunction. The elision was that difficulty fathering a child somehow impugned a man's virility.

The physician opened the session for patient questions. The men present were encouraged to "share" their diagnoses and "open up about" their feelings about their infertility and were praised in advance for having the courage to speak up. The physician recognized one man who raised his hand and took the microphone from a nurse. The man seemed to be white, in his forties or fifties, and considerably older than the Asian American woman he was seated with. He explained that he had recently remarried, that his new and younger wife, sitting by his side, was eager to have children, but that he had had a vasectomy after the two children from his first marriage had been born. The physician repeated the options for vasectomy reversal that had been covered in the preamble and then asked the man for his question. The man asked about his chances of success. A nurse who was sitting with me whispered that this information had also been covered in the introduction and speculated that the questioner meant to separate himself from the others in the room—to announce that his lack of potency was the result of his hypermasculine prior life course and should not detract from his masculinity. He had previously needed to be prevented from fathering too many children and was now starting over again with a young second wife. On other occasions, I had seen women patients with secondary infertility (women who had already had one or more live births when they turned to ARTs) use subtle means to exempt themselves from other infertile women. This slightly disrupted group solidarity and demarcated that primary infertility was indeed a matter of stigma. This man seemed to need to explain why he could not get his current wife pregnant and not to prove that he was already a father, whereas women with secondary infertility seemed to need to let others know that they were already mothers.

Staff members, including at least some female staff members, sometimes appeared to link imagined virility and male fertility. During a surgical vasectomy-reversal procedure that I observed, a nurse taped the man's flaccid penis to one side, covered his body with sterile blue sheets, and positioned a small square opening over his testicles, where the operation would occur. As she completed this preparation, she commented that the large size of the patient's testicles was a sign that the surgeon probably would be able to "milk" some live sperm from the patient's vas deferens, posterior to the vasectomy site, that could be frozen for subsequent fertilization in case the reversal did not work. While quantity of semen is correlated with size of testicles, the presence of live sperm and the quality and concentration of the sperm are apparently poorly correlated with size of testicles, and yet are both essential to fertilization. The fact that he had fathered two children fifteen and thirteen years ago probably was a good sign that live sperm would be found. But there was no reason to be especially optimistic about his future chances of fathering a child based on the size of his genitals. Nonetheless, it must have been a fairly automatic thing to say in passing, for no one contested it.

In talking to patients and practitioners during the breaks, and from observing the dynamic in the group, it seemed that there was some agreement on a hierarchy of male-factor diagnoses, stretching from less to more compromised manliness. Patients who were seeking vasectomy reversal were at one end of the spectrum; they usually were attempting to start second families. Those with congenital azoospermia were at the other end of the spectrum; the complete absence of sperm meant that they would never become fathers by impregnating their partners. I overheard a woman confess to another woman that she had looked furtively at patients who identified themselves with primary failure of spermatogenesis (those who made no sperm at all) to see if they lacked other secondary sexual characteristics such as facial hair or if they had feminine hands or visible gynecomastia (breast tissue development). In a somewhat parallel way, female infertility can be associated with what gets called "virilization," such as the elevated androgen levels found in polycystic ovarian disease, for example, or hirsutism. I sometimes heard women patients discuss their fears that these diagnoses connoted masculinization and a compromised femininity. But women with more (hyper)feminine bodies or sexual histories did not separate themselves off from others in an infertility setting to show their gender superiority. If anything, infertility resulting from a previous need for contraception (abortion, pelvic adhesions, or tubal ligation) connoted promiscuity, rather than a female equivalent of virility. The relevance of feminization to the stigma of male infertility reinforced the sense that the infertility stigma for men included compromised virility and was not simply a matter of the compromised ability to have children, from which all the men in this group suffered at this point. The men with diagnoses that were not completely without hope often took great solace in the extent to which they still had fertility potential. In one ICSI case I followed, the availability of a very small number of sperm for treatment prompted the wife to comment to me that her husband had felt enormously relieved by the live sperm and comfortable about undergoing treatment from that point on.

Good Patient/Husband/Father

The second category of masculinity that I observed in these clinics is the complex of norms that arose around appropriate comportment. These were highly relational and quasi-official norms. Male patients were to behave as good patients, which meant showing themselves to be compliant by being supportive and committed husbands. Being a good husband and being a good patient indexed worthiness to be a father. Decisions about who should be treated and who would receive insurance coverage were couched in terms of the best interests of a future child or children conceived with the help of the technologies, and these interests were gauged by the nature and quality of the commitment between the patient couple.

When there was no male-factor diagnosis, men were expected to accompany their partners at appointments and at several clinics were praised for solicitous behavior, as long as it didn't dominate the clinical encounter. This was less true for couples with male-factor diagnoses because of the sense in which the male partner became the primary patient. Most treatments, including those for male-factor diagnosis, however, directly involve men only in an ejaculatory role, which consigns them, as well as frees them, to support their partners' more intensive role. Staff, especially clinic psychologists, would frequently ask women patients how and whether they perceived their partners to be committed and supportive, and patients themselves would bring up the topic if they felt their partner was insufficiently so. Patients vary a good deal in the extent to which they socialize in and with the site. Some women patients liked time for "the girls" (the woman patient and the female staff) to talk without their partner around. A certain amount of male bonding between the usually male physician and the male patients was considered appropriate, too, although again many men did not do this. Overt disagreements between partners made everyone uncomfortable and were seen as bad omens (see chapter 3).

The importance of this complex of norms was brought home to me in the breach. One couple came to the clinic knowing that they needed to use IVF if they were to have children together because the woman's fallopian tubes had been surgically removed after an ectopic pregnancy. She was in her midthirties and had had a child before the loss of her tubes, so it was known that some things worked and some things did not work. The couple was thus atypical in that the standard diagnostic work was not needed to determine the reasons for the infertility and no instruction was required in what to expect from pregnancy. They were eager to begin IVF before any time was lost and were reluctant to go through expensive and time-consuming preliminary procedures. The husband submitted to the standard semen analysis, which came back normal on all parameters. He was then asked to undergo another test to determine the ability of his sperm to fertilize an egg. This procedure required that the man show up at the hospital at 6:30 a.m. in the morning. The husband, who declared himself "not a morning person," decided that there was no particular reason to do this test because the odds of having a normal semen analysis but trouble with fertilization were slim, especially given that they already had one child and they already knew the cause of their infertility. He also declined to show up for his wife's monitoring exams and various information sessions, justifying his decision on the basis of the routine nature of the procedures and the demands of child care and jobs at home.

The couple began a cycle of IVF. At the point at which the husband's sperm was added to the wife's eggs for fertilization, however, the doctors decided to perform ICSI instead of leaving the fertilization process to occur of its own accord. ICSI was covered on the consent forms if in the judgment of the physicians it was deemed necessary. The couple was told after the fact that ICSI had been done because the physician and technician did not know if the husband's sperm was capable of fertilizing eggs as he had not submitted to the required test. One egg was damaged by the ICSI process itself, and the woman did not get pregnant with the two remaining embryos. On a second cycle, the husband explicitly declined ICSI (still without doing the fertilization test), the eggs fertilized and grew well, and the wife got pregnant. Although the husband was in some sense free to decide the extent to which he was supportive and compliant, the staff and physicians complained about his relative lack of support for his wife and his relative noncompliance, and their criticisms made the unannounced ICSI procedure seem somewhat like a punishment. The norm that male patients ought to manifest as committed and supportive partners was not only present but prescriptive in the site.

Norming and Performing Gender

The ontology of masculinity and gender as performed in these sites relates to other key theoretical debates. On the one hand, the above ethnographic examples relate to the structuralist tradition that has argued that masculinity is predominantly "positional" and that femininity is predominantly "relational." On the other hand, the ontological choreography of gender that is described in this chapter can be linked to core science and technology studies territory in that it concerns relations between biological and social repertoires of gender normativity. Together these questions address the role of gender in the political economy and in the social and technical innovations of infertility medicine.

The examples given above of masculinity in contemporary U.S. infertility clinics indicate that men are expected to act in ways that display their committed compliance as patients and their support for their partners. On occasion, these qualities had real consequences for treatment. They were also used as the inferential grounds by patients and practitioners of particular men's likelihood of making good fathers. Men's roles, then, were prescribed by idealized versions of husband and wife and of father and child relations. The abstract nature of this framing was particularly evident in the evocation of the father and child dyad, which was a hoped-for result and usually not a present condition in the clinics. At the same time, men's inability to father a child, especially when there was a diagnosis of male-factor infertility, constantly threatened to undermine men's sexual prowess and manliness, the husband and wife frame, and the father and child frame. The pervasive threats to manliness were marked by background assumptions, humor, relief, pride, and episodes of repair initiated by men themselves, by women patients, by doctors, and by nurses, as described above. This need to profess and repair manliness highlighted a significant difference between men and their relation to fatherhood in the clinics, and women and their relation to motherhood in the clinics. While both desired to be parents, women often expressed the sentiment (as often to rebut it as to claim it) that "You can't really call yourself a woman until you are a mother." Men, on the other hand, found that their infertility threatened to compromise their masculinity in the clinical setting—first because it betokened a lack of virility and also because it stopped them from becoming fathers. Women did not have to keep repairing their threatened sexuality in the site, but men did. The equivalent aphorism for men might have been "If you can't father a child, then you are not really a man," with the emphasis on the impregnation and its implications of sexual prowess as well as of claiming the pregnant woman and family in question.

In 1974, Michelle Rosaldo and Louise Lamphere edited a collection called *Woman, Culture, and Society,* which contained several now classic early papers on the ethnographic study of gender. Most famous among them were Sherry Ortner's "Is Female to Male as Nature Is to Culture?" and Nancy Chodorow's "Family Structure and Feminine Personality."[27] The title of this chapter, "Is Man to Father as Woman Is to Mother?" echoes the title of Ortner's essay and the uses by feminist epistemologists and philosophers of science of the related "Is Sex to Gender as Nature Is to Culture?"[28] Several contributors to the volume share Ortner's opinion that "the secondary status of women in society is one of the true universals, a pan-cultural fact." Ortner's sources of evidence for her claim regarding the universal and systemic lower status of women are explicit devaluations of women; implicit devaluations of women, such as the association of (some) women (sometimes) with defilement; and the presence of structural barriers to inclusion of and participation by women in highly culturally valued activities. The paper urges anthropologists to try to explain "the universal fact and the cultural variation." Ortner argues that every culture ranks most highly its ability to socialize or civilize nature and that every culture associates women and nature and associates men and culture. In addition, she recalls Simone de Beauvoir's famous 1949 thesis that the female "is more enslaved to the species" than the male and more "biologically destined for the repetition of life"[29] and augments it with Chodorow's argument that women's body, social roles, and psychic structure stem from her unique procreative and mothering functions, which place her closer than men are to nature. In this well-known feminist tradition, to be socialized as female is to be made close to nature through one's "relational," concrete, and subjective modes of behavior and habits of thought. By contrast, to be male is what is sometimes called a "positional" subject position that is defined by modes of thought and action that are characteristically abstract, objective, and carried out from the position of and in the name of the man himself.[30]

Ortner has been critiqued by anthropologists such as Marilyn Strathern for assuming that all societies make a nature and culture distinction, particularly one like that found in the West in modernity. Many have argued that the empirical evidence does not support the existence of the universal subordination of women. Marxists and others have critiqued the theory for extending conditions pertaining to the gendered division of labor that is characteristic of capitalist patriarchy to all times and places. Still others have shown that the equation of women with nature and men with culture is not found in all cultures and is not consistently true within particular cultures. And many have tackled the assumption that women's deep sense of self is universally or even primarily relational.[31] The element that has been surprisingly resistant to criticism has been the idea of the positionality of "male" subjectivity. There do exist ethnographic accounts of masculinity that show it to be relational in practice if not in official ideology. These accounts, to my knowledge, deal mainly with non-Western men or men in explicitly socially marginal positions in society.[32] In these cases, it could be argued that the feminization that is associated with marginality or otherness is responsible for the relationality.[33]

The account of masculinity that is presented above has two components, one of which seems to map fairly well onto Ortner's notion of relationality—namely, the norm of the committed and supportive patient and husband. Male patients' role and corresponding subjectivity were predicated on their relations with partners and future children. The other component—virility and the repair of gender identity—seems more positional, though. That is, in the clinics it appeared to be carried out from the position of and in the name of the man himself. Indeed, the relation of fatherhood seemed to short-circuit at the manly properties or lack thereof of the man himself. For the men, difficulties of paternity compromised manliness, but paternity was not the manifestation of that manliness in the same way that for the women their potential to be mothers was the sign or condition of womanliness. Despite this evidence for male positionality, however, and counter to Ortner's argument, masculinity in infertility clinics was not only positional; it was also prescriptively relational, and the two reinforced each other. Getting pregnant is the goal of these sites, which often clinically reduced to impregnation for men, whereas it meant something much more long term and less linked to sexuality for women. This reduction of men to their ejaculatory roles, combined with the masculinist culture and stratification of several clinics, meant that injuries to men's self-identity were much more about their positional identities than women's. Man was not to father in these clinics as woman was to mother.

Despite the use of the structuralist dichotomy between positional and relational identities (albeit reworked, so that male positionality is coproduced with relationality, and vice versa), the ethnographic examples of masculinity in this chapter conceptualize it primarily as performative, with male patients having to negotiate, achieve, and protect their male identity and to act out and have imputed to them appropriate masculinity. This emphasizes the constructedness, situatedness, multiplicity, and fragility of masculinity in the infertility clinics. Even mainly elite male subjectivity in biomedical Western settings can be fragile and contingent and takes work, which is an important corrective to the tendency of feminist theory to assume that exactly this kind of subjectivity is taken for granted and is a marker of a power differential with subordinate subjectivities. Stressing the constructedness, however, risks making gender identity in the clinics seem much more fluid and contingent and much less tied into systemic gender difference, including inequality, than it actually is. In complicating masculinity, this account thus jeopardizes the very thing that the feminist assumptions about masculinity implicitly linked—namely, masculinity and other abstract social and political categories. The taken-for-granted, view-from-nowhere masculinity may have been mute about the coexisting relationality and vulnerabilities of positional subjectivity, but at least it assumed that the hierarchies and systems of power that maintained (a class of) male as the unmarked gender were also implicated in the perpetuation and justification of gender difference wherever it was found. Using the notion of ontological choreography to spell out connections between gender in the clinic and the very topic that Ortner connected to gender—the nature and culture binary—shows how gender is knitted into the site in a manner that is anything but free of gender patterns of which these clinics are just one iteration.

Recent theories of gender performativity have begun this task. Judith Butler responded to precisely this kind of criticism of the limits of performative notions of gender. She reformulated in *Bodies That Matter* the notion of gender performativity with which I began. She drew up a list of attributes that are essential to the reformulated view that included the tenet that "gender performativity cannot be theorized apart from the forcible and reiterative practice of regulatory sexual regimes."[34] Likewise, she argued that the performance of gender is not a voluntaristic act by a preexisting subject who is making free choices about his or her embodiment. Rather, gender performance is always what she calls a "citation" or reiteration of a "norm," which itself carries the framing and constraining historicity of "heterosexual hegemony." In this way, she recognized that the immanent and experiential or phenomenological aspects of social interaction are never de novo but are replete with the histories of the people, places, and things in which they are manifest and from which they spring. In the spirit of this historical understanding of performativity, the male and female identities that are enacted in the infertility clinics discussed above were not random or even patterned in a manner that was wholly specific to infertility medicine. Parodic husband roles had to be conformed to, compromised masculinity was repaired, and virility was claimed and attributed. Yet all these achievements were possible because virtually everyone in the clinics implicitly knew a lot about these places from other similar places that they had been to or lived in. That is, the exercising of agency regarding gender identity reproduced familiar ways of acting and familiar future scenarios to which to aspire. Other things being equal, people in the clinic also recognized each other as types whose relative status they marked by routine but subtle interactions and whose behavior they judged as appropriate or not, based on equally tacit but well-rehearsed projections. It was not simply that people bought preexisting gender stereotypes to bear in a preexisting field of practice, however. A lot of what happens in contemporary infertility clinics is new both technically and socially. Among other things, as I discuss in chapter 5, who counts as a mother and a father is underdetermined and sometimes needs to be resolved. The identity categories of male and female are themselves somewhat in flux and in ways that are strictly tied to the procedural innovations. It is necessary, therefore, to think of gender performativity as historicized and iterative but also to enable it to innovate both the identity categories themselves and the procedures they are tied to.

From Nature and Culture to Innovative and Conservative

Ortner's original argument posited women as more closely associated with nature and men as more closely associated with culture (with culture understood to be socialized or civilized nature). As the discussion of comportment above illustrated, the clinics are more gender differentiated in more stereotypical ways than this demographic of patients is likely to encounter in other more egalitarian spheres of their lives. Ideologies or norms of, anxieties about, and individual performances of gendered identities are also strongly interwoven with narratives about what is natural in infertility clinics. I did not, however, find a straightforward connection between women and nature or between men and culture. This would have been impossible because a salient characteristic of clinic life is the lack of a unique or stable distinction between nature and culture in the site. Indeed, the dividing line between what is natural and what not is constantly moving. This occurs both within infertility medicine at large and for individual patients as they progress to more interventionist or substitutional techniques for conception. A decade or more ago, the technologies were collectively referred to as "artificial reproductive technologies," and expressions like "artificial insemination" were standard within the field as a whole. The word *artificial* has fallen gradually out of use, and the *a* in some well-known acronyms like (ART) has come to stand for *assisted*, not *artificial.* Similarly, in vitro fertilization used to seem like a highly unnatural means of conception, but as success rates improved, as a large number of babies were born by IVF, and as techniques were developed for manual fertilization (like ICSI), in vitro fertilization came to seem to be fairly normal and to be a paragon of natural conception because the sperm fertilizes the egg without manual assistance. Efforts to get infertility procedures covered by insurance companies have also added to the pressure to naturalize the procedures. Insurance companies tend to shun anything experimental, elective, unsafe, or ineffective and will not cover conditions that they can get away with claiming are not diseases.

In addition to this trend within the field for procedures to become normalized and naturalized over time, patients go through their own changes over time about what seems natural and acceptable and what seems frighteningly or impossibly unnatural. For example, couples in the mid-1980s to mid-1990s would commonly enter infertility treatment feeling adamant that they would be prepared to try certain procedures but that they would not countenance IVF. In the early years, IVF initially struck many patients as being too technologically interventionist. As treatment progressed and in the face of having received a diagnosis of blocked fallopian tubes or certain other conditions that indicated the appropriateness of IVF, many couples changed their minds. Counselors, physicians, and staff were aware of this trajectory.

A significant way to normalize the newness of the techniques and the kinship relations and social interventions they represent is to naturalize them as much as possible. This naturalization often has a one-step-ahead dynamic, bringing into the realm of the acceptable each new procedure that a couple undergoes in their treatment trajectory. A successful way to make new things seem normal is to interpret them as new examples of old things. Reading new ways of getting pregnant and starting families onto a conventional model of the normative (in this country, nuclear) family is a strategy that is commonly used to achieve this.[35] Naturalization normalizes and domesticates procedures, making them seem like appropriate ways of building a family rather than monstrous innovations. Stereotypically gendered identities—projected as virility in men, for example, when the masturbator was applauded, the wives were happy with whatever their partners got "off on," and the man who beamed with pride at his high sperm count—and strongly gender-divergent roles for the mother- and father-to-be enact an exaggerated version of the respective roles in the ideal nuclear family. This paraodic performance produces a peculiar mixture of conservative and innovative, in which conventional understandings of gender differences and roles are deployed to domesticate and legitimate the new. The performance and reproduction of gender stereotypes are thus important in bringing order to these novel sociotechnical settings.

Masculinity in modern capitalist societies is often theorized as being both patriarchal and rational individualist. Performative, parodic masculinity in the clinics showed this same interdependence of positionality and relationality, in the ontological choreography by which it restabilized preassigned but disrupted gender identities and the ontology of gender and kinship.

Although the nature and culture boundary is constantly moving in this site, it remains important for patients and physicians in understanding the procedures. As argued, one way to stabilize shifting notions of what is natural and normal around reproduction is to compensate with extremely conservative or stereotypical—parodic—understandings of sex, gender, and kinship. If ability to reproduce or if a claim to being a mother or a father is in question (even if only in one's own eyes) or is underdetermined by the biological facts (see chapter 5), one can reclaim that status by acting in what is likely to be read by others as a feminine or masculine or a motherly or fatherly way. Not surprisingly, this behavior is often more stereotypically gendered than might be typical for the patients in question. Paradoxically, it is invoked at precisely those times of most technical innovation. This means that my argument departs from the assumptions in Ortner's and many other theorists' work that the configurations of male and female and of nature and culture binaries can be read one from another. It suggests instead that broadly used classificatory binaries interact in a more complicated way. Their "looping" interaction brings about a mixture of reproducing the same old social order and yet being something truly novel that strikes many observers as especially salient about the new reproductive technologies.[36]

Judith Butler's, Donna Haraway's, and other theorists' deconstructions of the sex and gender and the nature and culture dichotomies and their connections to one another are crucial correctives to the structuralist assumption that the categories are fixed and connected to each another in a fixed manner.[37] There is not an isomorphism but a different choreography between sex/gender and nature/culture. As Haraway expresses it, "feminist theories of gender" should be "hatched in odd siblingship with contradictory, hostile, fruitful, inherited binary dualisms."[38] As I explored gender in infertility clinics, I found that supposedly natural gender dimorphism is often invoked most strongly at those times when the natural is unstable or poorly integrated into patients' lives. Patients and practitioners retrench into hyperconventional understandings of some of these sorting binaries to stabilize and domesticate others and remove stigma. It is not fully accurate, then, to posit that movements in male/female and nature/culture march in lock step, whether in a contingent and constructed way or a more structuralist way.

The examples from assisted reproductive technology highlight the ways that these repertoires of masculinity can coexist. Together they qualify patients for access to the site as would-be worthy fathers, and they enable men and others to repair threatened male gender identity and remove stigma. They also make it possible to take impregnation and fatherhood into the realm of the technological and consumerist. This is because it is possible to normalize and domesticate the social and technical novelty of reproductive technologies by reference to exactly the same set of norms and counternorms as those that characterize the somewhat contradictory role of men in late capitalist or turn-of-the-century U.S. society. Relying on fixed social gender roles helps secure the underdetermined or threatened biological sex, and biological markers of gender help repair damaged social gender. Scripted, performative repertoires of gender in this site function to enable order and normalize innovation, especially innovations in the ontology of sex, gender, and kinship and in their implications for society.

5

Strategic Naturalizing: Kinship, Race, and Ethnicity

Having looked at the reproduction of the clinical environment (including access, etiquette, and epistemics) in chapter 3 and at the work done by norming and performing gender in chapter 4, I turn now to an investigation of the strategies for demarcating affiliation in the clinics. What might assisted reproductive technology clinics tell us about contemporary kinship? This chapter analyzes the work done inside clinics during gestational surrogacy and egg donation to establish and disambiguate kin relations. Patients, practitioners, and third-party reproducers (egg and sperm donors and surrogates)—with the help of medical techniques, lab standards, body parts, psychological screening, and rapidly evolving laws—take on part of this work. The examples here show that the clinic is a site where certain bases of kin differentiation are foregrounded and recrafted while others are minimalized to make the couples who seek and pay for infertility treatment—the intended parents—come out through legitimate and intact chains of descent as the real parents.[1] All other parties to the reproduction, human and nonhuman, are rendered sufficiently prosthetic in the reproduction to prevent (if all goes well) contests over who the child's parents are. The alignment of procreative intent and biological kinship is achieved over time through a mixed bag of everyday and more formal strategies for naturalizing and socializing particular traits, substances, precedents, and behaviors.[2] Given the history of U.S. kinship, where under slavery the ownership of another's reproduction was a means of denying kinship, the significance of money in establishing procreational intent and thus claiming biological kin is especially important.[3] Its unintended as well as foreseen effects—especially in overturning or superseding laws that are enacted to prevent a return to possession of rights in the body or reproduction of another—should be carefully watched.

This chapter is inspired by the writings of Marilyn Strathern, Sarah Franklin, and their colleagues on the role of reproductive technologies in complicating naturalized linear cognatic descent and in reinvigorating the study of Western kinship.[4] The cases analyzed here elaborate Strathern's and Franklin's counterintuitive discovery of the underdeterminacy of biogenetic ways of determining kinship. I attempt to show that biology can nonetheless be mobilized to differentiate ambiguous kinship. In the process, the meaning of biological motherhood is somewhat transformed; in particular, biological motherhood is becoming something that can be partial. Lawyer and anthropologist Janet Dolgin has argued that recent legal cases involving contested parentage have played a part in moving contemporary U.S. conceptions of parentage from something that is socially fixed and biologically natural toward something that is more voluntaristic and enforceable through contracts expressing procreational intent. She has documented this gradual shift from family law to contract law, despite frequent attempts to protect an older idea of family relationships where children are concerned. As she expresses it,

> U.S. law has widely accepted the possibility that adults can negotiate the terms of relationships previously defined as nonnegotiable and has begun to recognize family members as fungible.... In cases in which children are concerned, the law has been slower to approve the amalgamation of family law with contract law.... But ironically, the judiciary's essentially conservative response, concerned to safeguard decent parents in proper families, itself contains the seeds of change.[5]

Like Dolgin and other legal scholars who are interested in "alternative reproduction," I am interested in these seeds of change despite conservative strategies that extend old concepts and understandings.[6] Legal cases cannot be decided without plausible narratives being pieced together, presented, and debated by interested parties. One major site of the production of these narratives for disambiguation is the ART clinic, where couples and others engage in a narrative construction as they contemplate treatment and as people, bodies, and techniques go through treatment. This chapter is thus about "doing" kinship in the contemporary United States as opposed to simply "being" a particular and fixed kind of kin as a cultural or natural fact of the matter. Race, ethnicity, friendship, contract, and their historical shadows are all powerful parts of this doing. ART success rates in the United States have been analyzed by race and ethnicity, and no significant differences have been identified between groups in success rates for different procedures, although incidences of some diagnoses differ, and access is not equal.[7] These findings and the use of race and ethnicity that is discussed below suggest that race is biologically irrelevant (same success rates on different procedures) and yet biologically significant (in diagnoses, access, and biological kin construction). This dilemma is familiar to anyone who has worked on the intersection of race and medicine in the United States since the formal decline of "scientific racism."[8]

I also draw on the literature in science and technology studies on the interconnections between or coproduction of politics and ontology, nature and culture, as described in chapter 1. I continue the strategy that is begun in chapter 4: rather than simply observe the dissolution of the boundary between nature and culture, I specify the means by which the facts and practices of biomedicine and the social meanings of kinship are used to generate and substantiate each other in specific cases. The analytic emphasis on naturalization is indebted to scholars who have shown the ubiquitous and multiple means by which identities get naturalized.[9] The emphasis on the strategies for enforcing procreative intent resonates with studies showing that reproduction is always already stratified.[10] Again, I attempt to contribute to these exciting areas of analysis by providing an account of some of the processes whereby the ontology of naturalization and the politics of stratification occur.

ARTs, by definition, involve "assisting" human reproduction by using many of the techniques of modern biology that have enabled the predominant systems of Western kinship reckoning to be mapped onto the science of genetics and the telos of evolutionary biology. Reproductive technology clinics would be expected to reveal connections between relatedness as determined by biological practice and socially meaningful answers to questions about who is related to whom. The science would help to hone or perfect our understanding of such terms as *mother, father,* and *child.* Tracking biomedical interventions in infertility medicine reveals something altogether different, however. Rather than revealing the essence of the natural ground to our social categories, this tracking reveals a number of disruptions of the categories of relatedness (especially, parent and child, but also sibling, aunt, uncle, and grandparent). In particular, the connections between the biological elements that are taken to be relevant to kinship and the socially meaningful kinship categories are underdetermined. Keeping biological and social accounts aligned and retaining biology as a resource for understanding the latter take work. Norms governing the family, laws regulating reproductive technologies, custody, and descent, the medical technologies themselves, and the financial dynamics of third-party medically assisted reproduction are the feed stocks and products of this kinship work. "Procreative intent," a more processual and less transparent notion than it appears, is made manifest and followed when kinship is sufficiently disambiguated to preempt conflict. Infertility patients, third parties, and practitioners routinely, and both formally and informally, do this kinship work; they are practical metaphysicians. In documenting kinship negotiations in infertility clinics, this chapter is therefore an empirical argument about metaphysics.

The cases discussed here involve two technically identical procedures that lead to different kinds of kinship configurations—gestational surrogacy and IVF with ovum donation—as they arose in contemporary ART clinics. The two procedures draw on substance and genes as natural resources for making parents and children, but they distribute the elements of identity and personhood differently. They allow a map to be drawn of what is rendered relevant to establishing parenthood (what I call a "relational" stage in the process of conceiving and bearing a child because it can support claims of parenthood) and what is irrelevant (what I call a "custodial" stage because it is a stage of the procedure that involves caring for the gametes or embryos as ends, not means, but cannot sustain parental claims). From this mapping exercise, some things that underlie the use of biology to configure kinship become apparent. The examples chosen are complicated cases that involve friendship egg donation, intergenerational but intrafamilial egg donation, and family-member surrogates. Of the six cases, one involves a familial egg donor and a commercial surrogate; the other cases involve friend or relative third-party reproduction where disambiguating kinship can be especially difficult. In these cases, the need to rule out the possibility of incest and adultery meant that the parties to the reproduction were explicit about how the correct kinship relations were being created and maintained. The fieldwork in this chapter is all drawn from California ART clinics, where sperm and egg donation, conventional surrogacy, gestational surrogacy, and commercial gestational surrogacy are legal.

I attempt to show the choreography and innovativeness of processes of naturalization. To do this, I argue for three things that are taken up again at the end of the chapter. First, drawing on the different ways in which these gestational surrogacies and egg donations distribute what is natural, I argue against a fixed or unique natural base for the relevant categories of kinship.[11] I also contend that high-tech interventions in reproduction are not necessarily dehumanizing or antithetical to the production of kinship and identity, as some critics of the procedures (including, for example, mainstream Catholic theologians and some feminists) have maintained.[12] Indeed, in the clinical setting, gestational surrogacy and donor egg IVF are a means, albeit a fallible means, through which patients exercise agency and claim or disown bonds of ancestry and descent, blood and genes, nation and ethnicity.[13] Third, for reasons that will turn out to be intricately related to the first two arguments, I argue that the innovations that are offered by these technologies do not intrinsically provide only new ways of drawing these fundamental distinctions, nor do they simply reinforce old ways of claiming identity.[14] The technologies contain both elements.

Substance and Genes Come Apart: Donor-Egg versus Gestational Surrogacy

Most Americans are used to thinking that there are just two biological parents and that they both donate genetic material—a bilateral or cognatic kinship pattern that is inscribed in our understanding of biogenetics. A baby is the product of the fusion of the mother's and the father's genetic material. Kin are divided into blood relations and non-blood relations, and blood relations are usually assumed to share biological substance with one another in a manner that simply reflects genetic relationship. By the turn of the century, however, it was already a commonplace in the United States that a woman can share bodily substance with a fetus to whom she is not genetically related. As a result of donor-egg in vitro fertilization and gestational surrogacy, the overlapping biological idioms of shared bodily substance and genes come apart. The maternal genetic material, including the determinants of the fetus's blood type and characteristics, is contributed by the egg, which is derived from the ovaries of one woman. Nonetheless, the embryo grows in and out of the substance of another woman's body; the fetus is fed by and takes form from the gestational woman's blood, oxygen, and placenta. It is not unreasonable to accord the gestating mother a biological claim to motherhood. Indeed, some have suggested that shared substance is a much more intimate biological connection than shared genetics and is more uniquely characteristic of motherhood, as genes are shared between many different kinds of relations. This lies behind the claim made by some feminists (and implying an ejaculatory role that is embraced by some lesbian couples opting for so-called lesbian IVF)[15] that egg donors should be thought of as second fathers, while the gestating woman should be considered the mother. It is also supported in the United States by a legal tradition that is exemplified by the California Uniform Parentage Act of 1975, which states that giving birth is proof of maternity. The law was enacted to protect nonmarital children and was invoked in a landmark contested-custody case involving gestational surrogacy (table 5.1). Where gestational motherhood arises from egg donation rather than surrogacy and is thus backed up by procreative intent, financial transaction, and laws that are sympathetic to a birth mother's claim to be a child's legal mother, the case for considering the woman who gestated the fetus as the "natural" mother can become almost unassailable. The cases from the field recounted below illustrate the in situ constructions of a case for natural motherhood as it arose in infertility clinics where I worked.

Gestational surrogacy means that eggs from one woman are fertilized with her partner's sperm in vitro (occasionally donor eggs or donor sperm are used in place of the gametes of one partner from the paying patient couple) and then are transferred to the uterus of a different woman, who gestates the pregnancy. The woman who gestates the pregnancy is known as a *gestational surrogate.* In some states, the woman from whom the eggs were derived and her partner have custody of the child and are the parents at the child's birth. In other states, although the laws are rapidly changing, the genetic mother and her husband have to adopt the baby at birth, and the surrogate's name goes on the birth certificate. If donor eggs from a third woman were used, the woman from the paying couple is still the mother, and she adopts the baby at birth, as in conventional surrogacy. A combination of intent, financial transaction, and genetics trace maternity through the various bodies producing the baby in commercial gestational surrogacy. In noncommercial gestational surrogacy, an emotional or a familial commitment takes the place of a financial transaction.

Gestational surrogacy is procedurally identical to donor-egg in vitro fertilization: eggs from one woman are fertilized in vitro and then gestated in the uterus of another woman. Two things make donor-egg in vitro fertilization and gestational surrogacy different from each other. First, in the case of donor-egg IVF, the sperm with which the eggs are fertilized comes from the gestational woman's partner (or a

Table 5.1
Legal and Regulatory Landmarks, 1973 to 2003

Griswold v. Connecticut, 381 U.S. 479 (1965) Held that a state could not ban a married couple from using contraceptives.

Eisenstadt v. Baird, 405 U.S. 438 (1972) Extended *Griswold* to invalidate a law forbidding the distribution of contraceptives to unmarried people.

Roe v. Wade, 410 U.S. 113 (1973) Held that there is a constitutional right to have an abortion.

Uniform Parentage Act, part of Senate Bill No. 347 (1975) Was subsequently incorporated into the California Civil Code as sections 7000–7021. These were repealed and replaced by equivalent provisions in the California Civil Family Code as sections 7600–7650, effective January 1, 1994. The legislation's purpose was to eliminate the legal distinction between legitimate and illegitimate children, but it was used in *Johnson v. Calvert* (see below) when considering maternity. Civil Code section 7003 provides that a natural mother and child relationship may be established by proof of having given birth to a child. Civil Code section 7004 specifies that a man can establish his paternity on the basis of a blood test. Nineteen states enacted uniform parentage acts.

Del Zio v. Columbia-Presbyterian Medical Center, No. 74-3558 (S.D.N.Y. 1978) Ruled for the plaintiffs in 1978 that the destruction of their in vitro embryo in 1973 had been malicious and had prevented them from having a child. This occurred many years before IVF had resulted in a live birth and was the result of the destruction of the couple's embryo by a doctor who thought that the procedure was illegal.

Indian Child Welfare Act (IWCA) (1978) A federal law that establishes tribal authority over all child-protective cases involving Native American children, including adoption, guardianship, termination of parental rights action, and voluntary placement of children. It covers all children who are members of a federally recognized tribe or who are eligible for membership in a tribe. The goal of the act when it passed was to strengthen and preserve Native American families and culture and to counteract the breakup of Indian families by nontribal agencies through the frequent removal of children from their homes. The ICWA may also have implications for egg and sperm donation and surrogacy by Native Americans.

In re Baby M, 525 A.2d 1128, *aff'd in part and rev'd in part,* 537 A.2d 1227 (1988) Mary Beth Whitehead was a conventional surrogate (that is, she became pregnant from one of her own eggs) who became inseminated with sperm from William Stern. She, William Stern, and Mary Beth's husband, Richard Whitehead, signed the surrogacy contract stating that Mary Beth Whitehead was contracting her ability to conceive and carry a baby to the Sterns. Richard Whitehead signed the contract to renounce his claims to paternity, and Elizabeth Stern did not sign the contract to avoid it seeming as if the contract was a form of baby selling. After the birth of so-called baby M, Mary

Table 5.1
(continued)

Beth Whitehead changed her mind about giving up the baby. After getting the baby back for what the Sterns thought would be a short while, Whitehead refused to return the baby to the Sterns as per the contract and fled to Florida. A Florida court and then a new Jersey trial court moved to terminate Whitehead's maternal rights. The New Jersey Supreme Court eventually reinstated Whitehead's legal maternity, granted her some visitation rights, but awarded custody to William Stern. The court was eager to avoid the appearance of motherhood being contracted and so sided with Whitehead to that extent. They also decided that the baby's best interests lay with the Sterns, who among other things, had submitted a brief describing their idyllic suburban home life.

Webster v. Reproductive Health Services, 492 U.S. 490 (1989) The American Fertility Society (subsequently the ASRM) filed an amicus brief arguing that the government should not come in between a patient and her physician's expertise regarding fetal viability.

Americans with Disabilities Act (ADA) of 1990 The ADA took effect in July 1992 and prohibits private employers, state and local governments, employment agencies, and labor unions from discriminating against qualified individuals with disabilities in job application procedures, hiring, firing, advancement, compensation, job training, and other terms, conditions, and privileges of employment. An individual with a disability is a person who has a physical or mental impairment that substantially limits one or more major life activities, has a record of such an impairment, or is regarded as having such an impairment. *Bragdon v. Abbott* (see below) made the ADA explicitly relevant to ARTs.

Johnson v. Calvert 851 p. 2d 776, 790 (cal. 1993) In 1990, a California couple, Mark and Crispina Calvert, paid Anna Johnson to be a gestational surrogate for them. Crispina Calvert had had a hysterectomy. Johnson is African American, Crispina Calvert is Filipina American, and Mark Calvert is Caucasian American. A contract was signed, the medical procedures undergone, and Johnson became pregnant with one of the Calverts' embryos. Relations between the Calverts and Johnson deteriorated late in the pregnancy, and Johnson decided to fight for custody of the baby. The California court system that heard the case (and its appeal) ruled that the baby boy could have only a single mother and—citing the intent and legality of the surrogacy contract and the genetic parentage of the child—ruled that Crispina Calvert was the natural mother. The ruling compared Anna Johnson to a wet nurse or foster parent and denied her visitation rights. Anna Johnson's defense cited in utero bonding and the California Uniform Parentage Act of 1975, which recognizes birth mothers as parents. The Uniform Parentage Act was ruled inappropriate as it had been enacted to limit discrimination against illegitimate children and not to cover cases arising from the new reproductive technologies. The Calverts appealed to the jury's sense of the primacy of their genetic tie over Johnson's gestational and birthing tie to the child by pointing to the child's blood type

Table 5.1
(continued)

and his racial resemblance to them and not to Johnson. There were also class-based speculations about Johnson's motives for suing for custody and about her fitness as a mother relative to the Calverts. The 1993 decision set the precedent for the primary consideration of procreative intent, rather than the best interests of the child, making ART reproduction more like "normal" reproduction and less like adoption.

Fertility Clinic Success Rate and Certification Act of 1992 (Pub. L. 102-493, 42 U.S.C. sections 263a-1 et seq.) (also known as the Wyden law) See figure 7.2 for clinic reporting obligations under the SART/CDC/RESOLVE program that runs the registry implementing the Wyden requirements.

Family Building Act (HR-1470, 1992) State of Illinois law mandating full infertility insurance coverage, including up to four cycles that progress to embryo transfer or further of IVF per couple.

Casey v. Planned Parenthood, 505 U.S. 833 (1992) Reaffirmed *Roe v. Wade.*

In re Andres, 156 Misc. 2d 65, 591 N.Y.S. 2d 946 (1993) In this gestational surrogacy case, the courts decided that the gestational and not the genetic woman was mother, although they granted custody to the commissioning couple and required the wife who provided the egg to adopt the child.

Family and Medical Leave Act (FMLA) of 1993 Federal act that grants most federal employees a total of up to 12 workweeks of unpaid leave during any 12-month period for the purposes of providing medical care to a spouse, parent, or child; in the event of the birth of a child; or for the purposes of adoption or foster care. Unclear whether leave can be sought under the act to pursue infertility treatment.

In re Marriage of Moschetta, 25 Cal. App. 4th 1218 (1994) A California case of traditional surrogacy that held that the intended father and the surrogate were the legal parents and shared legal custody after the separation of the intended father and his wife.

Multi-ethnic Placement Act (MEPA) of 1994, as Amended by the Interethnic Adoption Provisions (IEP) of 1996 This amended act, now part of the Social Security Act, established a "race-blind" federal adoption mandate for public adoptions in response to the perception that policies that considered race, color, or national origin in foster care and adoption were delaying the permanent placement and formation of family attachment of children of color. Does not apply to eligible children covered under the Indian Child Welfare Act (see above).

Jaycee B. v. Superior Court, 49 Cal. App. 4th 718 (1996) Jaycee was born with five potential parents—a commissioning couple, Luanne and John Buzzanca, a surrogate mother, and anonymous donors for both the egg and the sperm. John Buzzanca filed for divorce shortly before the child was born, and Luanne Buzzanca, who had physical custody of the child after her birth, sought child support from her soon to be former husband.

Table 5.1
(continued)

Washington v. Glukberg, 116 U.S. 2021 (1997) Held that under ordinary circumstances, there is no general right to physician-assisted suicide, thereby constituting an antiprivacy precedent in the area of medical life and death.

Bragdon v. Abbott, 524 U.S. 624 (1998) The U.S. Supreme Court, in a ruling that involved a claim of discrimination by an HIV-positive woman against a dentist who refused to treat her, held that under the Americans with Disabilities Act (ADA), reproduction is a "major life activity." This means that an individual is protected from discrimination if he or she has a physical or mental impairment that substantially limits reproduction.

Buzzanca v. Buzzanca, Sup. Ct. No. 95-DOO2992 (1998) Baby Jaycee was conceived by anonymous embryo donation & gestational surrogacy. The Buzzancas, who entered into the surrogacy agreement were deemed Jaycee's parents after the dissolution of their marriage for custody and child support purposes despite the lack of a genetic or gestational tie.

Dunkin v. Boskey, 82 Cal. App. 4th 171; 98 Cal. Rptr. 2d 44 (Cal. App. 1st Dist. 2000) A California case in which an intended father, having previously consented to his female domestic partner's artificial insemination, was held to have a breach of contract case for damages against his female domestic partner after they split up. The reasoning was that an unmarried couple should be treated the same as a married couple, possibly paving the way for a similar decision in same sex relationships.

Revised Uniform Parentage Act (UPA) of 2002 For children born as a result of ARTs, the revised UPA precludes sperm, egg, and embryo donors who do not intend to parent from claiming parental rights. It also allows single women and married and unmarried couples who are undergoing treatment with ARTs to establish parentage on the basis of their intention and consent to parent. The revised act allows for a determination that a child has more than one legal mother.

New Jersey Family Court (2003) Ruled that both women in a lesbian couple could have their names put on their baby's birth certificate as its parents, without either one having to adopt. One woman provided the egg and the other gestated the baby, and they used donor sperm and IVF. Both women share a financial obligation to the child, and either will have custody if the other dies.

Lawrence & Garner v. State of Texas 539 U.S. 558 (2003) A U.S. Supreme Court ruling of six to three that sodomy laws are unconstitutional. A neighbor had reported a "weapons disturbance" at the home of John G. Lawrence, and when police arrived, they found two men having sex. The presence of an African American man on the suburban property of a white man reportedly led the neighbor to think that a weapons disturbance was occurring. Lawrence and Tyron Garner were held overnight in jail and later fined $200 each for violating the state's homosexual conduct law. The neighbor was later convicted

Table 5.1
(continued)

of filing a false police report. The majority opinion was written by Kennedy, joined by Breyer, Souter, Ginsburg, and Stephens and held that intimate consensual sexual conduct was part of the liberty protected by substantive due process under the fourteenth amendment. *Lawrence* invalidates similar laws throughout the United States that apply to consenting adults acting in private. O'Connor concurred on equal protection grounds.

Note: In States with legal precedent for second-parent adoption for gay, lesbian, or transgender individuals, use of ARTs by these individuals may also be legally facilitated. See ⟨http://www.hrc.org/familynet/adoption_lawsasp⟩.

donor standing in for him and picked by that couple). In the case of gestational surrogacy, the sperm comes from the partner of the provider of the eggs (or a donor standing in for him). From a lab perspective, there is no difference: sperm collected by masturbation (or TESE) is prepared and added to retrieved eggs, and the embryos are incubated and transferred identically in both cases. The sperm comes from the person standing in the right sociolegal relationship to whichever of the women is designated as the mother-to-be. The identity of the intended mother depends on the person who came into the clinic for treatment for infertility, the various parties' reproductive history, and the source of payment. Where additional donors are used or where the egg donor or gestational surrogate is being contracted on a commercial basis, who is paying for the treatment strongly influences who the designated parents are. In cases where motherhood is contested, one or more of these aspects breaks down or falls out of alignment with the others. Both gestational surrogacy and donor-egg in vitro fertilization separate shared bodily substance and genes, but whereas donor-egg IVF traces motherhood through the substance half of this separation, gestational surrogacy traces it through the genetic half.

Cases from the Field

Cases 1 and 2: IVF with Donor Egg

1. Giovanna

Case summary: Giovanna will gestate embryos made from donor eggs from an Italian American friend and sperm from Giovanna's husband. Giovanna and her husband will be the parents if pregnancy ensues.

One afternoon, I was in an examination room with a patient who was waiting for a physician to give her an ultrasound scan. Giovanna described herself as an Italian American who was approaching age forty. She explained that she had tried but "failed" IVF before using her own eggs and her husband's sperm. Her response to the superovulatory drugs and the doctor's recommendation had persuaded Giovanna to try to get pregnant using donor eggs from a younger woman. Almost all clinics report better IVF implantation rates using donor eggs, which are retrieved from women under thirty-five years old, than using an older patient's eggs. Giovanna said that she had decided to use a donor who was a good friend rather than an anonymous donor.[16] Choosing a friend for a donor seemed to be an important part of reconfiguring the experience of pregnancy: if conception was not to occur inside her body or with her eggs, then she wanted to have the emotional attachments of friendship to and the ability to make corresponding demands on the woman who was to be her donor. Giovanna described her friend as also Italian American and said that she was excited and ready to help. She described the shared ethnic classification as being "enough genetic similarity." Further, Giovanna accorded her gestational role a rich biological significance: she said that the baby would grow inside her, nourished by her blood and made out of the stuff of her body all the way from a four-celled embryo to a fully formed baby.

Scholars have documented a gradual U.S. trend toward socializing gestation and making the pregnant woman susceptible to medical scrutiny and state regulation and intervention.[17] In these arenas, gestation is increasingly assimilated to the care that a parent provides to a child once it is born. When gestational surrogacy is uncontested, everything except for fertilization is equated to child care, despite its biological nature. This makes genetics the essential natural component that confers kinship and minimizes the role of gestation. Giovanna, however, cast her gestation as conferring kinship because of its biological nature, despite the absence of genetic connection. Against the socializing and genetic essentialist cultural trend, she renaturalized gestation.

Giovanna pried apart the natural, biological basis for specifying mother and child relations into separable components. In addition, she complicated the natural status of the genetic component that would be derived from her friend by socializing genetics. What mattered to her in genetic inheritance was that the donor share a similar history to her own. She and her friend were Italian American, came from the same kind of home, both had Italian mothers, and had grown up with the same cultural influences. Genes were coding for ethnicity, which Giovanna was expressing as a national and natural category of Italian Americanness. As in so many cases of contemporary biomedicine, genes have social categories built into them without which they would not make sense or be relevant. This is a reversal of what is often presumed to be the unidirectionality of genealogy. Genes figure in Giovanna's kinship reckoning because a chain of transactions between the natural and the cultural grounds the cultural in the natural and gives the natural its explanatory power by its links to culturally relevant categories.

Giovanna separates biology into shared bodily substance and genetics and formulates a meaning of genetics for her in the context of her procedure. She resists the scientistic impulse to assume that biology underwrites sociocultural potential and not the other way around and that biology is sufficient to account for sociocultural reality. For her, the reduction to genes is meaningful only because it codes back to sociocultural aspects of being Italian American (it is not unidirectional). Likewise, the ethnic category is not just a category that performs a transitional function between nature and culture. It is a category of elision that collects disparate elements and links them without any assumption that every one of the sociocultural aspects of having an Italian American mother, for example, needs to map back onto biology.

2. *Paula*

Case summary: Paula will gestate embryos made from donor eggs from an African American friend and sperm from Paula's husband. Paula and her husband will be the parents if pregnancy ensues.

In a related case, an African American patient, Paula, met me once at the clinic and spontaneously offered commentary on the kinship implications of her upcoming procedure. Paula had undergone "premature ovarian failure" and entered menopause in her early thirties before she had had any children. She and her husband had decided to try donor-egg IVF, and she was hoping to be able to carry a pregnancy. They had not yet chosen an egg donor, and Paula said that she would first ask her sister and a friend to be her egg donor. Paula expressed a strong preference for using a donor "from my community." She said that using a donor was not as strange as it might at first seem, as it was like something "we've been doing all along." She explained that in African American communities, it was not unusual for women to "mother" or "second mother" their sister's, daughter's, or friend's children.[18]

Paula's explanation suggests the possibility of legitimizing the natural "deviance" of her procedure by pointing to its social basis: it was OK to give birth to a child made with another woman's eggs because sharing child rearing was already a common social phenomenon. This is the reverse of the familiar strategy where naturalization can normalize social deviance, as, for example, when behavior is explained by genetics or by an underlying mental or medical problem. If using a donor to get pregnant is just one more way of doing something that is already a prevalent social phenomenon—dividing different aspects of mothering across generations, between friends, and between sisters—then it is not a radical departure from existing social practice. In presenting it like this, Paula was normalizing her reproductive options. Rather than being exploitative, using a donor is assimilated to other ways in which women help each other to lead livable lives.

Cases 3 and 4: Gestational Surrogacy with a Family-Member Surrogate

3. Rachel, Kay, and Michael

Case summary: Rachel will gestate embryos made from Kay's eggs and sperm from Kay's husband, Michael. Rachel will give birth, but Kay and Michael will be the parents. In addition, Rachel is Michael's sister.

Michael and Kay had a history of long-term infertility, including two unsuccessful attempts to get pregnant with IVF. They decided to maximize their chances on one last attempt at IVF by using Michael's sperm and Kay's eggs with a gestational surrogate, Michael's sister, Rachel. Several staff members referred to Rachel approvingly as an ideal surrogate during the weeks of treatment. Her own family of three children was already complete. She was actively compliant, making explicit references to how she would organize her life to do whatever was in the treatment's best interests. She was good humored about the long waits during appointments at the clinic. Her husband's job was lucrative enough that she could assume the risks associated with taking a leave of absence from work, and she was happy to take a break from work to spend more time at home with her children. The generic role of the "good patient," then, was masterfully deployed by Rachel to emphasize that her role was subordinate to those whose procreative intent they were all working to realize. I followed Rachel's treatment as a case study of gestational surrogacy.

An important activity during this couple's treatment was ruling out Rachel as a parent and counting in Kay because incest with Michael (her brother) would be implied if Rachel were the mother. Such conventional strategies as distinguishing between medium and information and between nature and nurture were used by the parties involved in the pregnancy to rule out incest and to negotiate descent and heredity. During one appointment, Rachel, the surrogate, turned to Kay, the mother-to-be, and said that it was lucky that she, Rachel, had had her tubes tied after her last child. Kay raised her eyebrows, seeming not to understand, and Rachel elaborated by saying that otherwise there might be a danger of one of her eggs being in her tubes or uterus and some spare sperm out of the petri dish from her brother fertilizing her egg. Kay understood, laughed, and agreed. I was struck at the time that a brother's and sister's sperm and egg would be incest but that gestating a baby with a brother's and sister-in-law's sperm and egg was not, despite the fact that her brother's baby would grow inside Rachel.

Five embryos resulted from the egg retrieval from Kay and the subsequent fertilization with Michael's sperm. On the day of the embryo transfer, the doctor tried to persuade Kay and Rachel to let him transfer three embryos into Rachel and two into Kay: "Are you sure you don't want to split the embryos?" Kay and Rachel stood firm. Later on, in the staff room, the doctor expressed disappointment that Kay and Rachel would not both have an embryo transfer and thereby enable him to "make history" as he called it. If they had both become pregnant, it would have been the first time that a single set of twins would have been born from different mothers.

Rachel's husband, referred to by the staff as an "awkward attorney," was not present for the embryo transfer and was conspicuously absent throughout the whole treatment cycle. Kay's husband, Michael (Rachel's brother), was in the room, though, as the father-to-be. Rachel's prosthetic role was heightened during the embryo transfer by the presence of her brother while she was lying with her legs in stirrups. After the embryo transfer, Rachel had to remain in a prone position for two hours. Michael left soon after the embryo transfer, but Rachel and Kay were "in this together," as Kay expressed it, and so Kay stayed by Rachel's side.

During the two-hour wait, Kay and Rachel discussed what the baby would look like if Rachel became pregnant. Rachel began by flattering Kay, saying that if the baby looked like a combination of Michael and

Kay, it would be good-looking. This fit the normal genetic reckoning that half the characteristics comes from the egg (Kay's) and half from the sperm (Michael's). Kay responded by saying that the baby might look more like Rachel and that if it came out looking like her nephews and niece (Rachel's children), she would be happy. Perhaps Kay was just returning the compliment. But if the baby could look like Rachel, then Rachel was not as merely custodial in this pregnancy as the logic of gestational surrogacy seems to need her to be to avoid being incestuous. Rachel took up the thread, asking rhetorically whether the baby could get anything of her from growing inside her. They joked about dogs looking like their owners and mentioned that identical twins can have different birth weights depending on how they fare during pregnancy, both classic cases of the effect of "nurture" on "nature." They resolved the question by accepting that if Rachel had an effect on what the baby looked like, it would be because she had provided a certain environment for the baby to grow in and not because she was part of the baby's "nature." Rachel's role in the pregnancy was thus returned safely to the realm of caregiver and provider of a nurturing environment. Their narrow geneticization of incest prevented biological embodiment of her brother's child from being incest. The genetic discourse of naturalization worked in opposition to the discourse of biological kinship through embodiment by *de*naturalizing the latter. The threat of incest was again avoided, leaving the logic of the procedure and both compliments intact.

Ten days later, Rachel's pregnancy test was positive. The embryologist recounted the story that Rachel told her: Rachel didn't phone Kay straight away but instead went out and bought a little teddy bear to which she attached a note saying "Your child(ren) are doing fine with Auntie Rachel. Can't wait to meet you in eight and a half months." She took the teddy around to Kay's house, left it on the doorstep, rang the doorbell, and hid. Apparently Kay opened the door, found the bear, and burst into tears. It was not until Rachel came in for her final infertility clinic scan that her husband, the awkward attorney, was mentioned in connection with the pregnancy. Rachel told the doctor that her husband was happy about the twins that she was carrying. Apparently he had suggested naming the twins after the doctor, calling one by the doctor's first name and one by his last name. In choosing names and the male doctor for eponymy, Rachel's husband was able to contribute to the babies' lineage. In building in quasi-patrilineal kinship through naming practices, Rachel's husband also illustrated the manner in which people in this site routinely used tropes (including conservative ones like male eponymy) with which they were familiar from other contexts and extended them to cover and disambiguate kinship in this novel setting.

4. *Jane*

Case summary: Jane will gestate embryos made from an infertile patient's eggs and sperm from that patient's husband. Jane will give birth, but the patient and her husband will be the parents. Jane is married to the infertile patient's brother, so she is the patient's sister-in-law by marriage.

While Rachel's treatment cycle was in progress, another noncommercial surrogacy treatment was ongoing at the same clinic. In this case, the surrogate, Jane, was related by marriage to the intended infertile couple. The surrogate was married to the brother of a woman and her husband who were providing the gametes and hoping to be parents. Staff members compared Jane unfavorably with Rachel from the beginning, complaining that her heart was not in it and that Jane and her husband's sister (the mother to be) were "passive aggressive" toward each other. When Jane failed to get pregnant after two attempts, the clinic's psychologist suggested at the weekly staff meeting that Jane's "unresolved feelings," deemed evident from her psychological evaluation, were getting in the way of implantation of the embryos. The psychologist presented a narrative that the demands of the emotional contract necessary to undertake a pregnancy on someone else's behalf could not be sustained unless the two had a very close relationship. She voiced a strong preference for close girl friends or sisters over relations by marriage when a noncommercial surrogate was needed. In Jane's case, then, reproductive failure was blamed on a relationship that was insufficiently biologically or socially grounded to sustain the biological demands of this particular form of custodial care.

Case 5: Intergenerational Donor-Egg IVF

5. *Flora*

Case summary: Flora will gestate embryos made from her daughter's eggs and sperm from Flora's second husband. Flora will give birth, and her second husband will be the father. Flora's daughter will be the baby's half-sister and not its mother.

A fifty-one-year-old woman, Flora, came in for treatment. She was perimenopausal and had five grown children from a previous marriage. She didn't fit the typical patient profile of the middle-class white woman who had postponed childbearing. Flora was Mexican and crossed the border from an affluent suburb of Tijuana where she lived to Southern California for her treatment. With five children, she already had what many would consider "too many" children. She had recently been remarried to a man many years her junior who had not yet had children. Flora was quite explicit about the gender, age, and financial relations between herself and her new younger husband and her desire to, as she put it, "give him a child."

Because of her age, Flora was told that if she wanted any significant chance of getting pregnant, she should find an egg donor. The donor eggs would be inseminated by Flora's husband's sperm, and resulting embryos would be transferred to Flora's uterus or frozen for use in subsequent cycles. Flora read widely in the medical and popular literature and frequently made suggestions about or fined-tuned her own protocol. She also picked her own egg donor: one of her daughters in her early twenties who already had children.[19] The mother's and daughter's cycles were synchronized, and the daughter was given superovulatory drugs to stimulate the simultaneous maturation of several preovulatory follicles. The daughter responded dramatically to the drugs, and the physician and embryologist removed sixty-five eggs from her ovaries (five to ten eggs is considered normal). The eggs were inseminated with Flora's husband's sperm, forty-five fertilized, and five fresh embryos were transferred to Flora's uterus that cycle. Flora did not get pregnant in the fresh cycle or in the first two frozen cycles but did in the third.

Flora did not seem overly perturbed by the intergenerational confusion of a mother giving birth to her own "grandchildren" and to her daughter's "daughter and sister." I never heard her mention that her sixth child would be genetically related to her current husband as well as to the father of her first five children.[20] Instead, she discussed her daughter's genetic similarity to her. Like Paula, the African American woman mentioned earlier, Flora also assimilated her case to existing social practice, in this instance to the prevalence of generation-skipping parenting (where a grandparent parents a child socially and legally) in communities with which she was familiar. Nonetheless, Flora signaled some ambivalence on the part of the daughter. When asked who her donor was, she replied, "My daughter," adding that her daughter was "not exactly excited, but she doesn't mind doing it." The daughter herself said when the mother was out of the room that she didn't mind helping them have a single baby but that the huge number of embryos stored away was unsettling. After all, she added, her mother already had a family: "She doesn't need to start a whole new family. One baby is one thing but . . . !"

The daughter's reluctance to see her mother having "a whole new family" might have been due in part to a distrust or disapproval of her mother's relations to her new husband or with a reluctance to have the grandmother (Flora) of her children being a mother of babies again. The daughter's anxiety about the stock of frozen embryos, however, seemed to be at least partly an anxiety about the existence of unaccounted for embryos using her eggs and her stepfather's sperm. Using one of her eggs to help initiate a pregnancy that was clearly Flora's and her husband's was acceptable. It placed Flora between the daughter and her stepfather. The embryos in the freezer, however, were in limbo. If they were not used to initiate a pregnancy in Flora, then they existed as the conjoined gametes of the daughter and the stepfather. As in all the cases recounted here, the trajectory of treatment, as a proxy for procreative intent, was of paramount importance. Keeping biologized notions of descent in line with that intent and keeping incest and adultery at bay were clearly possible in the world of the clinic, which is structured around treatment trajectories. The status of the embryos in the freezer, however—even though technically owned by Flora and her husband—were off that trajectory and provoked Flora's daughter to express anxiety about inappropriate kinship.

Case 6: Intergenerational Gestational Surrogacy with Donor Egg

6. Vanessa and Ute

Case summary: Vanessa has contracted commercially to be a surrogate for Ute and her husband, using the husband's sperm and donor eggs from Ute's daughter.

A final case concerns Vanessa, who began a surrogacy agency shortly after she served as a commercial gestational surrogate and gave birth to a baby for another couple. Vanessa had seen a program on television in which she noticed the "joy in the mother's eyes" when the baby was handed over by the surrogate. Vanessa's family was in some financial difficulty after their small-scale manufacturing operation closed, and the approximately $12,000 that she might make as a successful surrogate was attractive. She expressed the decision to try surrogacy in a religious idiom, as a chance offered by God simultaneously to do good and to make a fresh start. Vanessa was introduced to "her" couple, a "German woman of about forty" (Ute) and an "Asian man."

On the first treatment cycle, Ute ovulated before the physician took her to surgery for ovum pickup, and only one egg was retrieved at surgery. The one egg was successfully fertilized with Ute's husband's sperm and transferred to Vanessa, but Vanessa did not get pregnant. For the second treatment cycle, Ute and her husband decided to combine gestational surrogacy with donor-egg IVF. Ute had an adult daughter from a previous marriage who agreed (as Flora's daughter had) to be her mother's and stepfather's donor. Ute, her daughter, and her husband decided that the use of the daughter as donor would be kept secret from everyone outside the clinic setting. Ute explained that using her daughter's egg was the next best thing to using her own eggs because of their genetic similarity. The daughter was safeguarded from being considered the biological mother in the pregnancy by making the relevant fact be the genetic relation of her eggs to Ute. This reduction to the genetics of the eggs and the protection that it afforded the daughter were both reinforced by removing the daughter and keeping her role secret. The logic of gestational surrogacy—essentializing the genetic aspect of biological kin and making the blood and shared bodily substance of gestation custodial rather than relational—was also maintained. Ute was providing the essential genetic component, even though that genetic material had traveled a circuitous route from her, down a generation to her daughter, and back up to her again. This required reversing the usual temporal direction of genealogy, but naturalizing kinship to genetics to a large extent removed the need to mark kinship by descent and so removed the temporal direction from the kinship reckoning.

Vanessa did not get pregnant on the fresh IVF cycle using the daughter's eggs, but sufficient embryos had been frozen to allow for a subsequent attempt. On the frozen cycle, Ute's daughter's and husband's remaining embryos were thawed and transferred to Vanessa's uterus, and Vanessa became pregnant with a singleton. The pregnancy, unlike her four previous ones, was not easy for Vanessa. Under her surrogacy contract, she was not allowed to make her own medical decisions while pregnant, and she had to consult with the recipient couple and the doctor before taking any medications or changing her agreed on routine in any way. The recipient couple and the physicians took over jurisdiction of her body for the twelve months of treatment and pregnancy, disciplining her body across class lines. Nonetheless, Vanessa described the pregnancy almost wistfully, explaining how exhilarating the intimacy with the couple was and how spoilt she had been. During the pregnancy, Ute and her husband took Vanessa out, bought her fancy maternity clothes, and temporarily conferred on her their privileged social status. Vanessa underlined the ambivalence of the "kin-or-not?" relationship between the surrogate and the recipient couple in commercial surrogacy arrangements with the oxymoronic observation that they are "the couple you're going to be a relative with for a year and a half."

Vanessa seemed surprised by the severing of the ties of relationship between herself and the recipient couple after the baby was born and handed over. She said that the couple stopped contacting her and that when she called to find out how they and the baby were, the couple made excuses and hung up quickly. Vanessa's relationship to the couple for the year and a half was enacted because she was prosthetically embodying their germ plasm and growing their child, but at no point was she related to the recipient couple or the baby in her own right. Unlike the cases described above, Vanessa was commercially contracted, and her reproduction was classically "alienated" labor. The commissioning couple honored the capitalist contract; they paid her and appropriated the surplus value of her reproductive labor—the baby. The genius of capitalism is sometimes said to be that the fruits of one's labor can be exchanged for money, without setting up a chain of reciprocal obligation. Thus, once the baby was born, Vanessa was in many ways just like any other instrumental intermediary that had been involved in establishing the pregnancy, such as the embryologist or even the petri dish. Because she cared for the couple and the baby, had good reasons to feel connected to them, and did things like make phone calls meant perhaps that she needed some postnatal management to be kept in the background. But because she had been commercially contracted, the logic of disconnection was the same as for other intermediaries in the recipient couples' reproduction. The irony was that this particular naturalization was heightened by the deeply entrenched conventions of capitalism.

Making Kinship: Relational and Custodial

I do not know whether a pregnancy was established in every case described above (cases 1, 2, and 4). Here I try to isolate some of the strategies that were used in the clinical setting in each case for delineating who the mother was for each child. These strategies are not exhaustive and cannot be expected to be invariant in different arenas of the patients' lives either. For example, legal and familial constraints bring their own sources of plasticity and relative invariance that powerfully help determine kin. But the clinic is one significant site of negotiation of kinship and is of particular interest because it articulates between the public and the private and because it illustrates flexibility in biological and scientific practice. I emphasize the mothers' relatedness because in the cases I have chosen, the procedures raise a challenge to biological essentialism through the separability of egg, gestation, and biological mother. The cases also challenge biologically essentialist understandings of daughter, husband, father, grandmother, aunt, and child. For each case, I distinguish different significant intermediaries in establishing the pregnancy. I then sort the intermediaries by whether they are relational or custodial in the determination of who counts as the mother. I also ask where and under what conditions the ways of designating the mother are liable to break down or contestation.

In ART clinics, clinicians, patients, family members, surrogates, and others together embody the intent and effort that are crucial to reproduction in these settings. Procreative intent inscribes only some of these parties (usually a paying patient couple) as kin of the anticipated child(ren), however. For analytical purposes, I call any stage in the establishment of a pregnancy *relational* if it implicates kinship. Here, I am primarily interested in working out who the mother is, so a relational stage in this determination would be a stage that implicates one (or more than one) mother. As the cases showed, many resources for making a stage relational are not necessarily well differentiated. Biology and nature are resources; so are a wide range of legal, socioeconomic, and familial factors that make up procreative intent, such as who is paying for treatment, who owns gametes and embryos, whose partner is providing the sperm, and who is projected to have future financial and "nuclear-family" responsibility for the child. A stage can be made securely relational by separating but bringing into coordination biological and social accounts of the relationship. Depending on the kind of parenting in question, different kinds of coordination are appropriate. By contrast, I call a stage *custodial* if it enables relatedness but does not itself thereby get configured in the web of kinship relations.[21] A woman is custodial in conceiving and bearing a child if she is biologically involved in bringing the child into the world and yet is not (or not primarily) implicated as its mother. A custodial stage is somewhat like an instrumental intermediary, then, and yet custody is not merely instrumental. Both the custodial woman and the child-to-be are ends in themselves and not mere means.[22] The degrees of custodianship range from one that is strictly limited to one that implies long-term commitment but not parenthood. The cases described above exhibited different strategies for achieving kinship implication of some people and not others. Breakdown (from the point of view of clinics or designated recipient couples), contestation, and prohibition usually occur when an attempt by at least some of the actors to render one or more stages as custodial rather than relational is contested or fails.

Significant Intermediate Stages in Designating Motherhood in Each Procedure

In the first case, that of Giovanna, the Italian American woman who was planning to undergo a donor egg procedure with the eggs of an Italian American friend of hers, the two potential candidates for motherhood were the donor friend and Giovanna herself. The friend was made appropriately custodial using three strategies. First, the friend's contribution of the eggs was biologically minimized by stressing the small percentage of the pregnancy that would be spent at the gamete and embryo stage versus the length of time that the fetus would grow inside Giovanna. Second, genetics were redeployed, so that the friend's genes were figured as deriving from a common ethnic gene pool (Italian American) of which Giovanna was a member and so in which she was also represented. Third, the friend was secured as a custodial intermediary in the pregnancy by stressing her other bonds to Giovanna of friendship and mutual obligation. Despite her custodial role of the eggs destined to make Giovanna's baby, her relation to the baby could be reregistered as enhancing her already significant cultural and personal connections to Giovanna. To further disambiguate who should be considered the mother, Giovanna put forward strategies through which she could assert her own relationality. Giovanna's claim to motherhood was to come from her gestation of the baby and her provision of the bodily substance and bodily functioning out of which the baby would grow and be given life. Giovanna stressed the significance of the gestational component of reproduction and emphasized the importance of the experiential aspects of being pregnant and giving birth in designating motherhood. Further, Giovanna was married to and would parent with the provider of the sperm.

In Paula's case (the African American woman who was going to use an African American friend or one of her sisters as her donor), the strategies of separating relationality and custodianship were similar to those employed by Giovanna, but there were also interesting differences. The most striking difference was that Paula drew a more tenuous line between who was a kin relation and who was a custodian, being content to leave more ambiguity in the designation of motherhood. She took legitimacy for the procedure from the fact that shared parenting was commonplace among people she identified with racially, and she also commented on the natural, biological confirmation of these social patterns that her upcoming procedure would provide. Socially and biologically her motherhood—including its ambiguity—would be recognizable and legitimate. Her motherhood would be sufficiently relational, and her donor sufficiently custodial without either being wholly so. Like Giovanna, then, Paula gave an ethnic or racial interpretation to the genes such that by getting genetic African Americanness from her donor the baby would share genetic racial sameness with Paula.

The familiar trope (underlying, for example, the Human Genome Diversity Project)—that genes code for racial distinctions, group inclusion and exclusion, and ethnic purity—seemed to be readily available to patients like Giovanna, Paula, and Flora, for whom group racial or ethnonational identity was culturally significant. The equally prevalent trope (exemplified in the Human Genome Project)—that genes function as the thing that provides the definitive mark of individuality, "the DNA fingerprint," which is passed down cognatically from a mother's and a father's individual contribution—was the template used by the patients of unmarked or hegemonic racial and ethnic origin. Depending on whether genetics was socialized or individualized, different naturalization strategies were available. As in all the kinship innovations described here, these naturalization strategies drew on deeply rooted and familiar forms of forming and claiming kin and simultaneously extended the reference of the kinship terms being disambiguated through the strategies. Of the patients whose cases are discussed in this chapter, the African American Paula seemed to be the most comfortable with deindividualizing genetics. The Italian American Giovanna, the Mexican Flora, and the German Ute all also mobilized the idea of genetic ethnic or individual (depending on the case) similarity with their gamete donors to assert genetic connection to their offspring, but unlike Paula they were all reasonably strongly invested in making their claims to motherhood individual and exclusive.

Michael, Kay, and Rachel (the gestational surrogacy between a brother, his wife, and his sister, respectively) deployed yet other familiar resources for ensuring the custodial role of Rachel and the relationality of Kay. Because of the need to protect Michael and Rachel from incest, the negotiations over custody versus relationality were explicit and repeated at different phases of the treatment. Although the opportunities for ambiguity were legion, there was a "zero-tolerance" standard for ambiguity in the designation of Kay and not Rachel as the mother. Kay was the infertile paying patient, receiving treatment and hoping to parent. She also provided the eggs that contained the genetic information and was married to and intended to parent with the provider of the sperm. These elements of her claim to motherhood were stabilized by emphasizing the unidirectional, individualized genetic basis of heredity and a mirroring between biological kinship and this understanding of heredity. When a threat arose from the uniqueness of this basis of heredity (when Kay said to Rachel that she wouldn't mind if the children looked like Rachel and the ensuing conversation), the threat was removed by assimilating that kind of acquisition of characteristics to environmental custodial factors. The provision by Rachel of the bodily substance and site for fetal development was made custodial and not parental or incestuous through two strategies. First, Rachel was assimilated at all points by her active compliance, by the kinds of boundary discussions she initiated, such as how lucky it was that her tubes were tied, and by a model on which pregnancy and birth were instrumental phases of Kay's and Michael's reproduction, with which she was helping. Second, Rachel's relatedness to Kay, Michael, and the children she was carrying was used to make her custodial assistance exactly what would be expected of her social role. As Rachel put it, the children were fine with their auntie (whose social role would be expected to be custodial) but couldn't wait to be reunited with their parents. The natural relation of aunt underwrote a social role of custodianship that in turn helped denaturalize Rachel as a potential mother to the children.

In the Mexican Flora's case (the perimenopausal woman whose daughter was the donor from whom sixty-five eggs were retrieved), a significant strategy in making her daughter custodial was to see the daughter as custodian rather than origin of the genetic material. The eggs used were not the daughters eggs per se but were retraced to Flora by "rewinding" genealogical time, such that they contained genetic material sourced from Flora. The daughter's gamete contribution was expressed as a detour to her mother Flora's genetic material that could no longer be accessed directly from Flora. Flora, perhaps more closely than any of the other donor-egg patients whose cases are described here, attempted to recapture in her own claims to motherhood the genetic idiom of linear descent. Flora's case was complicated by the intergenerational element, however, and her daughter's custodianship was threatened from at least two sources. Flora and her daughter discussed the similarity of what they were doing to other intergenerational parenting in which a grandparent can be the social and legal parent. Drawing an analogy with prevalent social practices, as Paula had done, Flora strengthened the legitimacy of the procedure and so stabilized her claim to being the mother of the child. In making this analogy, however, the daughter's role was in danger of being compromised because grandparenting often allows for the "real" parent to reclaim his or her parental jurisdiction. Unlike for Paula, this analogy to social practices did not loosen the designation of who was to count as mother but was meant to disambiguate it even beyond the norm for the social practices with which she was drawing the analogy. Both Flora and her daughter were adamant that there was to be no ambiguity in who was to be the mother: it would be Flora. Flora's marriage to the provider of the sperm and the incestuous implications of reckoning it otherwise reinforced the disanalogy with social grandparent parenting. The other threat to Flora's daughter's custodial role came from the unintended stockpile of embryos that were formed when the daughter's eggs were fertilized by her stepfather's sperm. The frozen embryos in the lab were only tenuously tied to the trajectory of Flora's reproduction and its social, legal, economic, and emotional umbrella of procreative intent. The embryo's quasi-independence left room for them to seem like the products of incest or agents of additional nonintended pregnancies, both of which were troubling to the daughter.

In Vanessa's pregnancy (which she carried as a commercial surrogate for the German Ute and her husband, from an embryo formed with eggs donated by Ute's daughter), three potential candidates could be designated as the mother. Vanessa was the commercial gestational surrogate; Ute was the intended mother, who was the wife in the paying patient couple; and Ute's daughter was the egg donor. As for Rachel and Jane, Vanessa was made custodial by assimilating her role in gestating the fetus to the provision of a temporary caring environment. Unlike Rachel's and Jane's cases of gestational surrogacy, however, Vanessa's custodial role was not elicited by the obligations of a prior relationship to the recipient couple. Instead, her services were contracted commercially, and Vanessa had no further claim on the child after the birth. This disconnection was underwritten by the assumption of contractual arrangements that both parties agree that recompense is satisfactory despite the possible incommensurability of the things being exchanged. Furthermore, contracts assume that the transaction itself is limiting and does not set up any subsequent relationship or further obligation. Vanessa's custodial role was threatened when she experienced the sudden severing of relations after the birth of the baby as baffling and troubling. Like Flora's and Ute's daughters, Vanessa was "dekinned," but whereas the former two desired this, Vanessa did not (or at least mourned the relationship). The contractual relation ensured the temporary relation of shared bodily substance with the recipient couple, but insofar as it was uncontested, it sustained no further relationality.

Ute's daughter, who was acting as the egg donor, was excluded from being the baby's mother in two ways. First, the genetic contribution contained in her eggs was described as being closely similar to her mother's genetic material. This is not quite the same argument as that made by Flora and her daughter, where the daughter became a vehicle or detour through which genetic material originally from Flora had passed. In this case, the mother and daughter simply alluded to the similarity of the mother's and daughter's genes. The daughter was further spared from being implicated in parenthood by the recipient couple's commitment to keeping her role a secret. Ute sustained her claim as the intended mother not by wholly capturing either of the predominant biological idioms (genetics as represented by providing the egg or the blood and shared bodily substance as represented by gestation). Instead, the genetic component from the daughter was deemed similar enough to stand in for her genetic contribution, and the blood component was made custodial in the contracting out of the gestation. Neither natural base was sufficiently strong to overwhelm her claim to be the mother, which she asserted through being married to the person who provided the sperm, through having a daughter comply with her desires, and through having the buying power to contract with Vanessa.

Prognosis

The genetic essentialism of gestational surrogacy in procedures such as Rachel's and Jane's seems to be faring very well, even as it extends and reconfigures preexisting ontologies. Some gestational surrogates have contested custody, but decisions have gone against them more often than for conventional surrogates who are genetically related to the child in question (see table 5.1). On the other hand, there have not been egg-donor contests of custody in the United States on either the model of sperm in DNA paternity cases or the model of surrogacy.[23] A New Jersey Family Court decision in March 2003 allowed both mothers (the partner who donated the egg and the partner who gestated the baby) to have their names on the birth certificate without having to adopt, in a case of lesbian IVF that suggests that both gestational and genetic idioms are alive and well regarding maternity. Several of the patients coded genetics back to socioeconomic factors and thereby in some sense deessentialized genetics in this setting just as they no doubt did in other areas of their lives. Eliding ethnonational and race- and class-based categories with natural grounds for designating kinship is a strategy fundamental to bilateral, blood-based kinship systems. Despite a tendency to favor genetic essentialism in determining motherhood, then, this has not been universal, and procreative intent (tagged by intention, payment, and often the role of class in the court's perception of a child's best interests) has been much more important in determining legal custody, regardless of the decision on motherhood.[24]

The donor-egg procedures seem to offer the potential somewhat to transform biological kinship in the directions indicated—for example, by Giovanna when she draws on the trope of blood connection and shared bodily substance without genetics. If an overwhelming predominance of nonclinical cultural contexts comes down on the side of genetic essentialism, donor-egg procedures might well become assimilated to adoption or artificial insemination with donor sperm. The bid to make gestation in donor-egg procedures a means of conferring biological kinship would have failed. A procedure that began to be used in the mid-1990s took the eggs of older patients but "revivified" them by injecting cytoplasmic material derived from the eggs of younger women into the older eggs before fertilization. This donor procedure preserves the genetic connection between the recipient mother and the fetus and so brings the genetic and blood idioms back into line. It thereby tightens the flexible ontological space for designating biological motherhood that had been opened by donor-egg procedures. The U.S. Food and Drug Administration ruled against this procedure, however (see chapter 8), leaving open both the gestational and genetic routes to motherhood. The meritocratizing and commercializing of the donor-egg market (much like the sperm market, only much more expensive) of the last few years has emphasized the extent to which egg donation is beginning to match the paternity template despite the differences just noted. On the one hand, the genes determine motherhood, like fatherhood; on the other hand, if a proxy is used for the intended mother, the gametes should encode socioeconomic mobility in the way the sperm market does.

Paula used a more mixed ontology for motherhood than the other cases recorded here. She was satisfied without an exclusive answer to the question of which woman was the mother, and she was happy to accept that in some ways there might be more than one mother. Likewise, because she raised the possibility of biologically enacting what she described as already prevalent social practices (shared mothering), there was less at stake in having gestation without genetics confer motherhood biologically. Shared mothering was not presented as necessarily involving a natural kinship rift; indeed it was presented as a practice that preserves racial identity and integrity. Conventional adoption has historically predominantly gone in one direction, with African American children being adopted into Caucasian families, arguably disrupting racial identification without disrupting racism. For Paula, having her procedure assimilated to some kinds of social parenting would still entitle her to appeal to notions of biological sameness because of the group understanding of genetics that was offered paternalistically to and taken up defensively by minority groups. The option to individualize or personalize genetic generative agency is only differentially available.

The likely stability of Flora's and Ute's claims to motherhood is hard to gauge. Ute's claim could have been challenged by Vanessa's desire for contact with the child after its birth, set against the purchasing power of the contracting couple. Flora is likely to encounter social censure, just as she did in the clinic, for her desire to bear a child "for" her younger husband. If Ute and Flora confess to using or are found to have used their daughters as donors, they may be condemned for coercion and putting their daughters through medical interventions to regain their own youth. If either daughter contests the circumstances of her role at a later date, the settlement of who is the designated mother might break down. Likewise, the relations of the daughters' own children to the children who are born from their eggs may become complicated. But it is also possible that parenting with donor eggs for perimenopausal and postmenopausal women will play a part in breaking down some of the more oppressive aspects for women of the "biological clock." If Flora and Ute can maintain their claims to motherhood, they will be cases to hold against the elision of women's identities, femininity, youth, and ovulation. If Flora is buying into the cult of youth to keep her husband, as she claims, she may yet be subverting the wider essentialist identification of women's identities with their youthful biologies, as she also maintains.

Practical Metaphysics and the Dynamics of Naturalizing

In 2003, the Ethics Committee of the American Society for Reproductive Medicine published a report, "Family Members as Gamete Donors and Surrogates," in which they sought to bring order to over a decade of empirical forays in the United States into disambiguating kinship in third-party reproduction. Table 5.2 summarizes the boundary maintenance around incest and adultery in intrafamilial so-called collaborative reproduction. The document makes fascinating reading, and many of the anxieties described in this chapter in empirical detail are given prescriptive form here. Strikingly absent, however, is any sense of family that is other than the linear, cognatic one. There is no discussion of the use of ethnoracially differentiated narratives or of the realities of kinship-denying and kinship-conferring practices on which patients draw. The tendency to naturalize and foreground this model of kinship means that at present the dangers and possibilities from other U.S. narratives of kinship have not yet been addressed.

The implosion or collapse of nature and culture are commonly discussed. All concepts of nature, including scientific ones, are claimed to be always already shaped, marked, interpenetrated with the imprimatur of culture, and (somewhat less frequently) all concepts of culture are claimed to invoke legitimizing natural grounds for their systems of classification. Critics of these postmodern sensibilities rightly distrust the looseness or voluntarism that this seems to imply. Ontologically

Table 5.2
Regulating Kinship in Familial Collaborative Reproduction

Collaborative reproduction usually involves anonymous or unrelated known individuals, but some couples prefer to involve a family member in the arrangement. This may occur *intragenerationally* between siblings or cousins of similar ages, such as a sister providing eggs for a sister or a brother donating sperm to a brother. It may also occur *intergenerationally*, as when a mother gestates her daughter's embryos or a father provides sperm to his infertile son. SART does not collect data specifically on familial collaborative reproduction.

Ethics committee guidelines:

1. *The use of family members as donors or surrogates is generally ethically acceptable.* It can shorten waiting time for treatment, lower treatment costs, and preserve some genetic or kinship relationship. Providers of assisted reproductive technology (ART) should pay special attention to issues of consanguinity, risks of undue influence on decisions to participate, and the chance that the arrangement in question will cause uncertainty about lineage and parenting relations.

2. *Brothers may donate sperm to brothers and sisters may donate eggs to sisters.* In practice, a 1998 survey of North American ART clinics found that 60% of clinics would accept sperm from brothers, whereas many more would accept sisters (90%) and friends (80%) as egg donors,[a] suggesting a gender difference in beliefs about the acceptability of family members as gamete donors. Care should be taken to avoid the impression of incest or adultery.

3. *Intergenerational gamete donation and surrogacy are especially challenging.* The following questions demand close attention in intergenerational cases: Can a donor or surrogate closely tied to and perhaps dependent on the recipient couple make a free and fully informed decision? What are the consequences of the unusual resulting relationships on the donor or surrogate, offspring, and rest of the family? What are the consequences of the creation of new genetic relationships that would be otherwise impossible? Care should be taken to avoid the appearance of incest or adultery.

4. *Consanguineous gamete donations from first-degree relatives are unacceptable. Incest* refers to sexual relations between two closely related individuals. *Consanguinity* refers to marriage and/or reproduction between individuals who are closely related genetically. In ART, you can have consanguinity without incest because gametes are combined outside the body without sexual relations and embryo transfer is by catheter. Brother/sister, father/daughter, and mother/son incest and marriage are prohibited in North America. Because they do not involve incest or marriage, however, there are no laws against brother to sister or father to daughter sperm donation, or against sister to brother or mother to son egg donation, which would all involve combining genetically consanguineous gametes. The ARSM declares that combining consanguineous gametes in ART should be prohibited.

Table 5.2
(continued)

5. Counseling is encouraged for all parties, including partners of donors and surrogates. The Ethics Committee feels that assessing and judging motivations such as love, devotion, loyalty, and duty within an intimate family are best left to the family members in question. Providers should make sure that all parties have received appropriate counseling to give informed consent and should make sure that all decision making is carried out free of coercion. Additionally, the interests of the child(ren) to be should be considered in counseling, consent, and decision making.
Action advocated: While following these guidelines is voluntary, the ASRM directs clinic personnel as follows: When an assessment of the proposed familial collaboration reveals consistent concerns about undue pressures on the prospective donor or surrogate, or about unhealthy family dynamics, the program should feel free to deny these procedures.

Source: Adapted from Ethics Committee, American Society for Reproductive Medicine (SART), "Family Members as Gamete Donors and Surrogates." 2003.
a. Stern et al. (2001).

they decry the violation of common sense: there is a real material world that is subject to more or less regular laws, and there is a distinction between truth and falsehood, science and myth. Politically they disparage the loss of an Enlightenment platform and the neglect of persistent stratification among people that are implied by the moral and ontological relativism of a nature and culture implosion. Talking about a nature and culture collapse in vague and programmatic terms is almost always counterproductive. And yet there is something extraordinarily important about the insight that nature and culture are coproduced and that the explanatory relations between them are eminently revisable.

In an attempt to rescue and further this insight in this chapter, I have discussed small and detailed examples of productive (not deconstructive) negotiations of the boundaries and explanatory relations between cultural and natural concepts in these ART clinics. If the chapter has been successful, it should make it clear that the coproductive thesis about nature and culture is no more a "culturalist" position than it is a "naturalist" one. It doesn't dismiss the reality of the realm of biology any more than it dismisses the realm of culture. Indeed, studying the strategies used to bring order to new reproductive and kinship situations convinced me that making distinctions between social and natural roles and facts and then using natural roles and facts to ground cultural categories are absolutely fundamental to meaning making in contemporary U.S. society. At the same time, however, the cases made a compelling argument for the absence of a unique biological ground for answering the question, "Who is the mother?" There was more than one possible answer to this question, and cultural dynamics—most predominantly the bundle of factors that make up procreative intent and its enforcement—propelled the sorting and classifying of some things and not others as the biological facts of relevance. The narratives of the trajectories of reproductive intent were choreographed between cultural and natural constraints at all points and were neither biologically nor culturally voluntaristic. The ambivalence that is evident in feminist theory about the relative virtues of naturalism and social constructivism is entirely appropriate: sometimes important political and ontological work is done by denaturalizing what has previously been taken to be natural and deterministic; sometimes the reverse is necessary. What cases like these show us is that the theorist does not need to champion one strategy rather than another. Modern medicine gives us many cases where ordinary people flexibly manage the choreography between the natural and the cultural (the practitioners and patients using the technologies in question).

As I focused on the dynamics of naturalization in the clinics, some aspects of the process of cultural change in the midst of technological innovation became evident. First, the cases suggested that high-tech interventions are not necessarily antithetical to the production of affiliation and identity, the claims of antimodern Luddites that technology is intrinsically dehumanizing notwithstanding. In this clinical setting, gestational surrogacy and donor-egg IVF were the means through which racial, ethnonational, familial, and individual desires and biologies were all promulgated. In the kinship innovations described here, the various naturalization strategies drew on deeply rooted and familiar ways of forming and claiming kin and simultaneously extending the reference of the kinship terms that were being disambiguated through the strategies. Given this slightly counterintuitive aspect of these high-technology sites that technological innovation and cultural history implicate each other so strongly, it is no wonder that progressive cultural critics cannot decide whether the new reproductive technologies are best judged as innovative ways of breaking free of bondage to old cultural categories of affiliation or whether they are best denounced as part of a hegemonic reification of the same old stultifying ways of classifying and valuing human beings. The technologies are fundamentally both. Technological change and cultural conservatism go hand in hand; the famous "lag" between technological innovation and cultural reaction is nothing of the sort. The lag simply refers to the looping catch-up requirements in social, legal and ethical spheres to organize and calibrate the coemergence of entities and relations that are produced in these extended cultural and natural biologies. In the next and final chapter of part 2, I turn to the other side of the argument that being a means is not necessarily opposed to being an end through a consideration of individual agency through objectification. In part 3, I lay out the amalgam between the public and the private that characterizes U.S. ARTs. This amalgam, if furthered and protected in the appropriate ways, contains the means to make sure that the incursion of the contract and objectification on reproduction does not repeat the reproductive abominations of U.S. history. Instead, at its best, it heralds a new era in the relations between science and society where the public becomes privately implicated in science and has not just a voice but a bodily say in its direction.

6

Agency through Objectification: Subjectivity and Technology

This chapter concludes the predominantly ethnographic part 2 by addressing questions of patient selfhood and agency—particularly women's agency—when people interact with assisted reproductive technologies.[1] It addresses these issues through the study of epistemic notions and practices in their embodied, cultural, and historical context in ART sites. This chapter examines the interaction between patients and the medical technology by using interview and ethnographic data to question the humanist argument that selves need to be protected from technological objectification to ensure agency and authenticity. I argue that objectification is only sometimes a reductive state in opposition to the presence or goals of a subject. In the various nonreductive manifestations of objectification, patients can manifest agency (and so enact their subjectivity) through their objectification. I also argue that when objectification is antithetical to personhood, it occurs in circumstances that can be traced and understood. This has implications for intervention.

In questioning the link between objectification and a loss of personhood, the circumstances under which people are objectified reveal the dynamics whereby objectification sometimes is opposed to personhood and sometimes is not. Infertility clinics involve procedures where the patients'—especially women patients'—self and body are objectified in many different ways, usually under the authority and expertise of others. Patients in infertility clinics are also typically highly invested personally (financially and emotionally in terms of their long-range desires for their lives) in what they are undergoing in infertility clinics. Infertility clinics are thus instructive places to look for the possible coexistence of objectification, agency, and subjectivity. The dynamics of actual cases of objectification are accessed through ethnographic data.

The position against which I am arguing, then, is that medical technology objectifies the patient, which deprives her of agency. There are two motivations for taking up the questions of personhood and objectification that have structured this chapter. First, the argument of the chapter takes on one element of a long tradition that sees objectification as alienating and that sees technology as always in imminent danger of usurping selfhood.[2] Where objectification is theorized as an important part of understanding modern personhood, it is still metaphysically opposed to subjectivity.[3] This tradition is exemplified in a number of writings that are directed at infertility medicine, most notably the many powerful critiques by feminist scholars of reproductive technologies for their objectification of women (especially in what I call "phase 1" in chapter 2). The aim of this chapter is not to deny the subjugating and disciplining effects of many technologies, including reproductive technologies, on those who work with or are monitored by technology but rather to question whether the various forms of objectification per se are antithetical to personhood.

Taking on this tradition is a political as well as a philosophical move. Philosophically, it contributes to other science and technology attempts to animate ontology and to explain rather than start with binary opposites such as subject and object. Politically, this chapter takes inspiration from the writings of a new generation of STS and feminist scholars of technoscience who do not reject science and technology but try to negotiate a politics in use and development, paying attention to the possibilities of places of scientific, technological, or medical practice for different women.[4] As Franklin has argued, in reproductive technologies, in particular, patients and practitioners are bound up with the technologies in question, so a politics of "just say no" is unconvincing.[5] Some interesting recent literature on gender and technology has combined the philosophical and political strands in historical and contemporary accounts of the coconstruction of the subject and the technology.[6] The argument presented here further illustrates that the components of a subject position and the power of technologies are negotiated in a heavily constrained manner.

The second motivation for taking up the questions of personhood and objectification is that they promise to contribute to the discussion of the relations between humans and things that is at the center of contemporary science and technology studies. The dependence of selves on technology has not received as much STS attention as its reverse, the dependence of science and technology on social, individual, and political factors. An ever-increasing body of work, however, is documenting the connections between technology and the self, propelled by a rise of scholarly interest in digital and medical technologies.[7] In recent years, there has also been an interdisciplinary revival of the Maussian anthropological inquiry into culturally specific configurations of the self.[8] The exploration in this work of the multiplicity of selves—the different kinds of faces or personas or social roles that we routinely switch among as we go about our daily lives—has opened up the possibilities of meaningful conceptions of the self that are not tied to the essential unity of the self. This chapter builds on these literatures that take the construction of the self and the connections between the self and technology seriously and attempts to link them further to STS questions.

Agency and Selves

For many philosophers, agency is just the power to act, and action and agency are almost indistinguishable. By contrast, *agency* here refers to actions that are attributed to people or claimed for oneself, that have definitions and attributions that make up the moral fabric of people's lives, and that have locally plausible and enforceable networks of accountability assigned to them.[9] In one philosophical literature, debates about self and personal identity revolve around the conditions for the unity of a life. On what does the continuity of a person from one time slice to the next depend? What are the necessary and sufficient physical and mental conditions for the persistence of selfhood?[10] In this chapter, I take the self to involve a much longer-range project than that implied by the Lockean and Humean tradition. Personal identity is here taken to have at least these elements—to have a long-term orientation to the good, to be essentially in movement, and to have an irreducibly narrative character.[11] For women infertility patients, whether they remain the same instance of the same sortal is not at stake, just as our own continued existence is only rarely the subject of our own or anyone else's doubt. In an investigation of agency, as I understand it here, being a subject changes in ways that are the result of and simultaneously proof of the person's agency. To understand this interdependency, it is necessary to look at the local achievement of identity, without deciding beforehand what may or what may not be an element in that achievement.

The defining aporia of current discussions of agency is, perhaps, as follows: the constructed nature of the unitary human subject is a datum for the social sciences, but agency cannot be accounted for without presupposing the unitary human subject.[12] This aporia is particularly salient in medical settings, where we have become accustomed to thinking of patients as disciplined subjects *par excellence*—anything but active agents—and yet where different descriptions of the patient are at play, at least some of which require an acknowledgement of agency.[13] This chapter aims to shed some light on this apparent tension around ARTs by showing that the subject is dependent on the constant ontological entwining between ourselves and our environments. This ontological choreography changes how many descriptions we fall under, how many parts we are built of, and how integrated we are or need to be. In this site, we cannot presuppose an ontology of the unified subject because a coherent self-narrative requires ontological heterogeneity.

ART clinics are ripe for the examination of agency and the ontological commitments that go with it. The clinics deal on a daily basis with human gametes and embryos, which function in this clinical setting as questionable persons, potential persons, or elements in the creation of persons (see chapter 8). Embryos, for example, can go from being a potential person (when they are part of the treatment process), to being in suspended animation (when they are frozen), to not being a potential person (when it has been decided that they will be discarded or donated to research), and even back again to being a potential person (when a couple has a change of heart and frozen embryos are defrosted for their own use or for embryo donation). As the literature on fetal personhood has shown, the rhetoric of the potentiality of life is predicated on a woman and would-be mother, who must literally come to embody the potential life attributed to the embryos.

Just as an in vitro fertilized embryo changes status dramatically in the site, patients also undergo significant ontological change. A patient who is going through a cycle of treatment is sometimes a person who juggles her work schedule to be at the clinic, sometimes a generic patient who sits in the waiting room, sometimes ovaries and follicles that appear on an ultrasound screen, sometimes an anesthetized body that lies on a surgical table, sometimes a patient with blocked tubes, and so on. The genius of the setting—its techniques—allows these ontological variations to be realized, to multiply, and to be coordinated. By passing through them, a patient embodies new options for her long-term self.

In this chapter, I look at patient testimony to track these ontological variations. I am not interested here anymore than elsewhere in the book in using patient testimony or experience to develop a pro or con analysis of ARTs. Such critical or valorizing enterprises assume the distinctness and stability of patients and technologies as a starting point and use one to praise or debunk the other. My purpose in including patient testimony is to understand how undergoing these technologies may or may not transform the women and how technology figures in and enables their self-making.

I am assuming that interpretive charity is appropriate in assessing patient utterances.[14] By this I mean that I assume that patients are broadly rational and coherent witnesses of their own treatment and that apparent inconsistencies are to be explained not as evidence of merely ad hoc, post hoc, ignorant, justificatory, or self-interested verbalizations but as important elements of context that vary from one utterance and one speaker to the next. The building of self-interest and justifications happens as patients go through changes during treatment. This use of a principle of charity provides a meaning that is highly context-sensitive. Context includes features of the world such as "where" and "when" to resolve the reference of indexicals and demonstratives. In this site context also includes changing features of the patient to resolve of what and whom things are predicated.

The implications of context-sensitive charity are that an utterance that a patient makes before she is pregnant might be incompatible with an utterance that she makes later when she is pregnant. She might say that a procedure objectified her or went wrong in a cycle in which she didn't get pregnant but might say very different things for an apparently identical procedure in a cycle in which she does get pregnant, for example. If the procedure did not vary, this difference could be viewed as a paradigm case of irrationality driven by self-interest or a need for justification. On the other hand, an account that is justificatory or self-interested can be viewed as moving and helping to fix the identity of the patient at the time she is speaking.[15] In this way, the self-interest or justification becomes one aspect of the content of what has been said, thus removing the apparent contradiction between the utterances. Neither a pure self-interest story nor a story that ignores interests or treats them as epiphenomenal is quite right. There is nothing mere or necessarily irrational to building self-interest or giving justifications if the interpretation of meaning sometimes requires taking into account the changing lives of the patients who provide testimony at various stages of treatment.

A final methodological point involves the recoverability of the constitutive elements of the socionatural order out of which, and in maintenance of which, purposive action arises. Stable features of the way that an activity is organized can be identified, and the factors that influence those features can be discerned, or ruptures in the normal way of going on can be found or provoked to reveal what has to be in place for normal conditions to pertain.[16] A drawback of the first of these approaches is that it misses the contingency of the organization of activity and in this case the mechanisms of attribution and exercising of agency. A drawback of the second approach is that there is not necessarily a good correlation between what is needed to patch together a breakdown and what normally holds things together.

Focusing on either stability or rupture are not, however, the only alternatives. If stable features are stable only at a certain level of abstraction, thinking of social worlds as presenting a smooth surface that can only either be studied as such or dug up becomes less compelling. By interrogating a setting from a different resolution, its normal workings probably can yield clues about its construction.[17] Using ethnographic data, I vary the resolution under which the treatment setting is examined and view the normal workings of the site through the changing metaphysics of clinic, patients, body parts, and instruments. Because technology, objectification, and subjectivity are exemplified around assisted reproductive technologies, talking about ARTs is an appropriate way of talking about subjectivity, and vice versa.

Narratives of Assisted Reproductive Technologies

Rhetorical conventions for "reading" assisted reproductive technologies have grown up surprisingly fast in the more than twenty years since ARTs began to be available in the United States. One convention is a picture of a happy (medically extended) family—a heterosexual couple, a white-coated physician, and a long-awaited miracle baby (or babies). Another convention (as described in chapter 2) involves the poor success rates of ARTs, the use of women's bodies as experimental sites, the ardors of being an infertility patient, the class effects of surrogacy, the exclusivity and expense of treatments, the risks of multiple pregnancies, the high miscarriage rates following ARTs, an increased pressure to have biological children, and the eugenic vistas into which the technologies so easily play. Still another convention involves negligence and malfeasance—the insemination of women patients by a doctor using his own sperm, patient deaths from hyperstimulation of the ovaries, and the mid-1990s case of egg and embryo swapping at Ricardo Asch's clinic at University of California, Irvine.[18] Taken to an extreme, these narratives inscribe a stereotypical infertility female patient[19] who is paradigmatic of the objectified patient—either helpless and saved by the technologies or victimized by them. If she is supposed to be helpless, then the technologies and the predominantly male physicians come to the rescue of the woman who is otherwise unable to achieve the pregnancy she desperately wants.[20] Since this patient has no agency, the value and virtue accrue to the doctors and the technology. If she is a victim, then she seeks exclusive and expensive medical intervention not to save her life but to conform to a norm (becoming a mother). She is turned into an object of study, experimented on, and reduced to a physical presence in the name of procedures that rarely work. This patient has no agency, and all the criticism and debunking accrue to the doctors and the technology. This patient has no voice in the shaping and application of the technologies but happens to benefit from her objectification in the clinic by being one of the lucky ones to get pregnant.

A different configuration of the relations among patients, practitioners, and techniques, however, results in a different narrative of agency and objectification. Paying attention to a patient's own narratives and to the detailed circumstances of the production of particular narratives suggests that a multiply objectified user entails being neither helpless nor a victim. The woman's objectification involves her active participation and is managed by herself as crucially as by the practitioners, procedures, and instruments.[21] Patient agency is not only not incompatible with objectification but sometimes *requires* periods of objectification.

I look at patient narratives, therefore, to draw possible ontological connections among techniques and patients' accounts and understandings of themselves.[22] The comments presented here, by women patients were offered spontaneously or elicited in response to my questions about why a given procedure had or hadn't worked. The excerpts are comments about success and unsuccessful treatment cycles by women who did and did not become pregnant.[23]

An Unsuccessful IVF Treatment Cycle with a Patient Still in Active Treatment

Everything went fine; he put back three embryos.... It's a numbers game,... but maybe I didn't make enough progesterone to support a pregnancy.... Maybe the hormones affected my body, and it wasn't ready for a pregnancy.

I had quite a few eggs—eleven—but they weren't all good, and only some of them were mature.... And then only two of the embryos looked good, so I guess something was wrong from the start.

My tubes are really damaged, so I wonder if I have a lot of scar tissue on the inside of my uterus too.... He said it was hard doing the egg retrieval.

If it doesn't work next time, we're going to go with a mixture with donor [using sperm from a donor as well as the husband's sperm during IVF to increase the chances of fertilization].... Jake [her husband] didn't do well at the Ham test [a test on the sperm to see if it is motile enough to penetrate the zona pellucida of a hamster ovum].[24] ... It's a lot of money and time to keep trying in vitro when we only get one or two eggs fertilizing.... They do micromanipulation [microsurgical techniques that drill an entrance into the egg through the surrounding zona to let the sperm in when spontaneous fertilization doesn't occur] at [another center], so we might try that if this doesn't work.

It's such a roller coaster. First, you go in for the scans and blood tests, and you feel on top of the world if your eggs develop OK and you get to go to surgery. Then you panic: will they get any eggs?... The doctor tells you in the OR [operating room] how many they got and if they looked good, and you don't know if you heard right because you're still half under.... Then there's a wait for the fertilization, and if that goes OK, it's whether there are any embryos to freeze. And you get excited again if you make it to the embryo transfer. That's sort of the last step. And you definitely have a chance if you get that far. Until then, everything can go fine, but you don't have a chance of being pregnant.... Then comes the long wait [around two weeks until the menstrual period comes or a pregnancy test can be done].... If you get your period, it was all for nothing. That's the lowest point of all.... You're back to ground zero—all that expense, everything we went through.

Dr. S. said at the beginning that in the ideal world we'd be candidates for IVF, but our insurance doesn't cover it. So we went with the AIH [artificial insemination with the husband's sperm] and Pergonal [a drug containing luteinizing hormone (LH) and follicle-stimulating hormone (FSH) that induces multiple-egg maturation in a single cycle instead of the usual one egg per month] for a few cycles. But now that we're doing IVF anyway, we wish we'd done that all along.... AIH wasn't really indicated in our case because I had an ectopic [a tubal pregnancy, where the embryo implants in a fallopian tube instead of the uterus; the embryo cannot survive, and ectopic pregnancies can be life-threatening to the woman] two years ago.

An Unsuccessful Previous IVF Treatment Cycle with a Patient Who Subsequently Becomes Pregnant as a Result of ARTs

Last time I think the reason it didn't work was the hormones weren't right for me.... You know, sometimes you go through all this, and it's not really you that's being treated.... You're just one more woman with her feet in straps getting the same drugs everyone else gets, going through all those tests. This time, though, I felt more like a person.... I was more informed about what was happening, and Dr. S changed some things because, you know, it didn't work last time.

This time it felt like he let my body do more of the work its own way—as if my body took on the work.... Last time it was as if they were all just doing things to me and my body wasn't cooperating.... Nothing was really different in terms of the numbers of eggs they got or the fertilization rate or anything.

At [a different IVF center], it's just a production line. They see so many patients a day it's not surprising it didn't work.... You never know what doctor you're going to see.... I heard one doctor joke to another woman, who was also waiting for the egg retrieval when I was, that she had no eggs. Then he laughed his seductive laugh and patted her thigh and said he was just kidding: she had seven fine-looking ones! The woman was still shaking after he went, although she managed a laugh and a smile because, you know, you have to be the doctor's friend, the best patient.... I don't call that humor.... He should have known the effect it would have on her.... And she's not likely to get pregnant if she's treated like that—like she's at his mercy.

A Successful IVF Treatment Cycle with a Patient Who Became Pregnant as a Result of ARTs

I guess I just got lucky—on the first time too! Now we want a sibling for her. I just hope it'll work again, but you keep telling yourself it might not happen because you know it only works one in five times or whatever.... But you don't really believe it in your own case because it's all you have experience of. You know, you think, "Maybe there was a reason I got pregnant the first time. Maybe I'm really fertile in these conditions."

Last time we had the pergonal, but this time we moved to Metrodin [an FSH-containing drug similar to Pergonal and made by the same company but with no LH], and the nurse coordinator says she thinks Metrodin has better success rates.

I felt really good this time. Dr. S knew how my body would respond, and although we got fewer eggs than last time, I just had a better feeling about it. Maybe we got fewer but better ones. After all, all you need is one good embryo.... I probably wouldn't be saying this if it hadn't worked!

They take a lot of care here. I don't think it was any one thing, but each stage is done so carefully compared to [another center].... I can't really pinpoint any one thing, but it all adds up.

Now it's happened I don't want to think about it.... It all seems kind of irrelevant, how we got this pregnancy.... It's our baby now, and we're just hoping the pregnancy will go OK.

Commentary: An Unsuccessful IVF Treatment Cycle with a Patient Still in Active Treatment and Not Yet Pregnant

When patients are still in active treatment but not yet pregnant, their comments about a previous cycle that was unsuccessful tend to isolate a specific phase of the treatment as a likely faulty step in the process. For example in the testimony above, progesterone, drugs, eggs, scar tissue, and sperm motility are mentioned as possible causes of failure. Conception is also presented as a series of hurdles that must be jumped. When a specific body part or phase of treatment is isolated in this testimony, it is not idiosyncratic to a patient couple. If in vitro fertilization is to lead to a pregnancy, every IVF patient couple must have sufficiently mature eggs to fertilize and implant, sufficiently motile sperm to fertilize the eggs, and little enough intrauterine scar tissue to allow for successful implantation. Furthermore, the faulty step that is isolated is strongly linked to the couple's diagnosis: sperm motility for male factor, scar tissue for tubal factor, egg quantity and quality for IVF patients, and so on. The diagnosis-related step that is used to try to understand why the treatment cycle was unsuccessful, however, is expressed very personally. The relevant step is suggested as the reason that this individual patient couple did not get pregnant on a previous treatment protocol. When these patients mention ambiance, subjectivity, or anything personal about themselves, it is not cited as a reason that the procedure did not work. Expense is mentioned ubiquitously and operates as an integral element in diagnostic reasoning. Insurance coverage for each procedure serves as a continuously narrowing function on the possible treatment options. The practitioners are referred to simply as sources of authority.

A part of the body or a phase of the treatment is singled out as the missing link in the process of getting pregnant, and surmounting it comes to represent the pregnancy and thus the woman herself in her capacity as a child bearer. This synecdochal aspect of diagnosis—the events through which a woman's body parts come to stand in for the woman—makes these accounts characteristically mechanistic.[25]

The procedure is not about the fallopian tubes rather than the woman, and nowhere do the patients make this kind of reductionist move and refer to themselves as fallopian tubes or whatever the diagnosis is. They are individuals who have a given diagnosis that might respond to a particular medical treatment. The promise of the site is not bypassing or fixing the fallopian tubes but in getting a woman pregnant, much as I might mend a punctured tire so that I can ride my bike and not so that I can have an unpunctured tire.[26]

The patient orients to herself as an object of study and intervention, to the physician as instrument and epistemic standard bearer, and to the instruments and material setting as appropriate technology for acting on her body. This nonreductive but synecdochally operationalized ontology (the specifically targeted medical intervention doing work for the whole patient) was unquestioned in patients that I talked with as long as the patient was willingly in active treatment.

Commentary: A Failed Cycle with a Patient Who Subsequently Became Pregnant as a Result of IVF

Patients who have a successful cycle are more likely to say that previous unsuccessful cycles were unsuccessful because the women lacked agency, insufficient attention was paid to her specific situation, or she was unable to serve as an expert witness about her own treatment: "They were all just doing things to me and my body wasn't cooperating"; "Let my body do more of the work its own way"; "The hormones weren't right for me"; "You're just one more woman with her feet in straps getting the same drugs everyone else gets." These factors are more marked when an unsuccessful treatment was undertaken at a previous clinic: "It's just a production line"; "You never know what doctor you're going to see"; "She's not likely to get pregnant if she's treated like that." Patients who get pregnant at a given clinic tend to assimilate all their treatment cycles at that clinic as preliminary phases of the same treatment and are thus much less critical of them. Comments made about failed procedures at previous clinics tend to be personal, critical, and denunciatory,[27] partly because of a sense of loyalty to the current site and a need to assert good reasons for switching allegiance.

Failed cycles that are not fully assimilated to a subsequent successful cycle provoke patient responses that portray their objectification as dehumanizing or against treatment interests. This makes sense if one thinks of the synecdochal relation as having failed. The relation works only if fixing the missing link really does repair the functioning whole (the bypassed tubes must refer back to a now pregnant or potentially pregnant woman). If, however, the tubes are bypassed by in vitro fertilization, for example, but the patient does not become pregnant, then the patient is alienated from her body by the treatment. She is stranded at the phase at which she underwent lots of procedures but has nothing to show for it. A loss of subjectivity and agency occurs after a procedure has failed or caused harm, which replaces the functioning synecdochal relation with an ontology in which there is a rift between the patient as subject and the patient as object. The objectification is indeed reductionist in these circumstances.

Insensitive doctors who joke about a patient's eggs (or who make inappropriate sexual innuendos, as some patients report) subvert the patient's own participation in her objectification. You have to be "the best patient," but if that involves a conversation in which a physician or someone else exploits that objectification and his or her authority, a deprivation of dignity and autonomy transcends the treatment situation.[28] It is as if the physician is playing in the wrong key, misunderstanding and exploiting the nature of the objectifications involved in treating that patient. This behavior is as highly condemned by other physicians and staff as by patients.

Commentary: A Successful IVF Treatment Cycle

The accounts of successful cycles are less causal (the comment about Metrodin is causal but is not specific to her diagnosis), less naturalistic ("I guess I just got lucky"), and less synecdochal ("I don't think it was any one thing, but each stage is done so carefully here"). Unlike the personal comments that are made by patients in active treatment but not yet pregnant, pregnant patients who describe successful or unsuccessful treatment cycles tend to make a different kind of personal comment. Their personal comments tend to be highly specific to themselves rather than to their diagnosis: something amorphous made the difference between a protocol on which a particular woman would get pregnant versus a generic protocol for other patients with her diagnosis.[29] The technical interventions that are so prevalent in the first group of quotes (Ham tests, exact numbers of eggs, embryo quality, scar tissue, micromanipulation, and operating rooms) drop out by becoming inevitable, invisible, or irrelevant: "Nothing was really different"; "I just a had a better feeling about it"; "I don't want to think about it.... It all seems kind of irrelevant." In talking about successful cycles, the synecdochal relation has done its work. The objectifications brought about the desired transformation of the woman herself, and the cause of this success was the coordinated functioning of the whole referential chain—the desire to get pregnant, the treated body parts, and the now pregnant woman. Responses are correspondingly oriented to the achieved pregnancy and to the factors that affected the whole procedure.

Patients' Active Participation in Their Objectification

In infertility clinics, patients willingly accept the role of being the object of the medical gaze and in fact actively participate in it. One nurse commented that almost all women patients flirt with the physicians (with both male and female physicians, although to a lesser extent with females, she felt). She said that many patients want to be thought to be attractive, to have their bodies seen as womanly despite their infertility, and to be compliant with and special to the physician—to be good candidates for the procedure. The suggestion was that the circumstances of the doctor and patient relationship are inherently seductive, combine elements of intimacy and authority, and are accepted by the patient in her active subordination to it. No patient described herself to me in this way, but several commented on how well they got on with their doctor. The woman quoted above said, "You have to be the doctor's friend, the best patient" when talking about another patient's compliance with a physician's insensitive and unsettling joke.[30]

When a procedure took a large toll on a patient for no result and a result was extremely unlikely on statistical grounds, I often wondered how the patient managed to perform with such control, civility, and seeming passivity. On the other hand, when a procedure had a chance of working, I expected the patient to take an active interest in her own presentation as an object of study. The parallel with the sociological phenomenon of intentional subordination—subordinating the will to the structural power of another person or organization to achieve some overarching goal—is informative.[31] The power is not something that simply resides in the physician or institution, however, as this notion tends to suggest. The physician is a link in the chain that mediates access to the techniques. The patients do not so much let themselves be treated like objects to comply with the physician as they comply with the physician to let themselves be treated like objects.

I turn next to a brief description of some of the most routine elements of ART treatments: the pelvic exam, the ultrasound, diagnostic surgery, and the manipulation of gametes and embryos in the lab. The aim is to show what is made to appear by the different equipment and procedures. Each procedure has its own physical setting and uses various methods to objectify (parts of) the patient. I use each example to focus on one crucial aspect of this ontological choreography.

The Pelvic Exam

By the time a woman patient enters an examination room for a pelvic exam, she has already passed all the specific and generic trials that are required before she is allowed to be there (and so have the room, the instruments, and personnel of the clinic). Of particular importance is what might be called her "anticipatory socionaturalization." Most infertility patients will have spent more than a year trying to get pregnant before they approach a specialist doctor. During this time, their thinking about their fertility will have changed dramatically. Several people reported to me that they always knew that they might not be able to get pregnant if and when they wanted to but that their predominant fertility concern previously had been how not to get pregnant.

One characteristic of the process of anticipatory socionaturalization that is undergone by patients prior to their arrival at the clinic is to be aware of the phases of the menstrual cycle and the things that can go wrong when trying to conceive. This awareness typically involves some or all of the following: talking to other people who are trying to get pregnant, buying or borrowing self-help books on infertility, watching their own bodies minutely for signs of ovulation and menstruation, timing intercourse, familiarizing themselves with the statistical arguments about the likelihood of conception in a given cycle, checking for ovulation with home test kits or by taking their temperature every morning, going to a nonspecialist physician, getting a sperm count for their male partners, deciding for or against various support networks (who to tell about the infertility, how and where to seek help), going to infertility clinic open houses, and forming hypotheses about what might be going wrong in their own cases based on personal health history and hunches. This anticipatory socionaturalization means that when patients come to the clinic, their bodies are already considerably "un-black-boxed." Where they once either tried or tried not to get pregnant and had only a vague understanding of eggs, hormones, fallopian tubes, temperatures, sperm counts, cervical mucous, endometriosis, and luteal phases, they now itinerize their own bodies in these terms. Coming to the clinic allows new access to these processes and body parts. It renders the parts visible and manipulable and subjects them to all sorts of tests so that they yield facts on which to base diagnoses and treatment.

The pelvic exam takes place in an examination room. The patient lies on the examination table with her legs placed on padded leg rests. At the relevant times in the exam routine, the nurse hands the physician cleaning swabs, the speculum, and any other pieces of equipment that are needed. The physician sits on a mobile stool at the foot of the table, and his or her gaze, gestures, and the speculum delineate the physical zone of treatment and study. The character of the conversation between the patient and the physician changes so that the patient's internal reproductive organs become the focus of attention. This change is choreographed by the patient's, the physician's, and the nurse's coordinated positionings, as well as by the swabbing, gloving, and placing of the speculum. These mundane steps render the body and the instruments compatible and are at the heart of objectification.

When the speculum is in place, one gloved hand examines the vagina and cervix, and the other hand palpates the pelvic region from on top of the abdomen. The physician is able to establish the gross anatomical normality of the woman's body. Fibroids are distinguished from uterine walls, cysts from ovulating ovaries. Patient reactions or reports of pain are assimilated into the exam. The physician may separately feel the breasts and check the body for "unusual" amounts of secondary hair.[32] As argued in chapter 3, the appropriate topographic knowledge is embodied in the physician's skilled gestures and in the alignment of the instruments and the patient's self-testimony. Together these elements render what is seen or felt as something that is an instance of this or that probable diagnosis. The patient is routinely called on as an expert witness in the process of rendering body and instruments compatible. She provides access to two kinds of information that wouldn't otherwise be present in the room: pain information and information about her fertility history. These two classic sources of skepticism—first-person sensations and the past—are fully integrated into the procedures of the examination room and so are made part of the ontology of the site by the self-witnessing of the patient. Her reactions and answers to questions during the procedure calibrate that otherwise unavailable information with the rest of the activity in the room.

An everyday metaphysics of the person that is dependent on but different in kind from her body presents the woman as dualistic: she has a desire to be pregnant, and she has a body that refuses to cooperate. This behind-the-scenes ontology of the body can no longer reconcile the long-term self-narrative with the physical body.[33] But once the woman is in the clinic, the ontology changes. The pelvic exam educes (draws out) and classifies some of the contents of the Pandora's box that is now the patient's body.[34] For example, a cold metal speculum is an inert and blind (and often loathed) commonplace artefact. But when it is incorporated in this procedure, it becomes part of the trail along which the patient's uterus, ovaries, vagina, and cervix are brought into contact with possible treatment. The body parts are not any more real than they were all along and have not suddenly become relevant because the body's black box has been opened, although there's something right in both these formulations. The body parts become more real only in the sense that they are enabled to display properties in their own right. And they are more relevant only in the sense that they are rendered as functional stages to which treatment can be applied. The diagnostic and treatment setting draws out the body parts into a new metaphysical zone that consists of many perceptible functional stages where treatment can be focused. What happens during a pelvic exam is that these body parts come onto the scene of action of the patient's future chances of conception by becoming connected to new and different things.

The Ultrasound

Women patients usually undergo several vaginal ultrasounds in any given treatment cycle. They monitor desired follicle growth and potential problems, such as impending spontaneous ovulation, hyperstimulation, cysts, tubal pregnancies, and intrauterine pregnancies. The endometrium, or lining of the womb, is also measured to assess readiness for implantation. Since the mid-1980s, vaginal ultrasound probes have typically replaced abdominal ultrasounds in assisted reproductive technology clinics. The ultrasonographer sheaths the ultrasound probe in a sterile condom, coats the condom with jelly, and inserts the probe in the woman's vagina. By rotating the probe in different directions, the ultrasonographer brings first one ovary and then the other ovary into view on the screen. Follicles can be measured and compared. If, for example, the woman is undergoing a cycle of IVF, the ultrasonographer will be looking for several simultaneously developing follicles of roughly equal size. Each follicle is measured, and Polaroid photos of the screen are often produced that record all the images and the associated follicle measurements. When the physician doesn't watch or perform the ultrasound, he or she bases the decision to continue treatment or not on the ultrasonographer's hand-collated list of numbers of ripening follicles and their sizes, the photos, and an endometrial lining measurement. The physician also decides whether to continue medication, whether to alter doses, and what and when the next phase of treatment should be on the basis of this information.

As in the case of the pelvic exam, the ultrasound brings out new entities in a treatment zone that is composed of trails of instruments, technicians, and objectified patient.[35] This manifestation of actants on the screen—the ultrasonographer's list and the Polaroid photo—is how the setting produces places at which to anchor treatments. One trail moves from the ultrasound probe in the vagina, to the ovaries' appearance as images on the monitor and in the photos, to the appearance of the follicles as numbers and sizes on a list, to the ovaries' and follicles' classification in the patient's file, to the physician's room, where the next phase of treatment is initiated. As long as these trails of actants-in-the-setting flow back to the patient, the synecdochal relation between the body parts and the patient is maintained. This material maintenance of synecdoche ensures that the objectification of the patient that consists in the eduction and itinerizing of her body parts is not opposed to her subjectivity.

Diagnostic Surgery

Diagnostic surgery is often justified to the patient on the grounds that it affords the physician a "proper look."[36] Most infertility surgery is done by laparoscope or hysteroscope, which enables surgeons to visualize the peritoneum and inside of the uterus, respectively. The laparoscope is inserted through a half-inch cut at the navel, and a video system is hooked up to the laparoscope. An image of everything in the line of sight of the laparoscope appears on two monitors placed on either side of the patient's body so that the surgeon and assistant can both guide their instruments without moving around. The surgeon talks everyone in the room through the pictures on the screen, pointing out the diagnosis. As soon as the public witnessing of her pelvic region is complete, and if all is deemed normal, the surgeon removes the laparoscope and sews the patient back up again. If something is found that is considered to be relevantly abnormal and to need treatment, two or more tiny incisions are made, through which instruments for cutting, cauterizing, or grabbing are introduced as needed. These instruments are operated by the surgeon and by one of the scrub nurses or an assistant surgeon.

The organs become a focus of repair and therapy with all the qualities of the classic specimen of study. The uterus, ovaries, and tubes are represented on the monitor, floating apart from the context of the rest of the body and the whole person.[37] The anesthesiologist and his or her instruments are crucial in this form of objectification, holding the patient's consciousness in abeyance. The surgeon does not usually need to worry about whether the patient will remain under or experience pain or whether the uterus, tubes, and ovaries are correctly "wired in" to the rest of the body to be able to treat them as if they were unto themselves.[38] For example, some left and right fallopian tubes were described as being "normal" so that "dye flowed through them easily." Others were "occluded," or the "fimbria were damaged or occluded." Some were "stunted and blocked," had "spots of endometriosis on them," or "were stuck down with adhesions." They could be "Novi-catheterized," or one or the other could be "missing." Sometimes the tubes contained things like an "ectopic embryo," a "reanastomosis site of tubal ligation reversal," or "polyps at the corneal junction of the uterus and tubes." Adhesions on them could be "filmy" or "vascularized," and the adhesions could be lysed or left alone. Ovaries were left and right. They could be missing, stuck down with adhesions, or spotted with endometriosis. They often had cysts, which were either "corpus luteum cysts" or "hemorrhagic cysts." The patient herself could be "just there for us to get a look," could be "ginger" and so in danger of bleeding, could be "married to a colleague," could have been "crying in the office yesterday," could have wanted her "fibroids removed only if it could be done laparoscopically," could be "difficult," could want to "go home the same day," or could be "in poor health."

Infertility surgery also requires porters, three operating room nurses, one or two surgeons, an anestheseologist, an embryologist, an IVF nurse, and an ultrasound technician for egg-retrieval surgeries. Sometimes a representative of an instrument company, a medical or technical student, and a researcher (me) might be in the room. There could be as many as eleven people in the operating room. The room also contained anesthesiology machines, trays of instruments, cameras, screens, the bleeding-control equipment, the cleaning and counting equipment and a lot of paper. These routine entities and gestures are required to bring out the singularity of body parts and to educe mechanistic properties in body parts so that they can be tested and fixed.

Gametes and Embryos in the Lab

The embryology laboratory is the site at which all semen is washed, spun, and separated before inseminations. (Some clinics have enough room to have separate laboratories for gamete treatment, embryo treatment, eggs, and sperm.) The freezers contain patients' embryos, patients' sperm, and donors' sperm, and eggs are brought here from the operating room for insemination. Regular activities—such as the making of new culture media, the quality-control testing of the media on mouse embryos, the filling of the liquid nitrogen canisters for the freezers, and the ordering and receiving of new equipment and supplies—ensure the self-reproduction of the lab. The embryology laboratory maintains an ontology of connectedness between patients and body parts during the time when the two are separated. This spatial separation between the patient and her gametes and embryos makes possible events that would not otherwise occur but also increases the work that is necessary to mark and maintain a potential pregnancy trajectory. The gametes become temporarily independent genetic emissaries enlisting a whole space—the lab—that enables human embryos to exist outside patients' bodies. This independence allows functional stages, such as a blocked epididymus or fallopian tube, to be bypassed, and it allows manipulations of egg and sperm to occur. It also allows eggs and sperm alike to be derived from donors, joining in the lab a process of conception that will end up with implantation in the birth mother. These are powerful innovations.

Labs typically exhibit a moral economy of care that is oriented to maintaining the ambiguity that is imposed by a hybrid ontology: techniques, instruments, and medical professionals, and not just the traditional parents, are all part of the trails that lead to the establishment of pregnancy in this site. For example, work on embryos or potential embryos is carried out in semidarkness, with carefully controlled pH, and carbondioxide concentrations. Lab technicians feel that the care taken to protect the eggs and embryos from potentially harmful exposure to the outside environment "makes sense" because it probably approximates in vivo conditions. This care is exercised whenever embryos or potential embryos are around because of their link to a possible pregnancy. As described in chapter 3, at one clinic, a baby isolette from the neonatal intensive care unit had been adapted for working with eggs and embryos. These standards of care stand in contrast to the spinning, washing, and freezing of sperm and the freezing of surplus embryos when the life potential of embryos or fragile eggs that do not form part of a pregnancy trail is on hold or not at stake.

During the fertilization and development of embryos in the lab, developmental details are recorded. These notes are used to discriminate good from bad embryos: which should be transferred to the patient's uterus, which should be frozen, and which can be treated as waste. Embryos used for research are deemed to be a subset of the category of waste embryos. Segregating the embryos in this way helps to manage the disposal or freezing of some embryos. These developmental criteria justify exempting certain embryos from the moral and legal standards that apply to embryos as potential sources of life in the lab. The gametes and embryos are tied back into trails leading to the woman, and this structures the lab equipment, the lab procedures, and the behavior of lab technicians. The technicians often talk about their responsibility for the life potential of the eggs and embryos and are always aware of the preciousness of the eggs and embryos to the primary patient couple. As elaborated on in chapters 7 and 8, extensive legislative and bureaucratic standards penetrate the lab. They inscribe every embryo and prescribe reproducible success rates. The political and legal registrations of the embryos when outside the body are a measure of their precarious connectedness during this time. When the egg doesn't fertilize or when an embryo fails to develop normally, it is removed from the trail leading back to the uterus, and the ontological bond is severed. Leaving the trail relaxes the legal, moral, maternal, and political registrations of the embryos.

Cycles of Objectification

Each patient features under many descriptions during the course of treatment and during visits to the clinic. His or her subjectivity is multiply configured, and her or his agency is consequently crafted out of different elements in the different sites. When a couple seeks a medical solution to infertility, they exercise agency in their active participation in a number of different kinds of objectification. Just as I did in chapter 3 for normalization, I have separated out some of the kinds of objectification. The patient's agency is structured by these forms of objectification. This section points to different patient objectifications and the kinds of work they do (places they are used when infertility is being treated medically) but does not define the notions. Because women are the primary patients and because women undergo so many more of the objectifying procedures, I have focused on women.

Turning infertility into a medical problem through the eduction or drawing out of body parts in the treatment zone is one form of objectification in this site. A kind of agency may be eroded by this objectification, depending on the extent to which a patient is an integrated body or a person who acts as a whole over time. This kind of objectification can occur during the un-black-boxing of the body in the patient's anticipatory socionaturalization, during the itinerizing of a patient's body parts in the pelvic exam, during the rendering visible of body parts by the ultrasound, and during their manipulating and public witnessing in diagnostic surgery. The woman is rendered into multiple body parts many times during a treatment cycle. Objectifying the patient herself is a second and related form of objectification in this site. The kind of agency threatened by this objectification consists in the patient's identities as a social actor. For example, during a routine visit involving a pelvic exam, the woman begins as someone who carries the usual identificational weight when she is in the waiting room, moves to a place where it is appropriate to undress and wear a gown,[39] becomes an object of study that can be viewed from unusual angles with the help of instruments, and ends as someone who gets dressed and interacts face-to-face. Many of the woman's multiple social roles are temporarily irrelevant while she is being examined. The temporary loss of social identity makes it possible to "offer up" her body parts for intervention.

Generic bureaucratization of the patient is another form of objectification in this site. One agency that is associated with this objectification is the extent to which the patient is interacting as either a unique individual or a generic patient. Typically, although it can work in other ways, a woman patient is highly bureaucratically generic at the beginning of the appointment, is highly specific when she is being examined, and is somewhere in between at the end of the exam. In the waiting room, she is someone who has chosen to come to this appointment rather than to take other available passage points that our culture provides for infertility, and she is already being assimilated to a diagnostic form and into the normal routine of the clinic. Only generic properties are relevant to the clinical setting at this stage: she needs to arrive on time, come when she's called or fetched, and behave within the normal parameters for the clinic. She is objectified in a nonspecific, bureaucratic sense and is a token of a generic patient. This potentially threatens a patient's individuality but enhances her flow through the clinic.

Epistemic disciplining of the patient is also a characteristic form of objectification in this site. The clinic staff manages carefully what the patient is told and shown, what things are appropriate background knowledge, when informed consent is formalized, and when the patient may have access to information. Hospital ethics committees and physicians are careful to inform the patient of the risks and benefits of the treatment, for their own as well as the patient's protection.[40] This informational ethic is particularly salient in ART clinics, where everyone—the umbrella professional organization, the drug companies, the patient-support groups—produces free informational booklets on every aspect of treatment. The clinic is saturated with information, and the epistemic capital acquired by women patients in their anticipatory socionaturalization is altered and expanded in their career as a patient. At the advertised public sessions to attract prospective patients, lectures are held about the prevalence and causes of infertility and the state-of-the-art treatments available at the clinic. In addition to the various informational booklets, the clinic environment contains newsletters, support-group and counseling literature, and regular meetings on various topics (such as how to administer subcutaneous hormone injections). The only contexts in which I heard patients expressing a lack of knowledge or understanding were idiopathic diagnoses (that is, no known reason why the patient was not getting pregnant), unexpected breakdowns in routines (such as schedules that were changed by the clinic for nontreatment-related reasons), or confusion that cleared up after the procedure was experienced (a phenomenon with which everyone is familiar).

Epistemic disciplining of patients threatens to turn them into dupes who think and act in the best interests of the purveyors of information—the powerful professional organizations, drug companies, and support-group networks. As in other circumstances where rational, informed citizens are produced, discipline in the form of education also enhances participation. Patients are better able to participate in their own care because they have been initiated into the epistemic environment of the clinic. Producing informed citizens also produces epistemic standards. Certain facts about bodies, about particular bodies, and about treatment options are learned or are expected to be known. This generation of epistemic standards in the clinic normalizes diagnoses, helps constitute the practitioners as experts and the procedures as reliable, and facilitates the flow of authority and accountability.

These different but related forms of objectification are all associated with a kind of agency as mentioned above. When do they enhance agency, and when do they deprive patients of agency? As long as the activities in question promise to lead to pregnancy, synecdoche between the objectified patient and her long-term self is both enabled and maintained. In these cases, she exercises agency through active participation in each of the forms of objectification. However, if the synecdochal relation fails and the ontology of the educed treatment trails is not sustained long enough to overcome the infertility, then the different dimensions of objectification come apart from their associated kinds of agency. In these cases, the objectification stands in opposition to aspects of personhood that are cared about deeply and guarded carefully. The oppositional tension between objectification and agency alienates us from technology: operationalization as medical renders us as mechanistic discrete body parts, objectification of the social patient turns us into objects of experimentation and manipulation, bureaucratization turns us into institutional cogs, and we are hoodwinked by our epistemic disciplining. The ubiquitous possibility of this alienation that results from synecdochal breakdown in part explains our customary ambivalence toward the benefits of technology.

Thus, the project of repositioning medical ethics starts once we have arrived at a point at which we can explain why sometimes personhood and technology stand in opposition and sometimes they stand as partners. At this level of descriptive resolution, we can see and talk about technological alienation, the personification of technology, and the technologization of personhood. Instead of taking the creation of rational, informed citizens as the centerpiece, medical ethics could examine the conditions for the maintenance of synecdoche, collective options created and closed by trails of activity, and who and what gets to blaze which trails.

Retrofitting Agency and Objectification

I have argued that it is possible to discern potential gains for the long-range self within different dimensions of objectification, even when a notion of agency is commonly opposed to the dimension in question. I also have argued that the dehumanizing effects of objectification are not mere expressions of resentment at failed procedures but—resentment or not—are occasioned by metaphysical ruptures between the long-range self and the entities that are deployed in objectifying a patient. Because the rupture may not be immediately apparent or may manifest itself further down the treatment line, seemingly contradictory things can be said about identical procedures in good faith and with truth.

One effect of the temporal extension of metaphysical rupture is that whether objectification and agency are oppositional or coconstitutive, a treatment episode—and consequently the identity of the patient as subject—is revisable retrospectively. This does not mean that a patient might at one point think that she actively participated in her own objectification and then in the light of new evidence come to think that she had in fact been alienated by her objectification, although this may sometimes happen. This also doesn't mean that there was a fact at time t but that an assessment of t at t' gives different truth conditions for the state of affairs at t. The first of these would be a model of revisability based on the propensity to err displayed by the patient. The second version of revisability would be a form of epistemic presentism: an assessment of the state of affairs at time t must always be carried out from a time t', and the characteristics of time t' can impact the truth conditions retrospectively of the state of affairs at time t.[41]

The problem with these models of revisability is that they take the temporal component of cycles of objectification to be essentially static, so that time is a sequence of discrete time slices. Given such an account, arguments for retrospective assignments of agency would have to maintain that an individuated time slice is porous to elements that are derived from a distinct individuated time slice. It is notoriously hard to argue even for the effect of present actions on future actions when time is seen like this, despite the high degree to which we organize our lives on the premise that the present is thus oriented to the future. Instead, we must view ourselves as epistemically situated in the present but not isolated in a time slice—not epistemically trapped in the present.

Viewing the semantics of time as extended and dynamic (unless rhetorical devices are used to restrict temporal reference) licenses the idea that objectification and agency inherently possess retrospective and prospective properties. It allows patient testimony and patients' constructions of self-interest to be understood. As prefigured in the section on narratives of ART and analyses of accounts of why procedures did or didn't work, the apparent contradiction in testimony is resolved because the subject of the discourse changes depending on the stage and success of treatment. Patients still in active treatment operate with a metaphysics of educed trails in which objectification and agency are coconstitutive. When the patient talks about a failed treatment cycle that she doesn't assimilate to any ongoing treatment, she often expresses herself as having been alienated or dehumanized. In this case, the trails of activity have petered out without reclaiming the long-range subject, leaving a dualistic metaphysics in which objectification and agency are oppositional. When speaking about a successful cycle, the trails of activity have led back to and transformed the long-range subject, and the heterogeneous or hybrid ontology of the treatment zone becomes irrelevant.

Ontological Choreography

It was argued above that a woman infertility patient objectifies her infertility so that she passes through a number of places that promise to bring about desired changes in her identity. It also was suggested that by rendering herself compatible with instruments, drugs, and material surroundings, she was allowing them the possibility of transforming her but that transformation was not inevitable. She is locally and temporarily drawn out into a series of bodily functions and parts, working in a way that forges a functional zone of compatibility with the means of medical intervention. The instruments, drugs, physician, gametes, and so on all take on some of her by standing in for the phases that are diagnosed as not working. These processes of the rendering of the infertile woman and infertility procedures as compatible do not guarantee a seamless and successful solution to the infertility. The objectified body must not lose its metonymic relation to the whole person, and neither must the instruments lose their acquired properties of personhood for fixing, bypassing, or standing in for stages in a woman becoming pregnant. I call this process of forging a functional zone of compatibility that maintains referential power between things of different kinds *ontological choreography*. The choreography explained in this chapter is the coordinated action of many ontologically heterogeneous things and people in the service of a long-range self. The treatment is a series of interventions that turn “Where is it broken?” into a well-formed way of asking “Why aren’t you pregnant?”

III

Economies

7

Sex, Drugs, and Money: The Public, Privacy, and the Monopoly of Desperation

Reproductive technologies are multinational. Despite enormous discrepancies in access to clinics, the numbers of assisted reproductive technology clinics on every continent are rapidly growing, and as for many medical specialties, clinical and basic science researchers from many different countries publish in the major Anglophone journals of the field and participate in the annual professional meetings (table 7.1). Furthermore, the clinical field innovates quickly, and because it is dependent on tacit skills for handling gametes and embryos, it relies to a high degree on a transnational network of imparting skills. Drug development, testing, and marketing also are multinational affairs. Nonetheless, ARTs in the United States are a national enterprise that at once reflects U.S. political and scientific culture and is currently in the forefront of changing it. As Sharon Traweek found for the world of high-energy physics, even intrinsically international, highly collaborative scientific enterprises are inflected in deeply nationalist ways.[1] Indeed, ARTs might even be especially American in their current role, given that they involve the interaction of both science and the law in ways that seem to be telling of the nation. Sheila Jasanoff, preeminent analyst of the interactions among and coproduction of U.S. political culture, science, and the law, sees both science and the law as fundamental to understanding this national setting:

> American political culture derives its distinctive flavor as much from its faith in scientific and technological progress as from a commitment—some might even say an addiction—to resolving social conflicts through law.... Much of the recent entanglement of law and science reflects the American public's determination to bring under control the darker side of technological mastery: risks to public health and the environment, to individual autonomy and privacy, and to community and moral values. From Tocqueville to the present, commentators on American culture have called attention to this country's particular penchant for resolving political controversies and achieving social order

Table 7.1
Selected Assisted Reproductive Technology Organizations and Journals

Selected Organizations
Advanced Reproductive Care (ARC) (888) 990-2727 (provides funding packages) ⟨http://www.arcfertility.com⟩
American Infertility Association (AIA) ⟨http://www.americaninfertility.org/⟩
American Society for Human Genetics ⟨http://www.faseb.org/genetics/ashg/ashgmenu.htm⟩
American Society for Reproductive Medicine (ASRM) 1209 Montgomery Highway Birmingham, AL 35216-2809 (205) 978-5000 ⟨http://www.asrm.org/⟩
Centers for Disease Control and Prevention (CDC) report on 2000 ART statistics ⟨http://www.cdc.gov/nccdphp/drh/art.htm⟩
FamilyNet, project of the Human Rights Campaign Foundation for gay, lesbian, bisexual, and transgender families ⟨http://www.hrc.org/familynet/⟩
Fertilitext ⟨http://www.fertilitext.org⟩
Genetics and In Vitro Fertilization (GIVF) Institute Norfolk, VA ⟨http://www.givf.com⟩ See also ⟨http://www.jonesinstitute.org⟩ (pioneers of cost-cutting IVF without drugs, using immature ovarian tissue)
InterNational Council on Infertility Information Dissemination (INCIID) ⟨http://www.inciid.org⟩
RESOLVE 1310 Broadway Somerville, MA 02144-1731 (617) 623-0744 ⟨http://www.resolve.org⟩
Society for Assisted Reproductive Technologies (SART) (205) 978-5000 ⟨http://www.sart.org/⟩

Table 7.1
(continued)

Major ART Journals
Human Reproduction (U.K.) and Fertility and Sterility (U.S.) Researchers and clinicians from every continent publish in these two journals. For example, out of 150 articles appearing in *Fertility and Sterility* in 1996, the following areas were represented:
North America (USA and Canada): 42.7% (64 total)
Western, Northern, and Southern Europe (United Kingdom, France, Spain, Italy, Greece, the Netherlands, Belgium, Finland, Sweden, Switzerland, Denmark): 34.7% (52 total)
Eastern Europe (Slovenia, Hungary): 1.3% (2 total)
Australia: 3.3% (5 total)
Middle East (Israel, Turkey): 7.3% (11 total)
Africa (Egypt, South Africa): 4% (6 total)
Asia (Japan, South Korea, Taiwan): 6% (9 total)
South America (Brazil): 0.7% (1 total)
Important ART articles also appear in prestigious, general medical journals, such as the *New England Journal of Medicine* and *Lancet*; in related area journals, such as the *Journal of Obstetrics and Gynecology* and the *Journal of Reproductive Medicine*; in technical or specialist journals, such as the *Journal of Assisted Reproductive Genetics* and the *Journal of Clinical Ultrasound*; and in medical journals from other parts of the world, such as the *Medical Journal of Australia*, and the *Journal of Korean Medical Science.*

through law. It is hardly surprising that in an age of anxiety about the products of science and technology the U.S. public has increasingly turned to law to reassert control over the processes of scientific and technological change or to seek recompense for the failed promises of technology.[2]

Dr. Robert Edwards, one of Louise Brown's British lab fathers, wrote in 1989 that "bio-ethics—or moral philosophy in biology, or whatever label is used—have irrevocably entered the public domain."[3] Exactly what the "public domain" of bioethics is—especially in the United States and especially as concerns ARTs—is hard to say. The British have a habit of putting prominent individuals from various domains—members of the proverbial "Great and the Good"—on committees that produce recommendations about matters of public concern that can lead to regulatory capacity.[4] And in Britain, it is still possible for a single prominent expert in the field to become an honored and well-known public spokesperson, like Lord (Robert) Winston. Winston, the British IVF and preimplantation diagnosis pioneer, was the narrator of a six-part BBC television series on reproductive technologies for which he wrote an accompanying book subtitled *A Personal View of IVF*, and he was also the author of another book on ARTs, subtitled *The Definitive Guide*.[5] The United States also sets up commissions, but they tend to be composed of experts rather than widely recognized individuals, and their relation to the general public and to various executive, judicial, and legislative bodies tends to be obscure. Even Howard and Georgeanna Jones, doyens of U.S. infertility medicine and founders of the Norfolk, Virginia, Genetics and In Vitro Fertilization (GIVF) clinic that has remained the flagship of U.S. reproductive technology, would probably not call their book "the definitive guide" or command such widespread public name recognition that they could become the de facto national spokespersons for the field whose "personal view" would be of high interest. There may be something like an imagined community in the United Kingdom that corresponds to the public domain of which Edwards spoke. Elucidating the nature of the public or the public domain in the U.S. case is more difficult.

Despite this, books and articles on bioethics dealing with reproductive technologies in the United States use the word *public* freely, particularly in the combinations *public debate* and *the public*. Authors typically lament the lack of the former by the latter. In the period covered by this book—the early 1980s to the early 2000s—two things are true about the public and reproductive technologies in the United States. First, the number of people who have used them or know people who have used them has grown from almost nil to hundreds of thousands.[6] Second, the number of people who are familiar with them and have engaged in thought experiments involving them has grown from a small number to an enormous number. Reproductive technologies have been the subject of plots of popular television shows like *Friends*, of several movies, of numerous novels, and of countless lowbrow magazine and higher-brow newspaper and magazine articles.[7] Several highly publicized court cases have dealt with reproductive technologies. And reproductive technology examples, especially the landmark court cases, have become obligatory thought experiments in moral-reasoning classes on university campuses nationwide, proving themselves remarkably susceptible to decontextualization and abbreviation. In some sense, then, the public is saturated with reproductive technologies. Conspicuously missing is anything approaching an agora of ideas where the public can openly and equally discuss the pros and cons of reproductive technologies. Indeed, it is hard even to imagine what the space, the technologies of dissemination, mitigation, and aggregation, and the rhetorics of this agora might be.

Does this matter? A common bioethical definition of what counts as public in pluralist cultures like the contemporary United States relies on reason:

> In a pluralist culture, it is important to know what will and will not count as public—that is, available to all intelligent, reasonable, and responsible members of that culture despite their otherwise crucial differences in belief and practice. A public realm assumes that there is the possibility of discussion (argument, conversation) among all participants. The only hope for such discussion in a radically pluralist culture is one based on reason.[8]

My contention in this chapter is that the public agora of reproductive technologies is provided not by reason but by the politically capacious notion of privacy, which collects assisted reproductive technology's publics, contours its contestations, and propels its innovation. Privacy is not a unified thing that is in turn a unifying cause behind which different constituents rally. Rather, in its U.S. incarnation, privacy is a range of rhetorics that cover a wide range of behaviors and beliefs and that have a rich history in the U.S. Constitution, its amendments, the routine practice of the law, everyday life, science, the market, and the bedroom. Privacy in the United States is always marked out in dynamic relation to the various meanings, branches, and spheres of the public, such as the law, the state, and the demos. The law—the classic interface between the private individual and the polity in the U.S. version of democracy—is especially important. Reproductive technologies have been among the arenas of action and legislation that have altered and extended the scope of privacy in recent years, and so what privacy means changes even as it allows disparate people, techniques, and beliefs to coordinate.

In this chapter, then, I explore connections between three kinds of privacy that are fundamental to postwar U.S. science and culture and that come together in a particularly dramatic way in the field of reproductive technologies. These three realms, which are characterized by different degrees of separation and protection from and regulation by the state, are reproductive privacy, autonomous scientific research, and the private sector, and I describe all three in action in the clinical context. Their confluence in contemporary U.S. reproductive technologies makes an important if somewhat paradoxical space for social and technical innovations. The space is driven by its own powerful quasi-market logic that in its extreme form collapses supply and demand in what I call a "monopoly of desperation" that it may well share with other areas of biomedical and biosocial innovation. I argue that this dynamic space has implications for gendered citizenship in that it gives actors, especially women, access to new embodied political risks and opportunities. This confluence of privacies suggests that the biomedical and life sciences need to be integrated into political philosophy in a way that goes beyond how science and technology are generally discussed in relationship to democracy, agency, and social order.

Reproductive Privacy

Reproductive privacy, one of the meanings of privacy that collect publics, is most in evidence in talking with patients because it deals with protections on access to and use of procedures. Reproductive privacy in the United States is epitomized by the U.S. Supreme Court decision that legalized abortion, *Roe v. Wade* (1973), and includes in its penumbra the rights to privacy that are generally interpreted as protected in the U.S. Constitution's Bill of Rights (1791) and the fourteenth amendment to the Constitution (1868), on which *Roe v. Wade* and much other legislation draws. This part of reproductive privacy is part of the constitutional right of privacy, which in turn has been interpreted as being part of the liberty that is protected by the due process clause of the fifth and fourteenth amendments. Cass Sunstein, in his futuristic mock cloning trial, succinctly summarizes this constitutional strand of reproductive privacy and its possible limits:

> There is an acknowledged constitutional right to some form of individual control over decisions involving reproduction. In *Griswold v. Connecticut*, 381 U.S. 479 (1965), [the] Court held that a state could not ban a married couple from using contraceptives. In the Court's view, a right of "privacy" forecloses state interference with that decision. In *Eisenstadt v. Baird*, 405 U.S. 438 (1972), the Court extended *Griswold* to invalidate a law forbidding the distribution of contraceptives to unmarried people. In the key passage of its opinion, the Court said, "If the right of privacy means anything, it is the right of the individual, married or single, to be free from unwarranted governmental intrusion into matters so affecting a person as the decision whether to bear or beget a child." And in *Roe v. Wade*, 410 U.S. 113 (1973), the Court held that there is a constitutional right to have an abortion. This decision the Court strongly reaffirmed in *Casey v. Planned Parenthood*, 505 U.S. 833 (1992).
>
> On the other hand, we have held that there is no general right against government interference with important private choices. Thus in *Washington v.*

Gluckberg, 116 U.S. 2021 (1997), we held that under ordinary circumstances, there is no general right to physician-assisted suicide.[9]

Lauren Berlant has added to this account of the history of reproductive-privacy protection in the United States by showing how important the jump from *Griswold* and *Roe v. Wade* to *Casey v. Planned Parenthood* actually is, in that it introduced privacy protection within marriage for women and thus broke the bounds of a hitherto banal and sentimental heterosexual intimacy as the basic imaginary of privacy.[10]

As well as being a constitutional and living legal matter, reproductive privacy became a key post—World War II political issue in the United States through the abortion debate. ARTs frequently involve the production of human embryos ex utero that do not always rejoin an in vivo trajectory. As is discussed in chapter 8, the "disposition" of embryos in infertility clinics has been a fast-evolving area of material practice, ethical and bureaucratic accounting, and political positioning. The presence of embryos whose disposition needs to be accounted for means that clinics are always somewhat implicated in the politics of the abortion debate. The pronatalist impetus that patients and practitioners alike express in seeking and providing assisted conception, however, means that infertility treatment is as easily incorporated into prolife perspectives as into reproductive-choice agendas. Bipartisan political alliances that promote the passing of legislation that mandates insurance coverage for infertility procedures have been successful in some states because ARTs seem both to offer reproductive choice and to be profamily.[11]

Reproductive privacy also invokes a person's or a family's private life, which interacts with legally protected privacy, especially in the realm of family law, and is to be differentiated from the customary distinction between the public and private spheres. As Martha Minnow, Harvard Law School professor of family law, rights, and difference, has expressed it,

The work of [early twentieth-century] scholars focused on the distinction between government and private property. Another historical distinction in this country's law, rather than drawing the boundary between the public and the private marketplace, divides private families from all other institutions, including market and state. Law has long regulated relationships not only between public and private realms but between children and state officials, parents and children, husbands and wives.[12]

In reproductive technology clinics, private body parts, private acts, private aspirations, shame, and stigma are prevalent and finely choreographed, as is described in part 2. Patients frequently express a desire to maintain a carefully guarded privacy, to keep their medical records and their condition from being disclosed, and to control the aspect of stigma management that is organized around disclosure (who to tell and who not to tell about their infertility). Privacy protections are also important where not just stigma but active state-sponsored or more implicit discrimination might be feared. I have talked to many patients who wish to have their sexual orientation or marital status kept private as far as it is possible under the law and under local clinical guidelines. And still other couples wish to have their personal political or religious beliefs respected and integrated into their procedures.

Hopes that privacy will be respected invoke fears of diffuse yet ominous breaches of that privacy, as when we fear that too many bureaucratic traces are being kept of our lives or that these traces are being taken out of context, sold, or somehow used to shame us or penalize our access to services or rewards.[13] Reproductive privacy, then, is about what one does in private and about what public apparatuses are used to maintain and protect privacy boundaries. This realm of privacy summons up intimacy, sexuality, and reproduction, all of which have historically been notoriously fragile in their privacy (a fragility that is immediately apparent when the word *illegal* or *unwanted* is placed before those terms). The expression "in the privacy of the bedroom" references a larger domestic or private sphere of which sexual and sometimes reproductive behaviors are parts. Feminists have long argued that the preservation of this private sphere is both essential and deeply problematic for women and that the boundary construction of this sphere is highly political.[14] Private lives in reproductive technology clinics are dialectically public in this way, just as elsewhere. Despite, or perhaps because of, so many different but interacting strains of reproductive privacy, patients refer to and draw on privacy to cover widely varying experiences with the technologies. The following eight examples give a sense of the widespread resource that different meanings of privacy provide to patients on the right and the left, prochoice and prolife, who make themselves into the public of these technologies.[15]

In the examples of Naomi and Jules and then of Beth and Pat, the long interpellation of privacy into feminist and queer theory and politics is demonstrated through the idioms of "choice" and of "private and consensual," respectively. Naomi and Jules appreciate the privacy that protects their access to parenting, and they extend the metaphor of choice into the arena of picking a sperm donor. Beth and Pat struggle with their own definition of infertility and with the way that they incorporate it into the heterosexual clinical and insurance definition to have access to procedures without compromising their privacy. They thereby experiment with extending "private and consensual" from sexuality to parenthood.

1. Naomi and Jules chose a sperm donor through a local commercial sperm bank. They both were professional women. Naomi was just about and Jules just over age forty when they bought sperm. Naomi was a professor, a rising star at an elite private university; Jules worked in industry. Naomi was, at least since the 2000 census, multiracial. Jules was white and Jewish. They had postponed childbearing for an additional year while Jules had treatment for a life-threatening illness. This illness left them with no doubt that Naomi would be the birth mother and thus the one inseminated. Both agreed that this gave her priority over certain decisions, in line with legally protected rights of pregnant women to "choose," and Naomi made the final choice about the donor. While they agreed that a Jewish donor would have been acceptable, Naomi chose a donor who matched her own predominant racial identity as she experienced it, which was African American. Jules supported her in this decision, speculating that it stemmed from wanting in part to preserve or pass on this identity visibly and in part to pass on their cultural advantages to a child of that racial minority. Mostly, however, Jules explained that she felt that it was none of her business why Naomi chose the way she did. Naomi was the nursing mother, and Jules was more of the primary care-giving mother, staying home while Naomi traveled or worked long hours. They joked, with more or less tension and in ways that are typical of dual-career professional couples, about the burdens that these gendered roles created. Jules was not on the daughter's birth certificate, but she legally adopted the girl when she was a little less than a year old. After the adoption, Jules expressed frustration at the difficulties that lesbian and gay couples had to deal with if they wanted to parent but delight at the outcome of the process in their particular case. She also praised sperm banks and the medicalization of conception for the relative anonymity of artificial insemination that had protected their privacy while also allowing them access to child bearing. In the course of their discussions with me, they explicitly referred to both the right to choose and the protection of their privacy.

2. Beth and Pat, a long-term cohabiting couple, had decided to try to have children. They researched transnational adoptions and pregnancy with sperm from a gay friend. Some of their lesbian friends told them about using IVF to allow one of them to get pregnant with an embryo made from the egg of the other mixed with donor sperm. Beth and Pat were intrigued by this possibility. They lived in a state that provided some insurance coverage for infertility, and the local clinic that they called had a Web site that claimed that the clinic respected the right of gay, lesbian, and single women patients not to be discriminated against in reproductive matters. When I spoke to them, they had decided to try a cycle of IVF, with Beth intending to gestate and Pat to provide the egg. They were excited as well as concerned about the upcoming procedure. They were anxious about the medicalization of their reproduction that IVF requires and the risks for both of them in that. But they also had encountered several hurdles in the preliminary phase that they thought had wider implications for lesbian relationships and families in the contemporary United States. They had been welcomed as self-paying patients but did not qualify for insurance coverage for IVF without evidence that they (or only the gestating partner; they weren't sure which) were infertile. The definition of infertile *that was used by clinics was interpreted as heterosexual and required uncontracepted exposure to ejaculatory coitus over a year or more. No one seemed sure how this criterion could be enforced or how infertility would be established for lesbians who were contemplating this procedure, however. Was it enough, as Beth suggested, that she had not gotten pregnant over the six years of hers and Pat's cohabitation, despite their desire to coparent and their lack of contraception? Or as Pat refined it, was it enough to qualify as infertile that they could not get pregnant with both of them as biological parents in any other way? In the end, Beth referred to previous attempts to conceive using fresh donor sperm and artificial insemination (AI) that had not resulted in a pregnancy, and this was accepted as evidence that she was infertile. Because these previous attempts were "turkey baster," or private, inseminations that had been carried out without the involvement of a medical facility, the clinic could not confirm whether they had actually occurred or how many attempts were involved. When I left the interview, I did not know whether Beth and Pat had actually tried this low-tech AI and failed to get pregnant or if they had pursued this option to a certain point and then decided that it did not solve their particular circumstances of childlessness given the available options. Their privacy on this matter was handled with delicacy by the clinic staff and caution by themselves.*

Beth and Pat were also anxious because their friend who had previously agreed to be the sperm donor had been turned away by the clinic because according to National Institutes of Health guidelines, his sexual orientation meant that he was at high risk for HIV. This meant that they had to pick from commercially available anonymous sperm donors. Instead of picking a donor to match certain culturally and phenotypically salient characteristics of the nongesta-

tional mother, as several of their friends had done who had used a sperm bank, they decided that in the new procedure they would "match" Beth, the gestational mother, because the egg was going to come from Pat. The final concerns that they raised were the lack of legal precedents, parental custody rights, and financial obligations. Their intention was to attempt to have both their names put on the birth certificate as natural mothers. As they entered treatment, they were eagerly following other cases like theirs in the United States.[16]

Some patients name religious freedom as an important part of their reproductive privacy. This means different things to different people. Janice and John, her husband, were officers in the Salvation Army, voted Democrat but considered themselves conservative on most social issues that relate to family. Their orientation to reproductive technologies involved privacy not in the sense of nondisclosure but in the sense of integrating the technologies into their own religious understandings of their infertility. They embraced the technologies as a useful means, despite their ineffectiveness in their case and despite the secular scientific clinical setting. Marie and Alex, an Orthodox Christian couple, interacted with the technologies in such a manner that no step in the procedure was a secular technical means that could be co-opted to a religious end. Each step had to be assessed, transformed, and justifiable as an end in itself and on its own terms to the satisfaction of their priest.

3. Janice, a stay-at-home mother in her early thirties with a daughter in elementary school, had been trying for years to get pregnant again. Both she and her husband, John, were officers in the Salvation Army, and both felt deeply committed to God and family. For the first few years after the birth of their daughter, they paid for medical insurance that included maternity benefits, expecting to have a second child. When it began to seem as though they might be having a problem, Janice scoured the local library and bookstores for books on fertility and discussed conception with everyone she knew. Her husband was supportive but quiet, almost never joining in these conversations. Janice exhibited remarkably few inhibitions about disclosing her failure to conceive, a fact that she put down to her religion. She felt that there was a meaning to all suffering and that it was an insult to God to feel ashamed by the trials that had been sent her way. Eventually they began to seek an insurance policy that would support infertility treatments. She adopted the same attitude toward treatment that she had displayed toward infertility, trusting in God, family, and friends to get her through it. Janice and John spent eight years trying to get pregnant, including the two IVF attempts that were allowed by their new insurance—one that didn't result in a pregnancy and one that ended with a devastating miscarriage. Eventually, they got pregnant "all by ourselves," as Janice wrote to me in a letter. Their little boy completed their family, and Janice concluded her letter with the words "isn't modern medicine wonderful?!"

4. An Orthodox Christian couple, Marie and Alex, began visiting an infertility clinic in their area to find out why they were not getting pregnant. Their hysterosalpingogram came back showing that the wife's fallopian tubes were closed from scarring from a previous pelvic infection. The doctor discussed this finding with the couple and told them that because they were young, the husband's semen analysis was normal, and the wife's hormonal indicators were all normal, a good chance of success lay with IVF. The couple told the doctor that they were not sure that their faith allowed them to do IVF. Among other things, they mentioned that masturbation was prohibited. The doctor informed them that sterile, nonspermicidal silastic condoms with a small pinprick hole in the end could be used and that one of their Roman Catholic patients had used them. Sperm could be collected in these during normal intercourse, and because the condoms had a hole and no spermicide, they were not considered contraceptive. A staff person would explain how to collect and deliver the semen-containing condom. Alex and Marie said that they would consult with their priest and get back to the doctor about whether they wished to proceed. This particular doctor was strikingly entrepreneurial, both technically and socially, and suggested immediately that the couple bring their priest with them to one of the clinic staff's weekly meetings so that everyone could talk and be as well informed as possible in deciding if this was feasible.

A couple of weeks later, the couple and their priest arrived for a clinic meeting. After introductions, the doctor presented the couple's diagnosis. He presented IVF as being extremely close to nature in that it used the husband's sperm and the wife's eggs but simply bypassed the blocked tubes, thereby restoring their natural function. The priest then explained that their faith, in common with Roman Catholicism, did not view having children as a right but as a privilege and so the means and not just the end had to accord in their details with the precepts of the religion. He was particularly worried, he said, by the idea of fertilization taking place outside the body. A lab technician suggested that perhaps they might adapt a version of IVF that had been developed for countries with little medical infrastructure that involved placing a test tube with the sperm and eggs in it in the vagina for fertilization, instead of allowing fertilization to occur in the lab. The priest brightened at this suggestion, and the physician beamed at the technician and later complimented her for her ingeniousness. They discussed this option some more, and the priest seemed inclined to give it his blessing until a second problem arose. The doctor felt obliged to comment that the medical community did not technically view the vagina as part of the inside of the body, which they consider to start at the cervix. The priest, however, felt that this medical definition did not pose a particular problem. If fertilization could occur inside the vagina, then it was clearly also inside the body. The lab technician then told the priest and the couple that the vagina is very embryo toxic and that the test tube containing the couple's gametes would have to be hermetically sealed if the embryos were to have any chance of surviving. The test tube would thus not be in biological contact with the inside of the woman's body during fertilization. Everyone looked anxiously at the priest, but he waived the objection aside. The meeting broke up merrily, with the treatment scheduled to begin shortly thereafter.

The next couple of weeks proved to be a tense period for the lab technician, who contacted other technicians with expertise in this kind of IVF. She carried out a trial run, terrified that something might go wrong. She hermetically sealed a test tube of culturing fluid and wore it inside her own vagina for the forty-eight hours that were required for fertilization. She wanted to check the seal and also feared that the test tube might come out of the vagina during defecation, perhaps during the very moment at which fertilization was occurring. Her trial was successful, and Marie and Alex began their cycle of IVF. The gametes were placed in the test tube, which was sealed and inserted in Marie's vagina like a tampon. After retrieving the test tube, the lab technicians found two embryos, one of which resulted in a successful singleton pregnancy.

It is relatively well known that some patients oppose the kind of privacy that is protected under reproductive choice because it permits the destruction of embryos in some circumstances. Less well known is that patients sometimes restrict themselves from certain treatments and deny themselves particular understandings of events because of their private views, including views in favor of reproductive choice and their own personal kinship circumstances. The cases of Agnes and Bart and of Clare and Simon are examples of this.

5. Agnes was a successful, forty-one-year-old businesswoman. Her first husband, considerably older than she, had three children from a previous marriage and with retirement on the horizon had not wanted to start a new family. Bart, her second husband had one child from his first marriage and was eager to have a child with Agnes. They began trying to conceive shortly after getting married, and after eighteen months of trying, they began infertility treatment. A cycle of IVF initiated a pregnancy that began normally. A fetal heartbeat was detected in early scans, but Agnes lost the fetus at the end of the first trimester. Her pro-choice beliefs made it hard for her to think of the fetus as a child and thus grieve the loss of a hoped-for baby. The existence of stepchildren from both marriages made the option of adoption less attractive than it might have been if the family had been childless. Autopsy results revealed that the fetus had a chromosomal anomaly, a common occurrence, she was told, at her age. Egg donation would have been a medically obvious solution to her body's response to treatment, but she felt like it would be yet another form of step-parenting because she would not be genetically related to the child but her husband would. Agnes resolved to try IVF with her own eggs again.

6. Simon and Clare are at risk for a genetic disease of which they are both carriers. They had suffered from unexplained infertility since their marriage five years previously. They discovered the genetic risk after a second IVF attempt resulted in a pregnancy that ended after four and a half months of gestation, when the fetus died in utero. It was found to have the fatal condition. Their mourning was difficult, although it was made more legitimate than Agnes's (above) by the relatively advanced age of the fetus at death and the opportunity that they had had to hold, name, and bury it. During this break in treatment, they debated how to proceed in their efforts to become parents and considered both adoption and further infertility treatment. Being able to conceive with IVF emboldened them to try again, but they were worried about the genetic disease. They decided to try another cycle of IVF but agreed with their physician that it made sense to use preimplantation genetic diagnosis this time. Preimplantation diagnosis involves checking each embryo to see if it carries the affected gene before transferring it to the uterus. One of their anxieties, however, was that Clare had had so few eggs on her previous cycles that no embryos would be free of the disease and survive the biopsy for implantation. Clare also discussed the possibility of trying IVF with donated embryos that were left over from another couple's successful IVF. This made sense to the couple because neither partner felt that it was absolutely necessary to have a genetic connection to their child (they were strong supporters of adoption) and because with embryo adoption they might be able to experience pregnancy and birth. Clare's ability to be pregnant had been established, and they felt that if they never had the chance to experience a successful pregnancy and birth, they would carry considerable feelings of loss with them. They were on the brink of embryo adoption as a backup to their own IVF should no embryos be available for transfer. After further research, however, the cooption of embryo donation by ultraconservatives gave them pause. During the stem-cell research debates of 2000 and 2001, the expression "embryo adoption" came to replace "embryo donation" in public opinion, the media, and the law and was rhetorically effective as young children, gestated by prolife women from donated embryos, were displayed to the nation. Simon and Clare are strongly prochoice Democrats. She works in literary publishing, and he is a public-interest lawyer.

Despite their openness to adoption, they couldn't participate in embryo donation because of its prolife associations and their sense that prolifers were promoting it to attack a woman's right to choose.

Chloe and Jonas wished to keep their marital status private but were ambivalent about the dichotomy between prolife and prochoice framings as they moved from the reproductive technology clinic to obstetrics. Finally, Leonie, a promarket Republican, and Rich illustrate the use of reproductive technologies and the private sector as models for each other to make sense of their infertility experience, both justified by a kind of social Darwinism.

7. Chloe began living with a man when she was in her midthirties. She already had two children from an early first marriage. Her new partner, Jonas, had been married before as well but did not have children. They decided that they wanted to have a child together but hesitated to get married because of their previous bad experiences. Chloe had had her fallopian tubes surgically removed after an infection, and so they knew that they would have to use IVF if they wanted to have a child. Because Chloe already had children, her insurance agreed to cover two cycles of IVF that got to the stage of egg retrieval or beyond instead of the more generous and more statistically realistic four cycles that the policy permitted for couples without other children. They discussed what they would do if Chloe did not get pregnant after the first two cycles and joked about applying for the insurance through Jonas instead of through Chloe so that they could claim the full four cycles. They decided, however, that the risk that they might be asked to prove that they were married outweighed the benefits of this option. They were grateful that their privacy had been respected and that no one had asked them whether they were in fact married. Their cohabitation and their presentation as a couple who desired a child was treated as sufficient proxy. A first cycle failed, but the second cycle was successful, so the question was moot. At midpregnancy, Chloe was offered an amniocentesis for the detection of certain genetic fetal anomalies, as she was now over thirty-five. The prolife ambiance of the reproductive technology clinic and the sense that the baby she was carrying was a "miracle baby" seemed to her to be negated by the simplistic framing of amniocentesis as a chance for a "therapeutic abortion." She expressed frustration that she could not find anyone to talk to who could bridge the infertility and obstetrics worlds and help her to interpret the prolife and prochoice elements in them.

8. Leonie works in public relations for a manufacturing company, and her husband, Rich, is a businessman. He, like a number of men in his family, is subfertile, with a low sperm count, low sperm motility, and high sperm abnormality.

They had been trying unsuccessfully to get pregnant for many years, including attempting cycles of AI, IVF alone, and IVF with ICSI. Despite their empirical inability to get pregnant, Leonie told me that they were "the kind of people who should be having babies," which she justified in terms of their good looks and the social and educational advantages that they could pass on. A slew of scientific publications debating the possibility of inheriting male subfertility if they got pregnant by a means like ICSI that bypassed the sperm problems, however, made Leonie anxious about the question of dysgenics. She took several months off treatment, learning as much as she could about the politics and ethics of infertility treatments and the possibility for dysgenic or eugenic consequences of using them. By the end of that period, she had reconciled her own beliefs that it would be eugenic for her and her husband to reproduce with the scientific literature that suggested that it might be dysgenic. If they did pass on their subfertility to a subsequent generation, they would also pass on the ability to deal with it culturally and financially. In addition, they would produce offspring who would be responsible about the numbers of children they bore, possibly even helped in their euceptive action by their subfertility. She also became convinced that arguments for selective access to reproductive technologies were not eugenic in what she called "the bad sense" and that as such the price structure of infertility treatment was justified. She made all these arguments through privacy. On the one hand, she argued that whether she and her husband reproduced was a matter of fundamental privacy, the core of their private lives. Simultaneously, she argued that as long as reproductive technologies were private and firmly in the private sector, it did not matter if they were eugenic in the sense of promoting the reproduction and well-being of more successful people; only things like government-mandated sterilization or population goals were problematic. Private eugenics was no worse because it was exactly the same as wealthy parents being able to give their kids music lessons or college prep classes. Additionally, she believed that reproductive technologies needed to stay in the private sector because that ensured that some restrictions on their use would be exercised without requiring the government to intervene explicitly in people's private lives. Despite having reconciled her beliefs and fears, however, she hesitated to resume treatment. She had found it surprisingly stressful, especially in the onslaught that it represented to her sense of her husband and herself as successful, and she decided to take some more time away from treatment to reestablish their confidence in their own status before recommencing.

As these presentations of several couples' treatment choices and circumstances suggest, many different strands of privacy are attached to the notion of reproductive privacy, and they intertwine in ways that are specific to individual circumstances. Evident in these narratives are negotiating abortion politics, keeping reproductive technologies and reproductive decision making apart from the government, respecting individuals' privacy (whether their condition, sexual orientation, or marital status), integrating private religious and political beliefs into treatment, and bringing private circumstances to bear on interpretations of individual kinship.

Autonomous Scientific Research

In addition to the multitude of ways in which privacy enables and mediates patients' experiences of assisted reproductive technologies, the ideal of the autonomy of scientific research has also been important to the development of U.S. reproductive technologies. The idea that scientific research should be autonomous has deep roots in postwar U.S. culture and lay behind the ultimately victorious Program for Postwar Scientific Research developed by Vannevar Bush, the director of the Office of Scientific Research and Development, which massively increased the size and funding of the National Institutes of Health.[17] As explored in chapter 1, the idea of autonomous science has two strands. On the one hand, scientific research should be carried out in an autonomous realm to promote a disinterested search for truth and to avoid the potential for political distortions that, for example, characterized National Socialist science. On the other hand, high levels of government support for (without government intervention in) basic research in science and technology are seen as powering the military and economic superiority of the United States.

Science pursued for political ends—especially that involving the participation of people (so-called human subjects)—had been revealed at the end of World War II to have involved the gravest violations of humanity, up to and including euthanasia.[18] The Nuremberg Code of 1947 established international standards for assessing the violation of human rights in scientific and medical research and also began to establish the idea that human subjects, regardless of their identity and of their capacity to speak for themselves in given circumstances, have rights that ought not to be violated by scientific experimentation. It took another jump, however, before it became clear that the core U.S. postwar ideal of science carried out "autonomously" (with democracy and good science securing each other) did not guarantee human-subject well-being. As David Rothman has written, American articles of the period that addressed the Nuremberg trial drew from it not the lesson that the state should regulate experimentation but quite the reverse—that the state should not interfere with medicine.... The logic of the argument was that the atrocities were the result of government interference in the conduct of research.... Science was pure—it was politics that was corrupting. Hence, state control over medicine through regulations that intruded in the private relationship between doctor and patient or investigator and subject were likely to pervert medicine.[19]

Henry Beecher, chair and Dorr Professor of Research of the Harvard Medical School's department of anesthesia, was the initial whistle blower in the late 1950s to mid-1960s who first alerted the public to U.S. cases of postwar violations of medical ethics. The ethical problems in U.S. medicine came first to Beecher's attention out of his concern that ethical breaches were subverting good science.[20] Eventually, considerable publicity was given to these revelations of systematic ethical breaches that were integral to the research establishment—the Tuskegee study and other studies that were carried out on conscripted soldiers, the mentally ill, and the elderly by prominent physicians and research scientists at top institutions and that were published in top peer-reviewed journals in the optimistic immediate postwar period. The focus then moved to the ethical lapses themselves rather than the effect of poor human-subject protocols on good science. At first, Beecher was motivated by a desire to protect the sense left by the Nuremburg trials' revelations and U.S. articulations of national science policy that good science and good ethics went together. The ethical lapses themselves were suddenly and widely reported to widespread shock and condemnation only after Beecher published an article, "Ethics and Clinical Research," in the *New England Journal of Medicine* in 1966 in which he reproduced the research protocols, without naming names or giving references, of twenty-two examples of research that he claimed had risked the health or life of research subjects.[21] What began as a call to protect good, autonomous, science was eclipsed by the context and details of the disregard of subjects' interests by the best and the brightest of scientists—the "real" story.[22] A corollary of this, whether acknowledged or not, was that autonomous science could no more intrinsically protect all human subjects than could politically motivated science. Even if good science and democracy were linked structurally through the various claims of the so-called Mertonian thesis (described in chapter 1) and even if truth, objectivity, transparency, and accountability were epistemological and political bedfellows, human subjects were not necessarily protected.

This transformation from Beecher's initial concerns about bad science to a public demand for a U.S. biomedical ethics that would afford human-subject protection took a decade to result in domestic legislation. The resulting laws—the National Research Act of 1974 and the associated National Commission for the Protection of Human Subjects of Biomedical and Behavioral Research—inform the basic structures for protecting human subjects in medical protocols to this day. Assisted reproductive technologies are direct descendants of this hybrid of postwar autonomous research with human-subject protocols to protect the growing scope of individual rights that risk being subsumed to the scientific bottom line. Robert Pippin has shown that the mechanisms of informed consent that comprise the backbone of what Charles Rosenberg has called "the bioethical enterprise" are based on an Enlightenment ideal of a sentient individual who cannot be harmed if he is informed of the presumed-knowable risks and willingly subjects himself to them. In short, the protocols require freedoms from various kinds of subjective and situational vulnerability that simply cannot be assumed in experimental procedures with suffering patients.[23] And Monica Casper has shown that assessments of risks and benefits—which are at the center of the National Commission for the Protection of Human Subjects of Biomedical and Behavioral Research; its principles of respect for persons, beneficence, and justice; distinctions between research and practice; and judgments about brutal or inhumane treatment—are grossly underdetermined and must be applied by each institutional review board for each decision.[24]

Assisted reproductive technologies have had a difficult time being seen as medically necessary, and their brief history has been fraught with ethical dilemmas, so in some sense they have taxed this system to its extreme. To an outsider, ARTs in the United States might have been an obvious case for taking a more European style of governmental regulatory control. ARTs have managed to keep a high degree of professional scientific autonomy free from direct state, corporate, or public control. This autonomy has been achieved despite (or indeed, because of) human-subject protection; political pressure and patient activism for accountability; the establishment of instruments of data collection and accountability that are run jointly by patient groups, the government, and the professional organization; profitable infertility drug manufacturers that push research in some and not other directions; ongoing fights to classify ARTs as treatment for diseases and thence qualify treatment for insurance coverage and insurers' efforts to resist these moves; and a growing body of activist legal precedent. This is a testament to the importance and strength of this model of biomedical research in contemporary U.S. society.

The initial regulatory impasse around ARTs began in 1979, after which federal funding for IVF-related research was not available. Instead, private-sector funds paid for research. Moreover, a professionalized bioethics combined with a typically American medical organization's desire to retain a high degree of independence from direct government intervention to produce privatized but relatively autonomous biomedical research in the field.[25] As Howard Jones, a prominent U.S. IVF researcher, expressed it,

> There does seem to be a growing consensus that something needs to be done. If this is so, it is probably in the best interests of all concerned that it be done in an orderly volunteer way by those who know most about the subject. If those who know most about it don't do it, consumer pressure will get the attention of the legislative process, which could botch it.[26]

Protocols for experimental procedures, informed consent, and institutional review boards have formed the basis of the rapid innovations in infertility treatments that have occurred over the last twenty-five years. This linking has allowed infertility specialists, researchers, and instrument and drug companies to try out new protocols directly on patients. Informed consent works despite the problems mentioned above because infertility continues to be considered to be an individual treatment rather than a public-health concern. It is seen as a matter primarily of private decisions about reproduction and not as a public-health issue, as discussed in the section above. This remains true even though epidemiological data as well as popular cultural lore suggest that distinct life-style-related patterns (such as postponed childbearing) indicate who might be at risk for this highly resource-intensive condition. It also remains true even though the state does in fact take a great deal of interest in who reproduces and how much.

Pierre Soupart, a Belgian researcher who moved to the United States and Vanderbilt University in 1972 (the year that he published the first documented in vitro fertilization), was the first IVF researcher to apply for a federal grant for IVF research. His 1979 protocol proposed perfecting in vitro fertilization techniques and then examining resulting human embryos for chromosomal anomalies. It did not propose implantation. Jimmy Carter's Secretary of Health, Education, and Welfare, Joseph Califano, responded to protests from antiabortion and religious groups by establishing an ethics board to consider the application. The ethics board eventually decided that such requests for federal funding were appropriate, but the board was allowed to lapse during Ronald Reagan's administration. The American Society for Reproductive Medicine (ASRM)—prompted in part by the regulatory vacuum, indeed a Catch-22 of a go-ahead decision from a lapsed ethics board—composed its own Ethics Committee in 1985.[27] Comprised of theologians, ethicists, lawyers, physicians, and researchers, the committee has continued to write regular reports on ethical issues, including a response to the Vatican's *Instruction on Respect for Human Life in Its Origin and on the Dignity of Procreation.*[28] The committee's reports are published as a supplement to *Fertility and Sterility* under the title "Ethical Considerations of the New Reproductive Technologies." In June 1993 Bill Clinton lifted the effective moratorium on federal funding for embryonic tissue research, and in 1994 the National Institutes of Health published the *Final Report of the Human Embryo Research Panel,* also known as the Muller Panel.[29] The de facto ban of federal funds for ART research has continued, however.

Concurrent to these governmental and organizational steps in the normalization and regulation of research involving human embryos around ART clinics, pressure was rising to regulate ARTs and, in particular, to standardize the ways in which success rates were presented and advertised. Because the issue of truth in advertising successfully combined many critiques of the procedures (the use of women's bodies as experimental sites in the face of bad odds, the technological imperative to try technologies despite poor odds, and the discrepancy between published success rates and the figure that patients cared about—the "take-home baby rate") and yet relied on tried and tested regulatory procedures, it proved to be a successful rallying point.[30] Representative Ron Wyden (D–Oregon), working with patient groups and the ASRM, sponsored the so-called Wyden law, which Congress passed as the Fertility Clinic Success Rate and Certification Act of 1992 (table 7.2; see also table 5.1) and which mandated the creation of a federal registry of data on pregnancy success rates from all ART centers and certification of embryo laboratories.[31] The act was not fully implemented, however, because it was not funded. Instead, voluntary data reporting continued, with professional organization members continuing to bear the costs. The Centers for Disease Control and Prevention (CDC) joined in 1997 in running the registry and first helped report clinic results from 1996 procedures. The act has been amended to incorporate the funding and other practicalities of collection and the verification of clinic success rates.

Table 7.2
Clinic Reporting and Certification Obligations

The Fertility Clinic Success Rate and Certification Act of 1992 requires all U.S. clinics that perform assisted reproductive technologies procedures to report data annually to the Centers for Disease Control and Prevention (CDC), through the Society for Assisted Reproduction for every ART procedure then initiate. ART is defined as any procedure in which both oocytes and sperm are handled outside the body.

The following information on procedures and outcomes must be collected for each cycle:

What ART procedure was involved?

What was the woman patient's age?

Did treatment involve a patient's own eggs or donor eggs?

Were any embryos that were transferred freshly fertilized or previously frozen?

Were embryos transferred into a gestational surrogate?

Was the procedure cancelled?

Was there a pregnancy?

Was there a live birth?

Was there a multiple pregnancy?

Who Is to Report ART Cycles?

The medical director of the Society for Reproductive Technology (SART) program ("practice") at the time that the reporting is due is responsible for reporting every cycle performed by that practice during the requested interval ("reporting year") to SART (and hence the CDC). In situations where different practices are involved with the ovarian stimulation, the oocyte retrieval, or the embryo transfer, the obligation in "fresh" cycles lies with the practice that accepted responsibility for the embryo culture.

Validation

Validation is the process whereby, through random sampling, the veracity of the entire dataset to be published is established. This process is performed by the Validation Committee in conjunction with the CDC. The Validation Committee is composed of fourteen professionals from both SART and non-SART member programs. Sites to be visited (currently 40) are randomly selected by the CDC, with site visits performed by teams of two Validation Committee members. Currently, about 20 variables are validated from 50 randomly selected cycles. Additionally, all live births that are reported by the clinic are validated. The data collected from site visits are compiled and reviewed jointly by the CDC and the Validation Committee. Programs are notified regarding the overall outcome of validation and their specific program results.

The Society for Assisted Reproductive Technologies (SART), in conjunction with the CDC and the national consumer infertility network RESOLVE, now runs the registry that implements the Wyden law. SART is a branch of the ASRM and started as special-interest group of its parent organization (it was originally named the IVF Special-Interest Group). On its renaming in 1987, it began its dual goals of implementing a voluntary registry of IVF clinics and their success rates nationwide (begun in 1988) and of exerting "a positive influence on state and federal funding and regulation of the new reproductive technologies."[32] Once the government, in the form of the CDC, was involved in implementing the registry, evaluation of the effectiveness of the registry and of the problems that it revealed began. Continued failures of standardization in reporting (and hence comparing) statistics stood out, as did high multiple-pregnancy rates and high and unfair treatment costs.[33]

From its inception, SART was envisaged as the means by which the precarious balance between the public and the private could be maintained in ARTs. On its Web page, SART aggressively defines itself in this manner. In answer to the first of several frequently asked questions ("What is SART?") the 2000 SART president, Philip McNamee, replies that it is "The governmental watchdog for ART—prospectively preventing governmental intrusion." Only much further down in his remarks does he mention the circular mechanism by which SART's regulatory and promotional goals combine to ensure that there a considerable pressure from patient groups and practitioners alike to belong to SART:

> Our 370+ practice members represent 95% of all IVF units in the USA. Insurance companies are noting that SART members publish their validated data and have accredited laboratories. Some major carriers have begun to require documentation of SART membership for reimbursement.

The requirements that must be met for SART membership are tough and serve to reassure patients and insurers and to fulfill the federal requirement that assisted reproductive technology results be published. SART members must report every ART cycle undertaken in every year, including all begun and cancelled cycles, as well as ongoing and live birth pregnancy rates. Each member clinic must also agree to have its reported data "validated" by two of a team of fourteen experts from the CDC or the SART members who make up the Validation Committee when randomly selected. All members must submit to embryo laboratory inspection and certification every two years, which is usually carried out by a team from the Joint Commission of Hospitals or the College of American Pathologists. They must also abide by all of SART's practice, laboratory, ethical, and advertising guidelines and, since 2000, must be headed by a medical director who is a board-certified reproductive endocrinologist.

SART has grown in numbers and in scope from its days as an ASRM special-interest group to an organization that blends regulation and privacy in the interests of patients and science. The ASRM published the statistics collected for the year 2000 in December 2002 (statistics run almost two years behind because data collection and analysis cannot begin until all ongoing cycles at year's end have been completed and any resultant births have been recorded).[34] In the 2002 bulletin, Robert Bryzski, president of SART, commented on the importance of the information-collecting system and the scale of trust and bureaucratic bookkeeping that it requires:

> We are very proud of our role in bringing this information to the public.... Infertility clinics in the United States are virtually the only medical facilities in the world that report their medical outcomes in this comprehensive way. Our members work very hard to collect this data, and use it to improve treatments for their patients. It is a major undertaking for the SART member clinics to assist us in collecting this information, and we appreciate all the hard work that goes into it.[35]

The Private Sector

The third meaning of the word *private* that is important to the development and practice of U.S. assisted reproductive technologies is the one that is evoked by the phrase *private sector*. It is opposed to the state-run and -controlled public sector. This sector, not a sphere, includes the part of a country's economy that is owned by companies and individuals, and it incorporates consumers (and the advertising impetus behind the media) in its scope (table 7.3). Research that is carried out by drug companies and their sponsorship of the field fall into this category. Similarly, to the extent that assisted reproductive technologies are not covered by insurance (and given that effectively no reproductive technologies are provided as a public service under Medicare), they qualify as consumer goods. The importance of drug companies, R&D, and the consumer aspect of ARTs is hard to overestimate (table 7.4). To explore this aspect of privacy, I review the recent development

Table 7.3
Internet Sites for Gamete Sales and Donations, Embryo Donations and Adoptions, and Gestational Surrogacy

Gametes
Rainbow Sperm Bank ⟨http://www.gayspermbank.com⟩ San Francisco sperm bank that actively recruits gay and bisexual donors.
Repository for Germinal Choice (now defunct) ⟨http://www.geniusspermbank.com⟩ Used to recruit donors by IQ and by educational and professional attainment.
Ron's Angels ⟨http://www.ronsangels.com⟩ Egg-selling program where the donors are models. Program holds egg auctions and recently began holding sperm auctions. Primarily a front for a pornography retail outlet.
Embryos
Embryo Donation ⟨http://www.reprot.com/⟩ Arranges long-term cryopreservation and embryo donations with "leftover" donated embryos.
Snowflakes Embryo Adoption ⟨http://www.snowflakes.org⟩ Christian organization that arranges embryo adoption with frozen embryos after an adoption home study has been completed and with the involvement and consent of the genetic "parents."
Surrogacy
Conceptual Options ⟨http://www.ConceptualOptions.com⟩ Permits nonmedical (vanity) surrogacy.
Growing Generations ⟨http://www.growinggenerations.com⟩ Gay and lesbian egg-donation and surrogacy services.

Table 7.4
Major Fertility Drug Manufacturers and Their Activities Supporting the Field

The two largest manufacturers of fertility drugs for the U.S. market are Serono and Organon.
Serono Laboratories, Inc. is the U.S. member of the Swiss developer and marketer of pharmaceuticals, Ares-Serono Group.
Organon, Inc. is the U.S. affiliate of N.V. Organon, a multinational pharmaceutical company.
Both companies received FDA clearance to release recombinant fertility drugs in the United States in 1997. Serono began marketing recombinant follicle stimulating hormone (FSH) as Gonal-F, and Organon, as Follistim. Ferti.net is supported by Serono. Organon runs a hotline for help regarding insurance coverage for their IVF products and procedures: (800) IVF-PALS.
Other pharmaceuticals companies that have U.S. approved fertility drugs: Ferring Pharmaceuticals, Fiyisawa, Hoechst Marion Roussel, Milex, Steris, Wyeth Ayerst.
Companies that regularly sponsor ASRM events: Asta Medica, Ferring Pharmaceuticals, Organon, Parke Davis, Pharmacia Upjohn, Searle, Serono Symposia U.S.A., Solvay Pharmaceuticals, TAP Pharmaceuticals, Unipath, Wyeth Ayerst.
Fertilitext is supported by Stadtlanders pharmacies and includes order forms on the site.

of recombinant drugs to control ovarian stimulation during IVF, as well as the pricing structure of treatments.

As has been documented by Adele Clarke and other historians, exogenous hormones had been extracted and prepared for use in the control of the female human reproductive cycle for half a century before anyone attempted clinical IVF.[36] The marriage between practitioners of infertility medicine and drug companies occurred very early because it was known that exogenous hormones could induce ovulation.[37] For the development of IVF, this knowledge promised two of the clinical ingredients that have been essential to the rapid spread of assisted reproductive technologies. The first was the possibility of regulating women's menstrual cycles so that the timing of ovum pickup for fertilization could be optimal. This aspect also helped make treatment cycles more amenable to manipulation within doctors' and patients' daily schedules. The second crucial clinical aspect was the possibility of inducing women's bodies to produce more than one mature egg per cycle so that clinicians would have a better chance of having at least one embryo to transfer and so that patients would have a greater chance of getting pregnant. This surfeit of eggs rapidly became a critical element in the development of embryo cryopreservation technologies and of procedures such as third-party reproduction and micromanipulation, where having more than one egg at a time was a huge asset. The companies that were making fertility drugs underwrote the clinical expansion of assisted reproductive technologies more or less since the birth of Louise Brown, with Bourne Hall (Steptoe and Edwards's ART facility in the United Kingdom, which was set up after the birth of Louise Brown) famously being funded by Ares Serono. The annual meeting of the ASRM displays the extent to which drug companies and instrument makers collaborate with practitioners in research and in many other aspects of American assisted reproductive technologies. The development of recombinant fertility drugs, after an acute shortage of naturally derived ovulation-induction drugs in the mid-1990s, marked a new level of standardization of ovarian stimulation in IVF and illustrates well this aspect of ARTs.

Human follicle-stimulating hormone, when administered to women patients exogenously, induces the development of multiple follicles and so is used for in vitro fertilization and other infertility procedures that aim for the simultaneous maturation of several eggs. Prior to the availability of recombinant FSH, urinary human menopausal FSH (u-hFSH) was the only substance that had been shown to work in controlled ovulatory stimulation. The first baby was born following controlled superovulation with a recombinant FSH in 1992 as part of the long-term, ongoing, research program at Organon to replicate and express complex human molecules by recombinant DNA technology. In the mid-1990s, multicenter phase 3 clinical trials—that is, trials involving actual patients who were undergoing treatment—of a recombinant form of human FSH were undertaken, and by 1995 results on their safety and efficacy were reported.[38] In 1997, peer-reviewed drug company research declared that recombinant gonadotropins actually worked better than urinary gonadotropins.[39] In the late 1990s, two of the major infertility drug companies, Ares Serono and Organon, began marketing recombinant human follicle-stimulating hormone (r-hFSH) products. Serono and Organon market u-hFSH.[40] At Serono, for example, the urine from which their u-hFSH is extracted is collected from healthy menopausal women volunteers in Italy, but the Serono spokesperson whom I interviewed declined to tell me any further details of consent, collection, or purification. Urinary FSH's clinical shortcomings stem from variations in its carbohydrate side chains and from the presence of urinary proteins and luteinizing hormones, which impede interbatch standardization and might interfere with the induction of folliculogenesis (inducing eggs to mature).[41] The amino acid sequence in r-hFSH is identical to that of u-hFSH. The recombinant products exist in various isoforms but have less "junk" attached to them that can disrupt the timing or extent of folliculogenesis.

Recombinant FSH was developed in a standard biotechnological interspecies fashion by transfecting nonhuman cell lines with genetic material that is capable of replicating identical amino acid sequences to human compounds. The immortal mammalian cell line that Serono uses for genetically engineered r-hFSH production is the Chinese hamster ovary, which is at this point a widely available biotechnological tool.[42] Serono reported that neither the Chinese hamster ovaries nor the r-hFSH provoked the production of antibodies.[43] They market r-hFSH as Gonal-F, which became one of the most widely used ovulation-induction drugs almost as soon as it hit the market. For Gonal-F to come to the U.S. market, however, recombinant FSH had to be tested for safety and efficacy, first on nonhuman and then human subjects, according to standard FDA procedures. The nonhuman tests were carried out at Serono's own labs in Geneva, Switzerland, on twelve of their own monkey population.[44] They tested subcutaneous, intramuscular, and intravenous methods for administering the recombinant FSH and measured its consequent effect on serum FSH concentration. They also measured its active life and compared all results to their previous standard of care—namely, the drug Metrodin, which they make from urinary FSH. Urinary and recombinant FSH performed similarly, except that the genetically engineered variety provoked a smaller immunological reaction.

A monopolistic demand for particular drugs is created by the domesticating and standardizing of the responses of different patients' bodies and the mandatory standardizing of a particular drug into treatment regimes because it is more effective and more predictable than other regimes. Drug companies stand to gain enormous financial incentives from establishing this kind of monopolistic dynamic, and so they also play the part of benefactor of research, running trials to demonstrate that indeed the drugs that they manufacture are the most effective. This aspect of ARTs in the United States is typical of many areas of contemporary biomedical practice. As mentioned in the discussion in chapter 3 about the fate of natural cycles, the development of recombinant gonadotropins has enabled practitioners to streamline and standardize ovulation induction to the extent that it is possible that there is little or no "going back" without a significant breakthrough in maturing immature eggs in vitro with its own associated drug culture. In any case, the dependence of the field on recombinant gonadotropins is solidified by their complicity in the demand for accountability, by improving both success rates and standardization. The profitability of the recombinant drugs also ploughs back into the field in the form of postgraduate training, the underwriting of professional meetings, and the sponsoring of every level of R&D.

The other side of this deep intertwining of the quasi-autonomous field of ARTs with the private sector is the fact that these technologies are services toward which patients are consumers. The bottom line of drug companies and medical facilities alike is profoundly affected by the ability of patients to get insurance coverage for or to pay for ARTs. Similarly, many practitioners experience the inequity of access as intrinsically wrong and against their training and precepts. This means that physicians' organizations and drug companies, as well as patients, have put their considerable research and analysis skills into tracking and critiquing payment options and into lobbying for greater insurance coverage for ARTs. Because the U.S. medical insurance system is also outside of the government or public sector in the United States, with the exception of Medicare and Medicaid (Medicaid, in a well-known piece of state-sponsored selective pronatalism, covers contraception but not infertility procedures, whereas state insurance laws do not require private insurers to offer coverage for contraception), most of the activity around pricing has occurred in the private sector. (See table 2.1 for a summary of the current approximate costs of typical ART procedures; table 2.2 shows the infertility insurance coverage by state as of 2001, including, where appropriate, ART coverage.)

A notorious 1994 paper calculated the cost of a successful IVF delivery in the United States as ranging from $67,000 for an average couple to over $800,000 for a woman over forty who had more than six cycles of treatment and whose husband was oligospermic.[45] These figures contained estimates of prenatal and perinatal costs that were associated with multiple gestation, although these costs are in fact passed on to obstetrics coverage despite resulting from ARTs and so tend not to show up in discussions about the costs of ARTs. Another well-known study, published in 1995, calculated the costs to all insured families under a given program that included coverage for ARTs.[46] The study's author put the costs that IVF would add to insurance policies per insured family at $2.79 per year for one cycle of IVF and at $8.30 per year for three cycles. Fees for psychological counseling and for legal help with such things as third-party ARTs are typically not included in these estimates of cost (see table 2.1 for typical quotes for these fees, however).

In 1998, Blue Cross and Blue Shield of Illinois calculated per pregnancy costs that compared health maintenance organization (HMO) and preferred provider organization (PPO) care costs for infertility treatment (of which 5 to 15 percent can be assumed to have involved ARTs).[47] The signing into law on January 1, 1992, of the Illinois Family Building Act mandated almost full infertility coverage, including up to four assisted reproductive technology cycles with embryo transfer for a first child. Illinois thus has some of the best data on ART costs to patients and insurance companies under different health-care plans. In the 1998 study, overall PPO costs per pregnancy were $24,763 versus $14,328 per pregnancy in HMOs. A large part of the price differential came from the higher levels of surgery in PPO care. Surgery has tended to be overperformed in infertility medicine because it is lucrative for hospitals and can often be billed under a general obstetrics diagnosis for patients, which most insurances cover.[48] In Massachusetts, another state with comprehensive infertility insurance coverage, including IVF, the cost per delivery for ART pregnancies with 1993 data was $59,484.[49]

Insurance companies' primary concern is to minimize per pregnancy (their primary outcome measure) cost to insurers. They also try to avoid short-term cost-saving events that end up being more expensive, such as high patient dropout rates, which can be deceptive:

> On first impulse, one would conclude that a patient's drop-out from infertility care saves cost for the insurance carrier. On further examination, this may, however, not be the case since many infertile couples will re-enter treatment at a later date, requiring costly repeats of diagnostic and therapeutic procedures, while facing decreasing success odds due to advancing age.... In the final analysis, the quick achievement of pregnancy may, therefore, be one of the most important factors in controlling costs during infertility therapy.[50]

In states where there is little or no coverage—the majority—payment is out of pocket for patients and results in gross stratification of the services. In the states where there is coverage, patient and physician activism has tended to be behind the legislation in question. Physicians in states that don't have comprehensive coverage have pioneered cost-cutting protocols. For example, a university ART program in Colorado found that in certain patient populations, preprogrammed, unmonitored ovarian stimulation reduced costs without compromising outcome.[51] This inequity has generated critiques as well as some imaginative ways of managing payment, some of which are described in chapter 3. Consumer groups as well as consultants in infertility health-care coverage insurance reimbursement have a dual message: never take the first no for an answer, and always advocate for appropriate coverage, with the available help:

> Patients need to fully understand the insurance contract's benefits which are available to them and to advocate for appropriate coverage. There are provider and manufacturer resources to assist the physician's business office staff and patient with identifying coverage.... Patients should not be discouraged if the initial response from the insurance company is unfavorable.[52]

Some clinics as well as some independent financing programs, including one associated with RESOLVE, offer flat costs, where patients pay a set fee, regardless of how many cycles it takes, or where they pay for three cycles and any more are free. Recently, the genetics and IVF program in Norfolk, Virginia, has been offering donor-egg procedures on the basis of payment if and only if a pregnancy is achieved. In practice, only certain patients qualify for these programs, which can, in any case, represent daunting up-front payments. Jairo Garcia, writing on behalf of SART, says that the fact that in most states "ART procedures are available only to upper-income-level patients and are out of reach for most individuals" represents "a social issue and introduces a certain bias for patient selection, pathology, and treatment."[53]

Monopoly of Desperation

Each of the areas of privacy examined above—reproductive privacy, autonomy of scientific research, and the private sector—constitute particular moral and technical economies for the implementation of ARTs in the United States. Each one of them also brokers relations with various branches of government, leaving ARTs substantially affected by yet not directly regulated by the state. Another feature of ARTs is closely related to and dependent on these "publics of privacy," which derives from the nature of its consumer base. There have been no American social movements to speak of against ARTs since they have become clinically available, despite their high media visibility. Patient advocacy for "health care coverage for infertility, medical research, adoption benefits and other family-building issues" is well organized and effective.[54] Individual issues, such as the stockpiling of embryos addressed in chapter 8, come up regularly. And a low rumble of antiabortion activity from one side and anticommodification of the child and the body on the other can be heard in the background.[55] But the kinds of social movements that have, for example, characterized the introduction of genetically modified organisms or that have accompanied the activist organization of patients with AIDS have been more or less absent for ARTs, especially since the implementation of the Wyden law.[56]

ART activism shares with AIDS activism the characteristic that patients feel that they cannot afford not to be involved in the medical science. To misquote Donna Haraway (see chapter 1), it is the only (or most powerful) game in town. In this regard, even though treatment with ARTs is elective, the option to "just say no" does not address the needs and interests of most patient activists in the same way that it can in activism that opposes genetically modified organisms (GMOs), for example. This means that although ARTs and GMOs both involve the so-called new genetics, their politics are actually quite different. As part of their greater similarity, ART and AIDS activists also share the phenomenon of lay involvement not just in political and personal issues that are related to access to and quality of care but also in the science itself and protocol development. This can be characterized as enhanced democracy (as Epstein guardedly does for the case of AIDS activism), and it has profound effects on the development of the field. The stratified nature of ART coverage has been a major focus of patient and physician activism, as recounted above, but its mere existence makes it hard to describe patient activism in the case of ART as particularly democratic.

ART and AIDS activists differ in their approaches, then, in that ARTs are much more in the private sphere and the private sector than AIDS medicine. And in general, ART patients are not sick or dying, and so there is not the same compelling public interest to treat and cure patients as there is for AIDS sufferers. In addition, ARTs have a rhetorical advantage that has been exploited by RESOLVE for advocacy purposes, which is that they can be cast in a robustly bipartisan manner through appeal on the one hand to "family-building" and on the other to "reproductive choice." The stigma of infertility in contemporary U.S. culture is much more personalized and privatized than the social stigma attached to AIDS and is correspondingly almost entirely invisible in the political face of activism. Finally, the cultural capital of professional literacy, affluence, and position that spurred much of the activism that Epstein described in American AIDS activism is augmented by a kind of "motherly activism" in ART activism.[57]

In recent years, scholars of the new social movements and feminist political scientists have put together case studies and theories showing how important a political role for "motherhood" can be to social change. Working especially with women organizers in Latin America but also in Sub-Saharan Africa, South Asia, and East Europe, scholars have shown that in societies with a traditional gendered separation of spheres, women's roles as mothers, health-care providers, and community organizers can bring about significant community betterment even (or especially) when a community's men are in marginal positions in a highly stratified public sphere. While it might not appear at first glance to have anything to do with the kind of activism found in the United States surrounding ARTs, some of the compulsions and organizations are similar. RESOLVE is the most important patient-advocacy organization for infertility and is an example of the combination of U.S. biomedical-style activism that is associated with the growth of AIDS activism in the 1980s and of motherly activism.

RESOLVE is a nationwide organization "that serves the needs of the infertile population and allied professionals with support, education, and advocacy."[58] Like many women's activist organizations, a single individual, Barbara Eck Manning, founded it in 1974. Twenty-five years later, it had fifty-nine local chapters and close to 30,000 contact people. Its organizational strengths lie in its combination of national reach and local support, and membership dues and other funds are split evenly between the national and local tiers of the organization to reflect this. RESOLVE has been especially successful because of its lobbying efforts, its linkages with every aspect of ARTs, its strong connections to physicians, and its maintenance of the SART database (coming from its original role in the passage of the Wyden bill). Members of RESOLVE have at their fingertips the means to be up to date on political and legal issues of relevance to their treatment as well as cutting-edge research and first-rate statistics. Locally, chapters hold support meetings that grow into friendship, support, and sharing of supplies and information. This is a mostly female space that is run on the basis of the drive to be a mother and the biosociality or understanding, care, and support that can be received from others in the same position. By no means all infertility patients join RESOLVE or go to local chapter meetings, but for those who do, it is a space where they can find respite from the fertile world. RESOLVE publishes booklets with titles like "Managing Family and Friends" and "Coping with the Holidays," and a typical conversation might poke fun at the unthinking and painful comments of the fertile: "Why don't you just adopt? There are so many children needing a home," or "The world is overpopulated anyway," which, if true, must apply to everyone and not just the infertile.

RESOLVE also helps participants to establish the perspective of having a right to participate in reproducing because it is a major life activity. This transformation of something that is surely historically and culturally contingent—certain kinds of pressure to have certain kinds of families—and also extremely personal and stigmatizing into a less stigmatizing bodily disease and the basis for an ahistorical, transcultural human right to reproduce is the basis of the local activist solidarity. At the national level, RESOLVE has managed to present its advocacy in a manner that performs the great bipartisan straddling trick. On the one hand, supporting research and treatment for the involuntarily childless is profamily or, in their phrase, "family-building." On the other hand and befitting the important patient demographic of liberal, postponed child-bearing professional women, it offers women greater control of and choice about their reproduction. The compulsion to motherhood and the broad support that the role in its ideal form garners have fueled the gendered activism of the field.

A prominent aspect of the experience of reproductive technologies for infertility is their "never enough" quality, which Margarete Sandelowski diagnosed as mutually reinforcing mandates from culture, technology, and nature that together make up the "therapeutic imperative" (and which I describe from an epistemological point of view in chapter 3).[59] An important part of this is the desperation that individuals or couples feel or have attributed to them, which Naomi Pfeffer documented with characteristic wit, and which Sarah Franklin examined in an early essay.[60] This desperate, private compulsion to motherhood doesn't just propel activism but also plays a role in the market of ARTs in the United States. That is, it imposes what I refer to as a "monopoly of desperation." ART procedures, drugs, and techniques are produced in the conditions of a monopoly, but the monopoly comes as much from the consumer as from the drug manufacturers described above. A monopoly is usually defined as a sole supplier or producer of a product or service. A monopolistic industry is one that has almost total control over the market for its product and is consequently able to determine output and prices, thereby manipulating demand as well as supply. Because monopolies are prohibitively anticompetitive and because they prevent consumers from exercising individual choice as the means of determining demand, they are regulated in most capitalist countries. Markets are "free" and competitive relative to a frame that must itself be stabilized, and economic theories propose different means for the most efficient means of doing this.[61] Economists and those who study markets pay less attention to the consumer side of the market than to the producer side. When they do attend to consumers, the issues tend to revolve around the conditions for and competencies of disembodied calculative agents that are able to maximize utility in the face of choice.[62]

The conditions of desperation—drawn from the intersection of private struggles, the imperatives of the private sector, and autonomous medical research that must perfect treatment—work together to produce extremely focused patient demand for treatment. In a sense, a monopoly that is imposed at least in part by the consumer is a very good market. It ensures demand and consumer loyalty, forces rapid innovation and efficiency, and pushes to internalize externalities such as the social and moral costs of unequal access to treatment. The motherly compulsion and the implication of women's bodies and lives as part of the very substrate of ARTs produce focused, because embodied, calculative consumers. The ART market, by being about all three kinds of privacy simultaneously, contains many checks and balances that reflect explicit and latent currents of interest and opinion of ARTs' publics in its own dynamic.

Bodies Talk

In *The Queen of America Goes to Washington City: Essays on Sex and Citizenship*, Lauren Berlant argues that there is no *public* sphere in the contemporary United States.[63] She attributes this to a collapse of the distinction between the private authentic selves that are developed and affirmed in the private sphere and the public selves or citizens that are developed through their relation to the state and a shared public sphere. This "intimacy of citizenship," she argues, occurred through a rise of a right-wing ideology that separated a nation that was made up of patriotic sentiment and the politics of private life from anything to do with government or the state, especially at the federal level. From this axiom about the lack of a public sphere, Berlant goes on to develop what she calls a "theory of infantile citizenship." It seems undeniable that a strong current in American culture postulates an infant, or at any rate an idiot savant, as the ideal political antidote to the excesses of government against which the founding documents are supposed to protect its citizens. Movies—*Mr. Smith Goes to Washington, Mr. Deeds, Being There, Forrest Gump, Dick,* and *Dave*—repeatedly use this masculine, infantile political figure. The simpleton cuts through government excess and excesses alike, witnessing but literally unable to understand or enter into elitism, corruption, and political intrigue and thereby restoring the populism and simplicity necessary for truth, justice, honesty, and decency.

As the cases described above show, though, many people with a wide range of political beliefs and socioeconomic status interact with these embodied technologies through rhetorics of privacy in gendered, patterned, political, and far from infantile ways. Rather than being opposed to the public sphere and the state, the private sphere, autonomous science, and the private sector broker relations between people and their government precisely because these privacies mediate, modify, and mitigate power and conflicts of interest. The motherly activism and the ineliminable embodiment mean that this kind of politics is the counterpart of the masculine infantile citizen. The different privacies of the sex, drugs, and money of reproductive technologies in the United States constitute a unique market. It transforms and defines individuals even while it crafts proxy relations between state and citizen. And it allows for tremendous social, technical, and entrepreneurial innovations, even while reproducing many of the conservative tropes and their material effects that underwrite ideologies of motherhood.

The tracing of an emerging U.S. public around assisted reproductive technologies over the last twenty-five years through a focus on privacy may sound counterintuitive given that private is frequently opposed to public, but as I have shown above, one or the other and sometimes several interacting meanings of privacy are used by patients, scientists, practitioners, lawyers, drug company representatives, and physicians alike. It is also somewhat strange to talk about privacy in connection with technology at a time when new technologies of surveillance seem to be geared precisely to an onslaught on privacy. But if privacy as a right needs to be protected in the face of technologies that threaten it, privacy as individual and political agency also needs to be recognized.

The public of reason that is raised at the beginning of this chapter as a possible agora for deciding issues in this field is compelling because it promises an arena in which the right and the good can ultimately predominate and because it assumes that there is such a thing as the right and the good. A public that engages in various practices under rubrics of privacy promises no such thing. It sets up a marketplace of practices where individual and group identities, technical and social innovations and conservatism, moral and capital economies intermingle without regard to scale and where reason is neither an arbiter nor a guarantor. Reason frustrates because it falters on the irreconcilability of incommensurable moral universes. Privacy, on the other hand, dispenses with commensurability in almost all settings and instead generates practices that over time develop "best-practice" moralities as products and input, both explicitly and implicitly.[64] In assisted reproductive technologies, success rates and technical prowess have improved, and patients are effective moral pioneers, in Rayna Rapp's term.[65] But such a marketplace of literally embodied practices uses rhetorical feed stocks—habits, tropes, and narratives of self, family, and nature that come with locally specific historical baggage of preexisting meanings and hierarchies and that is fed into new practices of differentiation and stratification, as well as empowerment.[66] As Judith Butler has said, "This is a position which insists *both* on the exclusionary effects of the modernist narrative and the revisable and rearticulable status of that narrative as a cultural resource that has serviced a collective project of extending and enhancing human freedoms."[67]

8

The Sacred and Profane Human Embryo: A Biomedical Mode of (Re)production?

Where physics had iconic status in relations between science and society throughout much of the twentieth century, biotechnology, biomedicine, and the information sciences have increasingly inherited that mantle. But twentieth-century physics and twenty-first-century biology are very different sciences in terms of how (and how directly) they implicate citizens, influence (and are influenced by) statecraft, and produce knowledge. Many observers have attempted to characterize the connections between science and society in the United States and elsewhere, especially those that have accompanied the rise of molecular biology, genetics, and biomedicine. Even before it began, this new century had been dubbed "the biotech century" as the genetic and digital revolutions converged and grew ever more powerful.[1] Evelyn Fox Keller, in her book *The Century of the Gene*, heralded a new "Cambrian period, only this time not in the realm of new forms of biological life but in new forms of biological thought."[2] Paul Rabinow's *French DNA: Trouble in Purgatory*, reflecting his Foucaultian perspective, suggests that relatively stable post–World War II relations between body, society, and ethics "are today, again, being remade, and the assemblages in which they functioned, disaggregated."[3] Charles Rosenberg has noted that even for those historians of medicine for whom meaning is ineluctably culturally and temporally situated, Ludwik Fleck's concept of a *Denkstil* (style of thinking) applies. Rosenberg and some of his colleagues have decisively drawn out the characteristics of a biomedical *Denkstil* involving the consolidation and bureaucratization of disease concepts and a professionalized bioethics.[4] Hans-Jorg Rheinberger has argued that molecular biology has transformed the Enlightenment narrative of overcoming nature with society to a point where the "natural" and the "social" "are no longer useful concepts to describe what is going on at the frontiers of the present 'culture of biomedicine.'"[5] Margaret Lock and Sarah Franklin have drawn our attention to a "range of new entities, from cloned sheep to transgenic mice, immortal cell lines, and brain dead bodies," "to modify the elementary components of life itself."[6] And Eduardo Viveiros de Castro and Annemarie Mol and her colleagues have begun to theorize this "multinaturalism."[7] Addressing political forms more closely, Sheila Jasanoff and her colleagues, in *States of Knowledge,* have begun to spell out some of these emergent assemblages between identity, state, and knowledge.[8] Outside the idioms of science and technology studies and history of science, Ulrich Beck has presented the preeminent analysis of the rise of risk and uncertainty.[9] And a number of prominent scholars of postindustrial societies have suggested that the rise of information, knowledge, and service economies has brought about dramatic changes in modernity and in organized capitalism of which the digital revolution has played a central part and in which biomedicine and genetics will be increasingly important.[10]

What is to be done? Some analysts have advanced normative arguments or displayed normative methodologies for questioning the changes. Marilyn Strathern, with typical prescience, started the conversation on ARTs and has developed modes of analysis that are based on partial connections.[11] Donna Haraway has long urged a posthumanist ethics of hybrid subjects and in her recent manifesto offers a companionate ethic of being that is instructed by significant otherness.[12] Helen Verran has begun to develop ways of getting on together across different, emergent ontologies.[13] Everett Mendelsohn and his colleagues have called for historical perspective and the courage to confront the hard ethical and political implications of the genetic revolution.[14] Bruno Latour has called on us to take the metaphor and rise of the precautionary principle seriously.[15] Paul Rabinow has cautioned us against epochism and metaphysics, while nonetheless citing radical changes.[16] We ought, he says, to be asking, "What forms (of social and bodily recombination) are emerging? What practices are embedding and embodying them? What shape are the political struggles taking? What space of ethics is present?"[17]

Over more than a decade during which I have worked in reproductive technology clinics and visited labs, doing research on reproductive and genetic sciences, I have seen many changes. Molecular biology and genetic technologies have entered the clinic. This has occurred on the side of the development of recombinant fertility drugs, as well as through such clinical practices as preimplantation diagnosis and the promises of stem-cell research and cloning. The growing convergence between "empirical" reproductive technologies and genetic technologies has echoes in many areas of biomedicine.[18] Kinship and the facts of life have been in flux, and a private form of motherly activism has many interesting properties and strengths. The practices that I have discussed in this book are a part of a broader landscape and thus ought to be instructive about the evolving roles of and relations among science, identity, and society in biomedicine. In this conclusion, I draw out some of these general characteristics as I have come to see them through the case of reproductive technologies. To do this, I focus on the part that is played by a major actor in the previous chapters to which I have so far given conspicuously little voice: the human embryo. Perhaps more than any other actors in this book—women and men patients, bureaucracy, technology, privacy—the embryo's tale conveys the sense that assisted reproductive technologies are part of something new and particularly indicative of time and place. The human embryo in the United States carries (in more legible form than most entities) active and latent meanings that are considerably more widespread and older than the technologies in question. The presence in the clinic of ex vivo human embryos raises in especially acute form enduring tensions between the sacred and the profane that characterize biomedicine. These characteristics make the embryo a good spokesperson (figure 8.1).

I argue that a consideration of the human embryo sheds light on patterns of interaction that together make up a "biomedical mode of reproduction."[19] Although for analytic clarity, I contrast some key aspects of this biomedical mode of reproduction in which the embryo is a signature presence with more typical ways of thinking about capitalism, I do not mean to imply that biomedicalism will replace capitalism. Biomedicine and, more broadly, biotechnology are but one part of the U.S. and global economy, and although they are likely to grow substantially, they will not dominate the economy. The tiny proportion of the field that deals with human embryos may increase dramatically if embryonic stem-cell research is allowed to develop, but even on the most bullish long-term projections, it would remain a small fraction of the overall economy. In addition, the phenomena that I discuss as evidence for a biomedical mode of reproduction appear to coexist comfortably with capitalist modes of production, even where the differences are striking.

My claims about a biomedical mode of reproduction, then, are not arguments about the end of capitalism or historical epochism and inevitability. Among other things, the causal antecedents are temporally messy. Some events that are important to the current period stem from

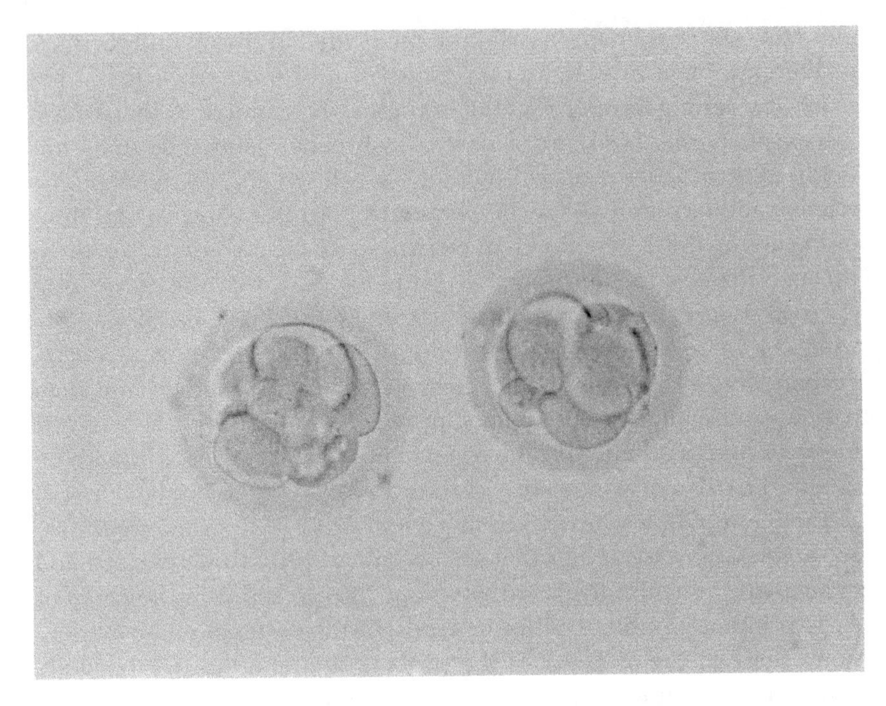

Figure 8.1
The Sacred and Profane Embryo in a Reproductive Technology Clinic. One of these embryos is now a child, and the other did not implant after embryo transfer. A third egg failed to fertilize normally and was discarded.

the immediate postwar period (including the Nuremberg trials and the rise of molecular biology starting in the 1950s), some come from specific biomedical innovations (like reproductive, recombinant, and digital technologies, which have punctuated the last fifty years), and still others have grown out of the burst of important 1970s legislation and commission reports (like the U.S. National Research Act of 1974, the 1979 Belmont Report, and social organizations like RESOLVE, which began in 1974).

Despite these antiepochal caveats, I argue that the biomedical mode of reproduction that I trace through the human embryo has its own characteristic systems of exchange and value, notions of the life course, epistemic norms, hegemonic political forms, security, and hierarchies and definitions of commodities and personhood.

I introduce key aspects of this mode of reproduction, using examples of embryos from my fieldwork[20] and divide these aspects into elements that relate to the economy, to science and society, and to

Table 8.1
Aspects of Capitalist Modes of Production and Their Possible Counterparts in a Biomedical Mode of Reproduction

Economy: Capitalism	*Economy: Biomedicine*
1. Production	Reproduction
2. Alienation from one's labor	Alienation from one's body parts
3. Capital accumulation	Capital promissory
4. Efficiency and productivity	Success and reproductivity
5. Waste disposal as a physical problem	Waste designation as an ethical problem
Science and Society: Capitalism	*Science and Society: Biomedicine*
1. Public understanding of science	Private implication in science
2. Moral responsibility of scientist	Bioethics
Identity and Kinship: Capitalism	*Identity and Kinship: Biomedicine*
1. Life course finite and descent linear	Loss of finitude and linearity in life course and descent
2. Too much homosex = homosexuality	Too much homosex = cloning
Too much heterosex = adultery; miscegenation	Too much heterosex = ooplasm transfer; xenotransplantation
3. Essentialism of natural kinds Social construction of social kinds	Strategic naturalization/ socialization of all kinds

identity and kinship (although as I argue throughout this book, they are intricately connected). Table 8.1 summarizes these points of contrast and emerging features of biomedicalism. I end by suggesting the implications of these dimensions for the integrity of the person, for agency, and for resistance in this mode of reproduction, pointing to counterintuitive but important consequences for human and nonhuman dignity and freedom.

The Human Embryo in the Lab

What was new about the new "artificial reproductive technologies" of the 1980s (which were subsequently renamed assisted reproductive technologies) was the array of procedures that enabled and required the routine presence of ova, embryos, and sperm in the clinic and outside the body. Human sperm has long been easy to obtain.[21] The extraction, nurturance, and manipulation of human eggs and the fertilization, nurturance, and manipulation of human embryos are difficult and require significant lab resources. The repercussions of this tension—between the sacred and the profane around the ex vivo human embryo—more than any single other thing make ARTs the proverbial microcosm of what I am calling the biomedical mode of reproduction in these sites. I use the ontological choreography of embryos in ARTs—what their ontological status is and how they coordinate very different kinds of time scales, matter, ideas, and organization—to describe the biomedical epoch that we have recently entered and to make projections about the future.

In the field of ARTs, embryos are treated as protopersons or even full persons by some people at some times and in some places, when they are maintained by certain kinds of equipment (see table 8.1). This is in line with the logic that increasingly makes fetuses subjects of legal and affective personhood as the gravid uterus has been rendered successively more transparent.[22] The expectation that an embryo produced in a clinic might lead to a pregnancy and become a baby is crucial in ART clinics, whether or not attributions of embryonic personhood are involved. For both patients and practitioners, clinical practice is based around maximizing the chances of establishing a pregnancy, which in turn involves calibrating the numbers of available embryos according to the procedures with the best outcomes, which routinely means that some or most of the embryos will not be involved in pregnancies. A trend over the last decade and a half has arrived at a point where no eggs or embryos can be left unaccounted for and their enumerated fates are fiercely debated (as I describe below and as table 8.2 shows). For the moment, there remains a parallel with "natural" pregnancy wastage rates that allows for embryo disposal or loss without sanction under certain circumstances. In these practices and emerging ontologies lie particular patterns that are representative of the biomedical mode of reproduction.

The Economy

From Production to Reproduction

The first economic characteristic of a biomedical mode of reproduction, which is exemplified by the sacred and profane status of the in vitro human embryo, is that U.S. biomedicine makes both production and reproduction parts of the economy.

Table 8.2
Gamete and Embryo "Disposition": Consent Forms and Disposition Options from a Typical Assisted Reproductive Technology Program

Sperm

Collection: "Instructions for Collection of a Semen Specimen"

Disposition of sperm: Not mentioned and not covered by a consent form

Eggs

Collection: Multiple procedures, coordinated in "IVF Cycle Calendar"

Disposition of eggs: "Consent for the Disposition of Unfertilized Eggs."

If fertilization is attempted, all resultant medically viable embryos must be intended for use in

- A current "fresh" cycle or
- Cryopreservation (freezing).

If eggs are not fertilized (intentionally or through failure to fertilize correctly), they can be

- Donated to research or
- Discarded.

Embryos

Conception: "IVF/ET Consent Form" and "Intracytoplasmic Sperm Injection Consent"

Disposition of embryos: "IVF/ET Consent Form" and "Embryo Cryopreservation Consent Form"

Fresh embryos can be

- Discarded if medically nonviable,
- Transferred in a current cycle, or
- Frozen for a future cycle.

Frozen embryos can be

- Stored frozen for a given period of time (often five or ten years and extendable if requested by patients),
- Thawed for potential transfer,
- Removed to another IVF facility,
- Donated for research,
- Donated to another couple (embryo donation or embryo "adoption" can be anonymous or open),
- Given frozen to patients for private disposal (thawed without intent to transfer), or
- Discarded at clinic (see below).

Table 8.2
(continued)

Frozen embryos can be discarded by clinic if • They do not survive thawing process, • The blood of husband or wife at time of donation tests positive for HIV, syphilis, hepatitis, etc., or • They are determined to have been "abandoned" (usually five years with no contact with patients despite due diligence).

Social theorists typically focus on production, whether in the service of the state or the market or both, to understand the social order and the motors of history. Even social anthropologists, who have made kinship central to their understanding of societies' economies, have theorized kinship as a system of production and exchange rather than of reproduction. Critics have pointed out that production also always involves reproduction. For example, the Marxist tradition from which I am loosely borrowing the central metaphor of this conclusion—mode of production—has been excoriated by feminists for ignoring the labor of reproduction and the reproduction of labor.[23] Equally, every theory or mode of production has its attendant theory or mode of destruction.[24] It is always informative to contrast different technologies and the social relations they entrain from the point of view of not just what they produce but what they reproduce and what they destroy.

While reproduction, production, and destruction are all always involved in all economic systems and in maintaining social order, one of these usually predominates. Capitalism is often defined, for example, in terms of the relation of different groups to the means of production. It uses the vertical hierarchical notion of class to differentiate the relations of various groups to the means of production. Nations that undergo chronic civil wars are an example of the kinds of social and economic disorder that can be described by relations to the means of destruction. The relevant notion of peoplehood here has tended in recent years to be the seemingly horizontal one of "ethnicity."[25] In economies and social worlds that are organized around certain biomedical conditions, including ARTs, I suggest that reproduction is becoming the predominant focus of value, exchange, emancipation, and oppression. The predominance of reproduction means that production and destruction are derivative. The notion of peoplehood that scholars of medical technologies have coined and begun using is "biosociality."[26]

What is entailed when a society moves from production to reproduction, which I am suggesting characterizes the biomedical mode of reproduction? From the perspective of the embryo, I consider a number of things this might mean and argue for what I think it does mean. First, as Karin Knorr Cetina and others have noted, molecular biology turns life into more or less standardized things that do work (such as phages, plasmids, clones, and cell lines). In other words, it makes reproduction productive in an industrial sense.[27] Although the so-called natural reproduction of organisms can be harnessed for productive use and turned into standard productive techniques, this definition fits biotech, agricultural, and animal products that are produced in a standardized form and from which other industrial products are made much better than it fits biomedical innovation.[28] In reproductive technology clinics over the last twenty years, there has been considerable standardization as well as innovation and constant pressure to perfect technique and improve results. The aim has not been the value-added harnessing of the productive power of the living tissues used, however, but reproduction itself. Embryos are tools and raw material in ART clinics, but they are increasingly rarely *mere* tools or raw materials. While they remain on the trajectory of a possible future pregnancy in infertility clinics, but their reproductive power is rarely if ever reduced to yet another form of labor. Exiting embryos from this trajectory is procedure-dependent. As I discuss below, there are specific places in the in vitro fertilization process, for example, where couples are asked to select among options for their excess embryos. And as described in chapters 3 and 7, this has become increasingly subject to "biographic" (giving embryos narrative trajectories) oversight. Deciding whether and when there can be such a thing as an "unused" embryo and deciding whether it should it be discarded, donated, sold, turned into spare parts, experimented on, or experimented with remain heavily contested.

The reverse of the idea that reproduction is productive—namely, that production is reproductive—can also be imagined. Feminists, as mentioned above, have made this point, and Sarah Franklin, Helena Ragoné, and their colleagues have shown the work that is undertaken to produce and reproduce reproduction.[29] This idea also lies behind one reading of Foucault's notion of "biopower," which he argued was "an indispensable element in the development of capitalism" and "would not have been possible without the controlled insertion of bodies into the machinery of production and the adjustment of the phenomena of population to economic processes."[30] Division of labor, disciplined and rationalized civic and work and private lives, and the rise of bureaucracy produce docile bodies, but these bodies simultaneously experience the highly differentiated and voluntaristic aspects of modern individuality. Production, in other words, makes subjects. But the same argument cannot be made about embryos in treatment trajectories. The reproductive, self-making aspects of their productive activity cannot be stressed because they do not engage in an independently validated system of production. Rather, in a reproductive technology clinic, the reproduction is the entire story.

Coming from a slightly different perspective, a feminist or an environmentalist model of production might demand that the cost of reproduction no longer be treated as an "externality," to borrow an economic idiom.[31] This would take a form much like the calls that are heard to make polluters pay for downstream environmental costs in industrial production by considering clean-up or pollution mitigation as part of production costs.[32] Paid wet nursing, child care, housework, and dependent care are key parts of capitalism that have remained feminized, raced, and underpaid because to the extent that they are not publicly funded or provided as corporate perks, they must find their own "market level." This assumes that they have their own market and does not see them as an integral part of the operating costs of industrial production.[33] A move from production to reproduction would, according to this model, be a radical move to reduce externalities—a move to frame productive markets so that the costs of reproducing the social and natural participants of those markets were included in the frame.[34] This position would be motivated by equity and social justice, and it has great appeal. It is not, however, what I am attempting to capture. Far from being a model of equity, the reproduction at the heart of the biomedical mode of reproduction is a core part of contemporary patterns of what Shellee Colen has called "stratified reproduction."[35]

Finally, the costs of reproduction could be based on a model from a preindustrial bucolic past or Luddite future postindustrial utopia where social order was predicated on kinship or community and thus on social reproduction rather than on reproduction that is in the service of the relations of production. Gayle Rubin and more recently Susan McKinnon have pointed out the patriarchal and exploitative nature of the exchange of women in preindustrial societies, however, making it unlikely that this constitutes a mode of production that is centered on reproduction per se.[36] Marilyn Strathern's pioneering work on reproductive technologies has shown the limits of thinking that capitalist means of production would disappear or be overthrown if humanity and relatedness were valued over capital. Indeed, she has argued that "culture is being enterprised up" in late capitalism, with reproductive technologies promising a perfect blend of consumerist choice and an assertion of natural relatedness.[37]

Instead of these important but different relations between production and reproduction, biomedicine gives us inklings of a more radical sense in which reproduction replaces production: the "intrinsic" and promissory value of reproduction is at stake. The embryo, a canonical biological entity in the biomedical mode of reproduction, is *constitutively promissory*, and its value stems from its life-creating potential. This reproductive capacity cannot be reduced to a market value (as for agricultural futures) or to the cost of the reproduction of human labor.[38] Contemporary biomedical objects and processes (like in vitro fertilization) are produced and exchanged in quasi-markets. For example, in the stratified commercial gamete (sperm and ova) markets with which U.S. citizens are familiar, consumers are charged for a chance to overcome infertility and to pass on the traits for which the donor was chosen. Nonetheless, the reproduction of a unique child, not a later satisfactory IQ (or whatever the relevant trait), determines success. Despite the commercial, service, and technological aspects of the treatment, the most ardent and sentimental of parental attachments to "the priceless child" is typically effected.[39] And unlike markets for other objects with high and supposedly intrinsic and unique value like priceless works of art, reproductive intrinsic value is self-validating and is not dependent on a community of experts and evaluation of skill.

Alienation from Labor to Alienation from Body Parts

The second economic characteristic of a biomedical mode of reproduction concerns the nature of alienation and exploitation: whereas under capitalism, workers risk being alienated from their labor, in the biomedical mode of reproduction, patients risk being alienated from their bodies or body parts.

The second striking aspect of the economy of biomedicine of which reproductive technology clinics form a part concerns the conditions under which exploitation and oppression occur. In the mode of reproduction, body parts rather than labor are at systematic risk of being alienated from the person. In capitalism, most workers exchange their labor for a wage or salary. This is more or less exploitative, depending on factors such as whether the wage is high enough to live on and how much the work enriches the owner of the business in question rather than the workers themselves. When neither the product nor the profit made from selling the product accrues to the people who labor, workers are alienated from their labor.

In reproductive technology clinics, body parts and the processes—gametes, embryos, and the gestational uterus, for example—that contribute to the reproduction goal of having a baby move in choreographed chains of custody that preserve and stabilize procreative intent.[40] Procreational intent and agreement about it are not assumed but emerge from the practice of reproductive technologies. Table 8.2 shows the paths that specify the possible fates of gametes and embryos in IVF without donors for a typical U.S. clinic. These paths rely for their coherence and enforcement on a closely coordinated but changing amalgam of laws, medical techniques, lab practices, payment structures, kinship determinations, and consent giving as described in the preceding chapters. All these aspects craft an active balance between the sacred life potential and profane materiel aspects of the body tissues and treatment prerogatives. A rupture in any of these paths potentially involves alienation of the body from the person. When these strands fail to come together or fail to give a unique answer about the identity of the child's parents or of the body parts that gave service, then there is alienation of body or body parts from at least one of the participants.[41]

Cases where people learn that their body parts have been exploitatively forced into chains of custody and exchange are increasingly common worldwide.[42] Contests over custody occur in a subset of ART cases. The most well-known U.S. examples involve surrogacy, and several widely publicized custody disputes have resulted from both traditional and gestational surrogate procedures (which are described in chapter 5).[43] Cases such as those involved in the embryo-switching scandal in California in the mid-1990s (discussed in chapter 6) and others where gametes have been mistakenly given to the wrong recipients have sparked custody disputes involving egg and embryo "donation." Feeling coerced or exploited at various specific points, if not overall, is common in both gamete donation and surrogacy.

Similarly, people can also be oppressed by the system in the straightforward technical sense of being barred from entry because they lack means, status, or citizenship or have the wrong citizenship. Charon

Asetoyer, executive director of the Native American Women's Health Education Resource Center in South Dakota, told me in March 2004 that "these technologies do not concern us because we do not have access to them. The Indian Health Service does not cover them, and we cannot afford them. More of our women are getting educated, and when women are older, sometimes they find it is harder to have children. So it would be good for them to have fertility services, but we don't have access." She also commented on the blood quantum regulations, the Indian Child Welfare Act, and the difficulty of having egg, sperm, or embryo donation make sense or be permissible in Native terms. Despite that, she thought that in an economy where work is hard to come by, acting as a gestational surrogate might be permissible work for Native women because the gestated baby would not be "an Indian baby."[44]

All these cases—being forced into or being taken out of stable treatment trajectories or being structurally kept out of them altogether—involve alienation from body parts.[45] Fernanda Vieira's work on reproductive technologies in Brazil has shown that it is possible to organize reproductive technologies themselves so that societal inequalities are internalized in the treatment. She has begun work on a program where infertile women who live in *favelas* are brought to clinics and given free infertility treatment in exchange for becoming egg donors to paying infertile couples. In related genetic, reproductive, and biomedical fields, scholars have researched the widespread nature of similar kinds of alienation from one's body—including Michael Lynch's work on failures of chains of custody of blood in criminal cases such as the O. J. Simpson trial, Veena Das's work on the Indian market for organs for transplant, Hannah Landecker's work on the severance and troubling remaining ties between subjects and bodily tissue in the history of the HeLa cell line, James Boyle's account of the *Moore v. UC Regents* case involving the tissue derived from the plaintiff's spleen, and Paul Rabinow's nationalist and entrepreneurial histories of French DNA and of PCR.[46]

In the biomedical mode of reproduction, stratified, value-added reproductive and other body parts produce value and create commodity. *Commodity* and *commodification* are, strictly speaking, the wrong words to use. The commodities thus created are not the Lukacsian or Marxist fetishistic commodities. Being constitutively reproductive, life saving or giving, and promissory, these commodities are not "things in themselves" that are autochthonous and autotelic, despite the tendency of popular culture to indulge in gene fetishism.[47] The life and reproduction that body parts and bodies hold in the biomedical mode of reproduction live in chains of custody and kinds of biological relatedness, not in commodities. The "mistake" of fetishism is replaced by custody disputes and mistaken parentage (it is always so with reproduction). Chains of custody can and do go wrong, get contested, and get broken.

From Accumulated to Constitutively Promissory Capital

The third economic characteristic of a biomedical mode of reproduction lies in the nature of capital: whereas in capitalism, capital is accumulated, in biomedical enterprise, capital is promissory.

Another characteristic of reproductive technology clinics and their gametes and embryos that has implications for a possible biomedical mode of reproduction concerns capital. For many reproductive technologies, capital should be theorized less as accumulated, as is customary in theories of capitalism, and more as what I call "constitutively promissory." The accumulated capital of conventional capitalism has long included futures, venture capital, and the stock-market risks that are associated with betting on future productivity. U.S. nonfinancial corporations have seen a steady rise of so-called intangible assets (such as patents and copyrights) in relation to tangible assets (such as real estate and equipment) over the last fifty years. But ultimate value still in theory depends on accumulation and measurable output and productivity. The biomedical mode of reproduction represents a change in the temporal dimension that is relevant to assigning economic value—a shift from the primary dimension being the past and present to its being something that unfolds over time in the future.[48] Likewise, it signals a shift away from production, productivity, and profit and toward knowledge, technologies of life, and promise.

Donor sperm and eggs have tended in certain striking ways to be treated like conventional capitalist natural resources in reproductive technology clinics in the United States. That is, they have been treated and valued according to their scarcity, the difficulty and expense of their extraction (low for sperm unless there is male-factor infertility and high for eggs), and the nontangible costs to those from whom the resources are extracted (usually, low for men and high for women). They have also been stratified on the egg and sperm donor market and priced accordingly, using increasingly codified and standardized conventions. Like the 4Cs of the diamond industry—carat, color, clarity, and cut—gametes are valued by eugenic criteria as well as by kinship criteria of genetic, phenotypic, and cultural "matches."[49] Just as they do for other natural resources, market values vary according to technical capacity, regulation, demand, and scarcity.[50] Nonetheless, there is a tendency to avoid paying for the actual gametes (which is evidenced by the use of the word *donor*), and that tendency is more marked for eggs than for sperm (see table 8.2). This is because the gamete market references future embryos and future children, whose potential existence determines their value.

By analogy with prohibitions on baby selling and owning a person other than oneself, which in the United States evoke fundamental antislavery prohibitions, embryos cannot be bought and sold. Embryos, despite being treated as material objects in the sense that they are routinely manipulated and stored, are treated as sacred in the sense of holding intrinsic, inviolable worth to the intended parents until or unless the embryos exit by defined paths of viability. This means that they need to be treated with extreme care while in anyone else's custody. As embryos only rarely remain viable if they are no longer precursors to an ongoing pregnancy, they usually cease being reproductive at the point at which they cease being sacred and vice versa. This means that their reproductive capacity rests on their promissory capacity of ineffable intrinsic worth, which will unfold over the future life of a child, if things go well. The exception to this occurs when embryos are designated for research, but according to the principles of vivisection, they must keep their hold on potential life to be able to model the very life processes that are being researched.

When the technology growth market collapsed in 2000, the stock prices of even profit-turning biotech companies plummeted. Worldwide recession and the events following September 11, 2001, have made it hard to say which part of the loss of value of biotech stocks is related to the sector itself and which part is recession linked. Even taking recession, terror, and war into account, a baffling disconnect seems to exist between product and profit, as they normally are calculated, for some reproductive biotech firms. Whether the prices of March 2000 were all "bubble" or not, a question remains about what exactly biotech companies are marketing and how that product should be traded and capitalized. The jury is still out as to whether value in the form of knowledge, promise, and the technologies of life in these sectors will somehow be translated into measurable accumulated profit and product or whether levels of value that remain constitutively promissory and future oriented can be exchanged in a capital market. A mature biotech market probably will develop a sense of product and profit that will stretch the notion of accumulation yet further, continuing a process that had already begun with the rise of service, knowledge, and health-care economies. At some point, however, the notion of accumulation may no longer stretch far enough. ARTs, by increasingly involving the manipulation of ex vivo human embryos that tie value to future unfolding of life, will perhaps remain the hardest case and most resistant to being valued as accumulated capital.

From Efficiency and Productivity to Success and Reproductivity

The fourth economic characteristic of a biomedical mode of reproduction is that improving success rates and growing reproductivity signify a healthy economy and are diagnostic of the economy itself. This contrasts to capitalism, where efficiency and productivity growth are diagnostic of capitalism and signify an expanding economy.

From Adam Smith to the present, theorists of capitalism have tended to emphasize the formal efficiency of a mode of production that is based in theory on a division of labor and a laissez-faire market. Even capitalism's critics, like Upton Sinclair, have been mesmerized by this efficiency. From the Taylor man to *Modern Times,* the mechanistic metaphors of capitalism signal the disciplined working body that maximizes productivity. By contrast, reproductive technologies and, perhaps more broadly, the biomedical mode of reproduction aim to improve success rather than efficiency. ART procedures involve so-called hyperstimulation of a woman's ovaries so that a large number of eggs can be matured at once, although the intended outcome is only a singleton pregnancy, and the most common outcome, as with "natural" efforts at human conception, is no pregnancy at all. Efforts at standardization and improvement are geared at improving results rather than increasing efficiency. When efficiency, as opposed to success, is increased, it presents a serious threat to the self-governance of the field as well as to claims to improving success rates. The two clearest examples of this are multiple gestations and the stockpiling of embryos in clinics.

The incidence of multiple births has increased dramatically since the advent of ARTs in the United States, and 5 percent of all babies who are born from ARTs that are carried out in the United States are still triplets or of higher orders. Triplets and higher-order pregnancies are in some sense very efficient—three for one, as it were—yet they are associated with high morbidity and mortality, including long-term effects for surviving children.[51] They also put high financial and emotional burdens on parents, from sheer numbers as well as from short- and long-term health effects. And the perinatal and longer-term health-care costs are typically substantial, raising questions about the distribution of scarce health-care resources. A decade ago, it was common practice in the United States to transfer five or more embryos per cycle, if they were available, and improvements in drug efficacy and ovarian stimulation protocol encouraged this. Yet increasing efficiency led to the problem of multiple gestations. While some patients found it acceptable to "selectively reduce" their pregnancies, many others (and not just highly religious patients) could not tolerate this idea. Selective reduction involves the termination of one or more fetuses while allowing for the continuation of the pregnancy. This procedure is used for multiple pregnancies that are the result of natural conceptions as well as fertility treatments, and it is associated with relatively high subsequent spontaneous miscarriage rates for any remaining fetuses.[52] Infertile patients are at a double disadvantage regarding this procedure. Patients who cannot easily get pregnant typically have enthusiasm for any kind of pregnancy, fantasize about the "instant family" that multiple births seem to promise, and often do not want to jinx themselves by making judgments about the best number of embryos to transfer. Likewise, they cannot easily get pregnant again, and so the high miscarriage rate associated with selective reduction is not acceptable to many.

Critics in the United States have called for the reform of or the imposition of federal regulations (similar to those found in many European countries) on the numbers of embryos that can be transferred to a woman's uterus in a given cycle of ART treatment. As part of its effort to remain self-regulating, the American Society of Reproductive Medicine (ASRM) has developed guidelines that are available on its Web site and that its member clinics are supposed to follow.[53] Over time, it has become increasingly rare for clinics to transfer more than three embryos at a time. For women in groups that are associated with high success rates, the numbers of embryos per attempt are now commonly restricted to two, and any leftover embryos are frozen. A successful outcome and not ever-increasing efficiency in establishing pregnancies is the goal. This is how the Society for Assisted Reproductive Technology (SART) describes this correction of efficiency on its Web site:

The SART Research Committee has prepared a large body of information about IVF including numbers of embryos transferred and multiple gestation rates from the SART dataset. Using this information, a CDC paper on multiple gestation and the SART/ASRM published guidelines on number of embryos to transfer, SART Executive Council members presented the USA experience at an international NIH/Bertarelli Foundation meeting in November 1999. This international meeting on how to decrease multiple gestation in ART was expected to be highly critical of the USA and its high multiple rates associated with ART. To the contrary, after hearing SART's point of view, our European colleagues were impressed with the progress that has been made and continues to be made in this country and agreed that our nongovernmentally regulated system is working in the right direction to substantially decrease the problem of high-order multiple gestations in ART.[54]

Similarly, more efficient production of embryos has led to the current crisis of stockpiling. When SART announced in spring 2003 the results of its national survey of frozen human embryos (carried out with the help of RAND), the number of currently cryopreserved embryos was put at approximately 400,000 in the United States.[55] This was far more than anyone had previously estimated. A government or even ASRM-wide decision on what to do with these embryos that would be equivalent, for example, to the one specified in the United Kingdom's Human Fertilisation and Embrylogy Authority (HFEA), would never be possible here.[56] In August 2001, George W. Bush limited federal funding of stem-cell research to that carried out on already immortalized stem-cell lines, so that no further human embryos would be "sacrificed." A decree to destroy the unused portion of this stockpile would be very unlikely to be proposed and politically impossible to bring about—embryo "adoption" agencies (see table 7.3) and anti-abortion groups in the United States and elsewhere have referred to this possibility as "genocide"—even if it didn't violate the fundamental privacy of ART decisions in the United States.[57] As the American Infertility Association, a major nonprofit infertility patient advocate group, expressed it on its Web site,

> It's a private issue gone very public. It's a complex web of personal philosophy, religious orientation, and social conscience about which everybody, and we mean everybody, has a strong opinion. But the fact is, and should be, what you do with the frozen embryos you don't use is your decision and yours alone.[58]

The study also, however, showed that almost 90 percent of these frozen embryos are accounted for by couples who wish to use them themselves and that another approximately 4 percent of them are slated for donation to research or to other couples. The embryos that are slated for use by other couples can enter a prolife trajectory—where they are "put up for adoption" as if they were already children looking for a home—as is practiced, for example, at the Fullerton, California Snowflakes Embryo Adoption Center. Or they can enter a prochoice trajectory and be "donated" rather than adopted, as they are at the Genesis Embryo Donation Center. Embryos that are designated through informed-consent forms for research enter the politics of what kinds of research are allowed, by whom, and with what funding and oversight. The remaining embryos are slated to be thawed without use or, if the intended parents can't be contacted, are classified as "abandoned." While many right-to-life groups object to all the last three categories—donation to research, thawing without intent to use, and abandonment—to many in the field the last group is the most worrying aspect of efficient embryo stockpiling because of the lack of accountability and clarity. Efforts are underway to reduce the numbers of embryos that get frozen in the first place and to tighten up the legal and accounting loopholes that allow embryos to enter the "abandoned" category.

Much of the activism and resulting legislation of the late 1980s and early 1990s, including the Wyden bill, was aimed at forcing clinics to report success rates. As described in chapter 7, the resulting centerpiece of the self-regulatory efforts of the American Society for Reproductive Medicine is the extensive database maintained by their Society for Assisted Reproductive Technologies in collaboration with RESOLVE and the Centers for Disease Control and Prevention (CDC). Clinics now voluntarily report their statistics annually according to standardized protocols to receive the ASRM stamp of approval, without which practice and business are likely to be poor. The professional organizations focus to a large extent on the collection and accountability of this information, combined with corrections to events such as multiple pregnancies and stockpiled embryos, where efficiency as a production goal has threatened to trump success as the reproduction goal.

From Physical Problems of Waste Disposal to Ethical Problems of Waste Designation

The fifth economic characteristic of a biomedical mode of reproduction concerns waste. Capitalist industry typically produces large amounts of material waste, and waste disposal is a constant political and logistical problem. In a biomedical mode of reproduction, the problem with waste is an ethical one of how to designate life material as waste.

The generation and disposal of waste are issues that are closely related to the question of efficiency. Because capitalism needs to renew market demand, it tends to produce specialized products and improved models of existing products in its value-added pursuit of profit margins, and these practices produce waste. This is true both because of the programmed obsolescence of products and because of the packaging required for the networks of transportation and distribution on which capitalist markets are based. Capitalism is intrinsically efficient, and yet it makes a lot of waste that must be disposed of. Along with the depletion of natural resources, another signature problem of capitalist societies (including their transnational hinterlands) is waste disposal. Sociologist Zzuze Gille has shown that different modes of production produce different kinds of waste and have different notions of waste disposal that are characteristic of their political economies.[59]

ARTs and their biomedical mode of reproduction also have a problem with by-products. Where in a capitalist mode of production, the principal problem with waste is its safe and long-term physical disposal, with ARTs the problem is the difficulty in designating anything as waste. Technicians and physicians cannot simply throw away embryos or eggs that are not used. Making an embryo into waste is an outcome and not a by-product. Correspondingly, the word *disposition* is used to describe the various trajectories for embryos, including options that end up with embryos being discarded, and the related notion of disposal is rarely used, let alone coupled as "waste disposal." There are thus only certain conditional outcomes by which embryos become waste, all of which involve conditions that cannot be known about before they occur, even if it is necessary to plan for their possibility. These conditions include the undesired but almost entirely uncontested "medical" reasons, where embryos die or a technician judges them abnormal after fertilization, division, or thawing. This category may grow, granting waste immunity to a wider range of embryos, as preimplantation genetic diagnosis becomes more widespread and reveals more abnormalities in apparently normal embryos. This medical designation of abnormality can also be used when technical errors in a procedure damage gametes or embryos. Patients can also choose to discard embryos under certain conditional circumstances, which also do not reveal themselves until after the creation of the embryos in question, such as a successful pregnancy with fresh embryos, rendering frozen embryos as surplus to a couple's reproductive intent.

As is summarized in table 8.2, couples have to decide when they begin infertility treatment whether any leftover embryos will be discarded, donated, frozen, or given to research, with the life trajectory and so life status of each option being different. The largely private questions on the status of embryos in the clinic (discussed in chapter 7) may well not be containable if stem-cell and other biomedical technologies break the tight fit that I described as a monopoly of desperation where physicians, patients, activists, and drug companies have managed to forge collective interests through interacting spheres of privacy. The principle source of embryonic stem cells is from so-called leftover embryos from reproductive technology clinics, and as mentioned above, these same embryos are already the subjects or objects of embryo adoption campaigns as well as regulatory battles over their use or disposal.

Science and Society

From the Public Understanding of Science to the Private Implication in Science

In capitalism, the public is called on to understand or be well informed about science to guarantee both a trusting public and reasonably accountable science and technology. In the biomedical mode of reproduction, the various publics are instead privately implicated in the science and technology in question.

A policy of high autonomy and high government handouts for basic research is taken to be a legacy of Vannevar Bush's eventually successful articulation of the ideal relations between the U.S. government and the research scientists who are held responsible for the Allied victory and the atomic bomb.[60] When the physical sciences dominated both national security and the cutting edge of the national scientific enterprise, this model promised a loyalty to government paymasters as well as the freedom to pursue science for its own sake. The scientific autonomy fit fairly well with security secrecy demands because both mandated freedom from outsider scrutiny. And the loyalty combined with science for science's sake underwrote the cold war ideology of national and political superiority that was expressed through scientific and technological superiority. The obligations of the government and of scientists to the public lay in keeping people well enough informed to be witnesses of this superiority so that they would continue to approve public funding of big science and provide a labor reserve of future scientists through public education in science. The model of public involvement is often summarized as "the public understanding of science."

As the life and biomedical sciences have moved to the forefront of national science (including national security), the relations between these sciences, the government, and the public have been changing. The physical sciences and their objects (like elementary particles) are tremendously needy, especially of real estate, computing power, and the formalities and probabilities of advanced mathematics. But these are very different demands and rights than those asked for by and owed to the organic objects of biomedical and life science. In biomedicine, people, their personal and collective identities, their bodies, and their body parts are materially and custodially involved and implicated in the science. The autonomy of science continues to be an ideal that informs research protocols and professional organizations, but it does not spell out the relations between biomedical sciences and the government or the public. Rather, ARTs' relations between science and its publics are played out in their interactions with the private spheres that are commonly opposed to both the public and the government. The production of enormous amounts of public information about ARTs by every interested actor is a key part of these technologies in the United States. But this is not geared at a disinterested public that needs minimal standards of scientific literary for attestation or affirmation. Neither is it geared at the public education system and a labor reserve of reproductive endocrinologists and geneticists. Information in the biomedical mode of reproduction is oriented to those whose bodies and identities are already implicated in the sciences in question, and it offers not just a passive model of understanding but an active model that combines features of the capture and the expression of agency of those involved in the technologies. Though there may be little to choose between in the excrescences of their respective acronyms, this can be characterized as a shift from the public understanding of science (PUS) to the private implication in science (PIS).

From the Moral Responsibility of the Scientist to Bioethics

The other core aspect of science-society relations in the biomedical mode of reproduction concerns the moral economy of science. Under capitalism, there is an emphasis on the moral responsibility of the individual scientist. In the biomedi-

cal mode of reproduction, the moral economy is the much more pluralistic and diffuse one of bioethics.

These two very different conceptions of scientific accountability require rather different scientific epistemologies and moral economies. The epistemology and morality of science in capitalism can be captured in its normative form by Merton's KUDOS norms. Its hallmarks are scientific method, certainty after skepticism, and the moral responsibility of the scientist. The biomedical mode of reproduction, on the other hand, involves an epistemology that is based more on plurality of interested parties, uncertainty of success, and risk to human subjects and that has instituted bioethics as its dominant moral economy (as described in chapter 7). Biomedicine has been characterized by the exposure of one set of ethical violations surrounding informed consent after another, from Nuremberg through Tuskegee, Project Whitecoat, and up to recent gene-therapy scandals such as the Jesse Gelsinger case and the finding that one of France's proclaimed cured "bubble boys" developed a new form of leukemia. The moral responsibility of the scientist is not enough, as the scientist cannot speak for the human subject. ARTs, through the monopoly of desperation, illustrate this connection between accountability and privacy. Other biotechniques such as DNA fingerprinting potentially pit the science and the human subject against one another (although the technique is used for exoneration as well as condemnation), in which case the need to address the technique from the point of view of accountability and privacy is even more imperative. Bioethics is necessary (even if it often wrong-headedly keeps calling for a single public agora) because of this irreducible, sometimes irreconcilable set of interests.

Identity and Kinship

Among several implications of the biomedical mode of reproduction for identity and kinship, the most striking is the potential to subvert the identity and kinship categories that biomedicine was initially designed to assert and repair. Thus efforts to rescue the life course contain the possibility of undermining that notion, efforts to make universally available the normative heterosexual nuclear family lead to the subversion of that family form, and technological solutions to highly intimate issues end up intertwining the technical and the personal in ways that displace the natural as the ground on which the social plays out, just as they make it necessary to attend to the social to disambiguate the natural. The first two features lie behind the sense that these technologies are both highly conservative and highly socially and technically innovative: they are. And the third point summarizes the ontological choreography that underwrites this combination of identity and change. In describing these concluding three characteristics of the biomedical mode of reproduction, I thus recapitulate and draw out the arguments in this book.

From a Finite and Linear Life Course to a Loss of Finitude and Linearity to Preserve the Life Course

Capitalist life courses are no longer brutish and short. They have phases such as childhood and old age, and they normatively involve reproducing in one's adult years. In the biomedical mode of reproduction, efforts to allow people to have children or to live to old age threaten to subvert the linearity and finitude of the capitalist life course that they attempt to rescue for patients.

I turn first to the threat that is posed to the integrity of the life course. ARTs, like most of biomedicine, are dedicated precisely to repairing and rescuing the life course. This is the impetus of the conservative streak of the sociotechnical innovation that I describe in chapter 4. The infertile are able to move on to conventional adult life phases and escape the limbo of what Margarete Sandelowski has called "childwaiting," just as in other areas of biomedicine, the chronically ill are promised relief from the time-lapse living that Kathy Charmaz has described as "Good days, bad days."[61] But the very techniques that enable physicians in some circumstances to reinstate the life course through the manipulation of the human embryo also hold the potential of subverting the inevitable and orderly, if deeply historically and culturally specific, procession through the stages of life. Other scholars have shown that both the technologies of birth and those of death are becoming increasingly dissociated from a linear, finite temporality.[62]

ARTs aim at "family building" and have as their fundamental rhetorical dynamic the correcting of deviance from the linear temporal life course. As I argue in part 2, people are extremely adept at narrating their lives according to these templates, whether they naturalize them or use them to naturalize their own perceived deviance, and the technical genius of reproductive technologies lies in their material connivance in that project.[63] In the effort to restore and repair the life course, frozen, or cryopreserved, gametes and embryos have been used to conceive children for the last two decades (and much longer for animals). Indeed, frozen sperm, in post-HIV, post-turkey-baster days, is the dominant donor sperm. Frozen embryos mean that families around the world have children who were conceived at the same time in the same petri dish but were born several years apart. These disruptions of birthing temporality are usually successfully subsumed in the United States into two other familiar temporal family templates that are drawn from the life course—namely, "family spacing" and "reproductive choice." Nonetheless, the widespread use of freezing of gametes and embryos leaves open the possibility of such suspended animation emerging at times and places that cannot be absorbed neatly into narratives of the life course.[64]

ARTs also suggest the possibility of the infinite extendibility of life as well as the possibility of suspended animation just discussed, through the physical and technical connections between in vitro human embryos and stem-cell research and cloning. The techniques of micromanipulation are key parts of cloning, and the laboratory and medical facilities in ART clinics make reproductive cloning seem within reach for those couples who are unable to reproduce any other way or for those wishing to avoid life-threatening genetic diseases that are carried by one partner or the other. "Human therapeutic cloning" (HTC), despite being a nonreproductive technology, promises to extend the lives of those who are already living by allowing them to grow replacement organs and tissues for failing body systems. HTC involves the production of cloned embryos using nuclear somatic DNA that ideally is derived from an affected patient, from which stem cells could be extracted. These stem cells could then be induced to differentiate into specific histocompatible replacement tissue and organs. The appeal of therapeutic cloning consists in being able to avoid the pitfalls of donated human organ shortages and the ethical conundrums of organ distribution; in greatly reducing the risk of tissue contamination that caused the blood scandals of the 1980s and the recent withdrawal from the market of contaminated lab-grown skin; in avoiding xenotransplantation; and in reducing or avoiding altogether the drug dependence and low success rates of therapies caused by rejection.[65] Therapeutic cloning, by suggesting a method by which almost anyone could grow their own custom organ and tissue replacements, hints at the possibility of the infinitely repairable body. While perhaps slightly more of a bricoleur's version of immortality than one might wish for, there are definite intimations of a greatly lengthened life course.

The problem with therapeutic cloning is its vivisection, inefficiency, and sacrifice, which it shares with stem-cell research in general as well as many areas of biomedicine. The 2001 presidential decision to allow research on preexisting immortal stem-cell lines was facilitated by the Mormons in Congress (Mormons, like most Jews, do not consider conception to have occurred until the embryo is implanted in a woman's womb) and by the fervor of those with bedside knowledge of the toll that is taken by disease and accident (epitomized by Christopher Reeve and by Nancy Reagan's split with George W. Bush over stem-cell research in the light of her experiences in caring for her husband). Ultimately, the use of reproductive cloning for immortality purposes as proposed by the Raelians or the use of therapeutic cloning for a dramatically lengthened life course will probably remain a fringe benefit of the wealthy, despite the fact that these are the points that the media has stressed. More disruptive to the life course in the general population might turn out to be the erosion of the distinctions between somatic cells and germ cells.

The first threat to the life course that is provided by somatic cell cloning occurs from the promise of ubiquitous cellular immortality. If some adult cells can behave in the same way as germ cells—that is, if mature, differentiated cells can themselves lead to reproduction—should all such cells be treated as potential life? Should every cell that can give rise to an embryo because it already has the full complement of DNA be sacred? Regardless of whether cellular DNA is in fact induced to display totipotentiality (as it would be in a fully fledged clone) or only pluripotentiality (as is proposed in human therapeutic cloning), the life potential is still there. If many cells can be reconstituted as totipotent, the moral and ontological separation between germ and somatic cells that has been important to our ideas of personhood comes under attack. Religious leaders, (Roman Catholics, in particular) have been important in this debate. Does it place any obligation on us to bring to life any particular cells? Do any of the abortion arguments apply for somatic cells as technical prowess increases? The United States has currently put this argument on hold by its blanket ban of federal support for all kinds of human cloning. But we are culturally unable to confront the threat from arguments about potential life posed by cloning.

ARTs' and biomedicine's efforts at salvaging the life course are moving into these realms of intended and unintended immortality just as the global politics of biomedicine are unlikely to halt the massive death toll from AIDS that is being experienced around the world. The connection between capitalism and modernity was in part forged over the idea of development and the orderly precession through the stages of demographic transition. One of the assumptions of demographic transition theory is that once a society has gone through demographic transition to a point where birth rates are low and life expectancy is high, life expectancy remains more or less constant and constantly high. The failure simultaneously to protect biomedical research and development and to provide AIDS medication at levels that is affordable to so-called developing countries means that this assumption no longer holds. While Western biomedicine is intimating immortality, a number of other countries are either failing to go through demographic transition or, more dramatically, are showing that the historical life course of modernity—once thought inevitably to involve the lowering of birth rates and higher life expectancy—is turning out to be nonlinear and reversible.

From Productive to Reproductive Notions of Homosex and Heterosex

Legitimate kinship in capitalist societies is bounded by prohibitions against too much sameness (such things as incest and homosexuality) and against too much difference (such things as adultery and illegitimacy), which together produce the nuclear family. In the biomedical mode of reproduction, legitimate kinship is bounded by prohibitions against too much sameness (such things as cloning and parthenogenesis) and against too much difference (such things as ooplasm transfer or xenotransplantation), which together allow the chains of custody of gametes and embryos to achieve legitimate reproduction.

My second point of comparison concerns kinship. Kinship is always determined relative to a mode of production (and vice versa), but for any given mode of production, there will be more and less fiercely policed methods for demarcating kinship. In particular, I claim that there are always notions of too much homosex and too much heterosex. In capitalism, these boundaries are tied to the basic unit of economic production, expenditure, and inheritance—the nuclear family. Correspondingly, too much homosex to confer kinship is perhaps iconically defined by homosexuality and incest in capitalism.[66] Too much heterosex, on the other hand, is defined by adultery and the associated illegitimacy, despite the recent gains in the ability of a mother to establish her relatedness to a child without naming a father.[67] In the biomedical mode of reproduction as exemplified in ARTs, the boundaries of kin escape the productive unit because they are attached to or internalized in the reproductive. When kinship determination becomes reproductive, too much homosex and too much heterosex should also become reproductive, and in fact these parallel notions of reproductive homo- and heterosex have indeed begun to surface. Reproductive cloning, were it to be authorized, threatens too much homosex for the logic of family building. Various forms of xenotransplanation (including the use of human DNA, where that DNA is determined to be feedstock rather than kinship generating, such as in ooplasm transfer) threaten too much heterosex.

Human reproductive cloning would involve the use of ARTs and cloning techniques to allow a baby to be born with the identical nuclear DNA to an already existing human. Much media attention has focused on reproductive human cloning, from fears of producing many Hitlers to the erosion of the individuality of ordinary folk. Commentators have countered this by emphasizing the importance of nurture in determining individuality, the natural occurrence of identical twins that are clones, and the curiously ordinary drive to have a child or to replace a dead child that is expressed by couples who are interested in using reproductive cloning to reproduce.[68] Despite the publicity that is granted to the Raelian cult, which is interested in cloning explicitly to achieve immortality, most people interested in reproductive cloning appear to be would-be parents struggling with infertility, parents eager to use the DNA of and thereby "replace" a dead child, or couples with one partner who has a serious heritable genetic disease. This ordinariness suggests that like users of other ARTs, many of those seeking reproductive cloning would incorporate the technologies into projects rescuing the life course. Nonetheless, reproductive cloning and its homosex or lack of heterosexual reproduction are unacceptable to the vast majority.

The practice of ooplasm transfer was forbidden by the U.S. Food and Drug Administration in 2001 but had already begun to have wide clinical application in the United States and elsewhere and has already resulted in the births of several children in the United States. Ooplasm transfer is an assisted reproductive technique that attempts to overcome the negative effects on pregnancy outcome of poor-quality eggs from the would-be mother. "Poor quality" is close to synonymous with "old" eggs, but the adjective *old* can be applied to a woman's eggs regardless of her chronological age. This biological age is frequently experienced by the women in question as just as real or even more real than their chronological age, and in an ontological move (described in chapter 3) that is characteristic of medicalization, it thereby constitutes a biological basis for an obsession with aging that is said to afflict a core constituency of those who use reproductive technologies.[69] To offset the quality of the egg, cytoplasmic material is extracted from an egg from another, usually much younger, woman. This ooplasmic material is injected into the eggs that are being fertilized along with a single sperm, through the technique known as intracytoplasmic sperm injection (ICSI). Although ooplasmic rejuvenation is not really understood, the mitochondrial DNA from the donor is thought to play a role in counteracting the impediments to implantation and subsequent development of the embryo of older eggs. While it was being offered, infertile women patients tended to prefer this means of fertilization to using entire nucleated donor eggs in place of their own eggs because with this procedure they remained genetically related to any resultant children. The role of ooplasmic DNA in the chromosomal profile and well-being of future children, if any, is not yet understood.

Ooplasm transfer has not (yet) been the basis for a maternity dispute between the cytoplasm donor and nuclear and gestational mother, and the fact that the FDA reviewed and rejected the procedure suggests that in many domains cytoplasmic material is seen as feedstock and not as kinship determining. The problem with this procedure, as far as the FDA is concerned, is its heterosex. That is, it introduces the DNA of three separate people into the embryo. Of course, the DNA in each gamete is itself the product of successful couplings of different sets of genes in the past, and if we are to believe geneticists, plenty of our DNA snuck in via viruses in our evolutionary past. Likewise, third-party reproduction of many kinds is already common. Nonetheless, ooplasm transfer is considered to be the first procedure to have combined the genetic material of three human gametes simultaneously to make a single embryo. As such, it has been seen as stepping over a line that has brought it into the stop-and-go regulatory territory of gene-therapy research rather than the private entrepreneurial clinical space of other third-party reproduction. If there is a tendency to assimilate other gamete donors and surrogates to social caring roles to minimize kinship claims, these ooplasm donors are reduced to genes as biological feedstock, thereby forestalling any claim to kinship. No one doubts that reproduction can occur using ooplasm transfer; it is the potential monstrous consequences of such heterosex that are feared. What troubles in this case cannot be treated as a custody dispute. The grammar of custody disputes is dismayingly complex where reproductive technologies are involved but is nonetheless articulated with reference to contenders to kinship. Ooplasmic transfer is treated more like xenotransplantation and is thus illicitly biological, unlike kinship conferring substance, which must be more than biological.

The importance of the movement of judgments of the boundaries of heterosex and homosex from production to reproduction lies in its potential to erode earlier boundaries. Ultimately, even though ARTs and other forms of biomedicine are driven in large part by desires to participate in and claim hegemonic family forms, they end up subverting those very forms. ARTs and the family-building rhetoric now work well, and in the United States their market and use are relatively free (thanks to the concatenations of privacy) not to limit their use only to heterosexual married infertile couples. Similarly, the literal forms of the productive nuclear family are not needed for reproduction with ARTs. Conception does not require coupling. When the frame is shifted from production to reproduction, questions of marital status and sexual orientation become increasingly irrelevant to the reproduction. ARTs have a huge potential to tolerate multiple family forms and to contribute to the demise of reproductive and kinship discrimination against single, gay, and lesbian would-be parents, despite barriers to access (discussed in chapter 3). This is part of the same set of movements over the last decade and a half that have led to practices of accounting for and caring for embryos as their disposition precludes wasting them or reducing them to an instrumental means. The rise of certain well-known ART facilities that are associated with both the left and the right in U.S. political culture (see table 7.3) as well as a growing body of law (see table 5.2) emphasizes this.

From Essentialism and Social Construction to Strategic Naturalization and Constructedness

In capitalism, social kinds usually required social, often constructivist explanations, and natural kinds required naturalist, often essentialist ones. A parent is a parent because of certain social roles, which may be thought of as constructed because they vary from place to place and over time. An embryo has certain biological properties, including a particular chromosomal complement. In the biomedical mode of reproduction, the notions of both parent and embryo require social and natural explanations, and at different times one repertoire of explanation must explain underdetermination in the other.

I turn finally to the interactions and coproduction of social and natural facts and scenarios in these clinics. As epitomized by the sacred and profane embryo, ARTs and biomedicine more generally are irreducibly both social and natural, and almost nobody contests this. This has implications for individual agency and collective identity (as I explore in depth in chapters 5 and 6). The biological facts of parenthood in a place where there are ex vivo gametes and embryos are underdetermined. Longstanding cultural and political narrative tropes as well as legal, medical, and familial conventions are deployed in the disambiguation of the facts of life. Similarly, biological facts are used to normalize and naturalize all sorts of social difference. Neither natural nor social kinds are essentialized in these sites, and neither is socially constructed. As implied by the metaphor of "family building," the ontological choreography that ties together disparate elements is constructed, and both nature and society are deployed strategically in this painstaking, highly constrained work of construction.

In terms of agency and resistance, this has important implications. Where we are used to thinking that resisting being commodified is important because it mistakes persons for property, people who turn to ARTs and to biomedicine in general sometimes seek commodification of their bodies. This might be to assert or retain a property interest in their body parts (such as embryos or gametes), to have access to treatment, or to assert kinship or the integrity of the life course (through procreational intent, for example). It might also be to help others attain kinship and restore the life course (as in surrogacy) and is evident in the combination of family law and contract law in ART cases that Janet Dolgin has documented.[70] Similarly, in theories of capitalism, there is a general ethical precept to resist being objectified, to never be treated as a thing rather than a person or a means rather than an end, and to avoid reductive essentialist categorization. Many ART patients actively seek out objectification, however, both to have access to treatment and also to be biologically essentialized to individual, race, or nation in contesting or creating kinship. These ethical "reversals" are an intrinsic part of the private implication in science and will be increasingly important in navigating changing relations between science and society.

Recapitulations

In this conclusion, I have attempted to point to some of the changing interrelations of the social and natural orders that are being ushered in by our forays into life itself. Like many other scholars, I have suggested that what is produced, how value is calculated, how research is undertaken and overseen, and how political order is negotiated are specific to the mode of (re)production. In addition, I have argued that modes of production themselves are determined relative to the content and nature of the (scientific) knowledge of the times. In the case of massive innovation in the life sciences, biotechnology, and biomedicine, I have argued for profound, if gradual and partial, changes that are consequent on the literal and rhetorical implications of the body or subject in the reproduction of certified knowledge about the social and natural worlds. In particular, I have explored the fields of meaning in which collective and individual identities forge their biographies. Without understanding these patterns that I have described together as a "biomedical mode of reproduction," it is hard to make sense of the moves that groups and individuals have made to assert and contest those identities. And without meaning making and decoding abilities, it is hard to intervene oneself. At this point, a certain expertise and standing is granted to all people just by being possessors of bodies, and all are able to benefit—either directly or indirectly—from the great enfranchising potential of the biomedical mode of reproduction.

Notes

Introduction

1. See Bartholet (1999) and Gill (2003).

2. King and Harrington-Meyer (1997); see chapter 3 below.

3. See, e.g., Clarke et al. (2003).

4. Lofland and Lofland (1971; 1984), Clarke (1990), Latour (1987). Lynch (1992), and Inhorn and Van Balen (2002) are my principle methodological resources.

5. See the glossary of terms for definitions of technical terms.

6. Nelson (2003, 1–2). This emphasis was also key to the framing of the April 2004 March for Women's Lives in Washington, DC.

Chapter 1

1. In contrast to the metaphor of promiscuity, antirealist philosopher of language John McDowell and feminist scholar of STS Donna Haraway have used the metaphor of modesty to describe their positions on realism and objectivity (Haraway 1997; McDowell, unpublished paper). Feminist philosopher and physicist Karen Barad's "agential realism" captures the catholic realism to which I am drawn (Barad 1999).

2. The so-called science wars, which involved a well-orchestrated (and much written about) stunt, appear to have petered out amid more pressing and worldly wars. Their relevance is generational, with younger science studies practitioners, myself included, feeling that it is not our battle. Some real issues behind the science wars continue to be worthy of public and specialist attention—such as the declining status of physics, the appropriate level of scientific oversight, the persistence of two cultures on academic campuses, and the role of objectivity in the politics of the old versus the new left. With some exceptions, however, these were not the topics that filled the pages of debate, and the name calling and mutual failures of comprehension are best

forgotten. For a selection of early salvoes and their echoes, see Gross and Levitt (1994), Sokal (1996a, 1996b), and the May 22, 1997, issue of *Nature* 387: 325 and 331–336. Three of the most helpful and significant works that have been written in the wake of the science wars are Latour (1999), Strum and Fedigan (2000), and Hacking (2000).

3. At an introductory level, see Collins and Pinch (1998, 1999). Anthologies include Jasanoff et al. (1995) and Biagioli (1999). Influential edited collections include Knorr-Cetina and Mulkay (1983), MacKenzie and Wajcman (1985, 1999), Lynch and Woolgar (1988), Pickering (1992), Clarke and Fujimura (1992), and Law and Mol (2002).

4. See Proctor (2000) for an argument that this thinking has prevented us from recognizing how much "good science," especially cancer research, the Nazis carried out.

5. See Merton (1942, 1973a, 1973b).

6. See Gieryn (1983), Gieryn and Figert (1990), and Mukerji (1989).

7. For two early anti-Mertonian pieces, see Barnes (1971) and Mulkay (1976). Those interested in pursuing SSK further are referred to Barnes and Shapin (1979), Barnes and Edge (1982), Shapin (1984, 1994), Barnes (1985), and Shapin and Schaffer (1985). Steven Shapin is also a master of the analytical review essay (Shapin 1982, 1992). See also Mulkay (1979), Pinch and Bijker (1984), Pinch (1985), and Collins (1985).

8. See Fleck (1935) and Mannheim (1929, 1936).

9. See Douglas (1966).

10. Wittgenstein (1953) and Winch (1958).

11. Most famously, Kuhn (1962, 1970). See also Kuhn (1977) and the philosophical writings of Willard Van Orman Quine (e.g., Quine 1953, 1969).

12. The Strong Programme had four theoretical and methodological principles for the examination of scientific practice: the account should capture causality, it should exhibit impartiality, it should be symmetric in the sense described in the text, and analysts should be reflexive as regards the status of their own truth claims. See Bloor (1991).

13. Bruno Latour (1999) has referred to this as the turn from science to research and from certainty to uncertainty.

14. The canonical statement of the great men, great science position is Weber (1948).

15. See Kuhn (1970).

16. See Shapin and Schaffer (1985), Shapin (1994), and Dear (1985) for early modern science and Schaffer (1992).

17. See Bijker et al. (1987).

18. See Schutz (1970).

19. See Garfinkel (1967) and Garfinkel et al. (1981).

20. See Lynch (1985a, 1985b, 1993).

21. See Strauss (1991) and Clarke (1991).

22. Shapin (1992).

23. For early modern science, see notes 7 and 16 above; see also Westman (1990). For nineteenth-century field sciences, see Rudwick (1985). For the history of modern biology, see Young (1986), Star and Griesemer (1989), and Clarke and Fujimura (1992). For the history of modern physics, see Galison (1997) and Gooding et al. (1989). For the history of technology, see Mackenzie (1990) and Hughes (1983). The history of medicine, an enormous field, intersects somewhat with STS. Much more literature can be found on contemporary medicine within STS (Brandt 1985; Epstein 1995, 1996; Rapp 1995a, 1999) than on the history of medicine, but key historical texts such as Foucault (1994) and Canguilhem (1991) inform STS practitioners who are working on contemporary and historical medicine.

24. See Zilsel (1942), Needham (1988), and Merton (1970).

25. Shapin (1992).

26. See Pickering (1992).

27. Biagioli (1990) and Findlen (1991) might be seen to be in this category.

28. See Hernstein-Smith (1997). The politics of identifying with the subjects and objects of a study are complicated, and the various sides in the science-wars context interpreted this as having more importance than perhaps it warranted in terms of the intellectual endeavors at hand.

29. See Goodman (1978), Cartwright (1983), Hacking (1983), Dupre (1993), and Galison and Stump (1995).

30. See Longino (1990), Wylie (1992), Harding (1986, 1991), Baier (1985), Code (1991), Collins (1990), and Butler (1990, 1993).

31. See Smith (1996), Cussins (1992), Berg (1997), and Star (1990).

32. See Latour and Woolgar (1986), Latour (1987, 1988, 1993, 1996, 1999, 2002), Callon (1986), and Law (1986, 1991, 1994).

33. See Pickering (1995).

34. Facts become unblack-boxed when they are contested and when people question the conditions and assumptions under which the data were obtained.

35. Latour (1999) uses the philosopher Alfred Whitehead to argue that scientists choreograph the critical ontological shift from eliciting properties or existences to discovering objects with essences.

36. Haraway (1991, 1992), Braidotti (1994), Braidotti and Lykke (1996), Davis (1999), Mol (2002), and Probyn (1995).

37. See, e.g., Jasanoff (1990, 1994, 1995, 1996a, 1996b, forthcoming).

38. See Dennis (1994), Irwin and Wynne (1996), Epstein (1996), Watson-Verran and Turnbull (1995), and Jasanoff (1996c).

39. Merton (1973, emphasis in original).

40. The geometry of inside and outside is misleading because it invokes hostility by its boundary drawing. In using this geometry, I refer to the diffuse differences between professional scientists and others.

41. See discussions of the infamous "Baltimore case" for a cautionary tale (Kevles 1998). Shapin (1999) makes clear why a bureaucratic accounting ideal of openness is not ideal.

42. Again, we all seek many kinds of outside notice, not only because we crave an audience and funding but also because we need to acknowledge our findings. Only some kinds of scrutiny seem dangerous. My argument is that, in the interests of society, this fear should be resisted as far as possible.

43. As Latour (1999) has argued, objects of science become important actors who have to be taken into account.

44. Philosopher of science Helen Longino, among others, has made this argument (see note 30).

Chapter 2

1. Becker (2000).

2. See Birke et al. (1990), Pfeffer and Woollet (1983), and Stolcke (1986).

3. Inhorn (1996) and Pfeffer (1993, 36). There is an analogy with the opposite situation, where a woman is told that she is infertile, forgoes contraception, and ends up resorting to abortion as birth control (Luker 1975, 63).

4. See Sandelowski (1993, 22–37) for a summary of historical approaches to writing about infertility. Naomi Pfeffer (1993) argues in her United Kingdom–based history of infertility and medicine against the common perception that the new reproductive technologies marked a watershed in concerns about or medicalization of infertility. Marsh and Ronner (1996) and Elaine May (1995) point out in their histories of infertility in America that thirty years' worth of technical and political work on in vitro fertilization preceded the birth of Louise Brown, even if her birth symbolically marked a medical breakthrough.

Several feminists (e.g., Adams 1994; Newman 1996; Squier 1994) have shown the prominence of feminist and other reproductive technological imaginings in the twentieth century.

5. Firestone (1970) and Spallone (1989). There is an interesting parallel here with feminists' attitudes to pornography. Radical feminists such as Catherine MacKinnon (1987) and Andrea Dworkin (1981) link coercive sexuality, pornography, and the status of women and thus are aligned with religious conservatives in opposing pornography, whereas most mainstream feminists are more tolerant of pornography than is the general population.

6. Canonical works include Arms (1975), Bernard (1974), Donnison (1977), Ehrenreich and English (1973), Homans (1986), Kitzinger (1978), Martin (1987), Oakley (1984), O'Brian (1981), Rich (1976), and Rothman (1982, 1986, 1989).

7. Kitzinger (1978, 74).

8. Hammonds (1995), Davis (1983), Ehrenreich and English (1978), and Jordanova (1989).

9. Ginsburg (1989), Gordon (1977), Luker (1984), and Petchesky (1984).

10. Terry (1989).

11. Bleier (1984), Bordo (1986), Cockburn (1982), Cowan (1983), Easlea (1981, 1983), Fausto-Sterling (1985), Fee (1979), Hubbard (1982), Hubbard et al. (1979), Keller (1985), Longino and Doel (1983), Martin (1991), Merchant (1980), Moscucci (1990), Rose (1982, 150–152), Sawicki (1991), Schiebinger (1989), and Wacjman (1991).

12. The points that follow can be found in these anthologies and monographs: Arditti et al. (1989), Corea (1985), Holmes (1992), Holmes et al. (1981), Klein (1989), McNeil et al. (1990), Raymond (1993), Rothman (1986), Rowland (1992), Scutt (1990), Spallone (1989), Spallone and Steinberg (1987), and Stanworth (1987). Where a single work is footnoted after a point is made, the reader's attention is drawn to an additional journal article or book chapter within one of the anthologies where the point in question is emphasized.

13. The use of the word *artificial* in conjunction with insemination has fallen out of favor, as it has in connection with most assisted reproductive technologies. New ways of doing things commonly are labeled artificial because they go against customary or natural(zed) practice, but eventually the practice comes to seem normal or natural enough, and the connotations of the word *artificial* begin to make people uncomfortable.

14. Sandelowski (1991).

15. Solomon (1989).

16. Klein and Rowland (1989).

17. For the French case, see Marcus-Steiff (1991). As elaborated in chapter 7, fears about the quality of care and the misrepresentation of statistics were central issues for infertile couples in the United States who worked with Representative Ron Wyden (D-Oregon) and the American Society for Reproductive Medicine (ASRM) on the passage of the Fertility Clinic Success Rate and Certification Act of 1992 (also known as the Wyden law), which requires clinics to collect and make public the results of their treatments. Consumer activism in conjunction with the ASRM and the Wyden law has also been important in reproductive tissue-donor screening and in the College of American Pathologists' Laboratory Accreditation Program. In the United Kingdom, regulation occurred at the government level, focusing on the embryo-experimentation debate but ultimately legislating clinic standards and reporting requirements into the Human Fertilisation and Embryology Act (1990). For descriptions of the regulatory situation in the United Kingdom and other countries, see Warnock (1985), Van Dyck (1995), Franklin (1997, 85–88), Pfeffer (1993, 164–169), Jones (1996), and Kirejczyk (1994).

18. Rapp (1987), Roggencamp (1989), Saxton (1989), and Spallone (1992).

19. Spallone and Steinberg (1987, 212).

20. Arditti, Klein, and Minden (1989, xxi).

21. Seal (1990, 21).

22. Arditti et al. (1989, 371–456). As AIDS screening of sperm became necessary, these people turned to a more high-tech subversion of establishment technologization, but they started by promoting home-based low-tech uses of insemination for "turkey-baster" babies. The recent development of so-called lesbian IVF (see chapter 7) shows that this strand has continued.

23. Franklin and MacNeil (1988).

24. Kaupen-Haas (1988), Patel (1987), and Williamson (1976).

25. Mies (1987).

26. Spallone and Steinberg (1987, 211–212).

27. Modell (1989) and Stacey (1992).

28. Crowe (1985).

29. Bartels et al. (1990), Birke et al. (1990), Lasker and Borg (1989), Rodin and Collins (1991), and Wymelenberg (1990).

30. Purdy (1989), Singer and Wells (1985), and Whiteford and Poland (1989).

31. Cohen and Taub (1989).

32. Edwards (1989). See Casper (1998) for a comparison with the almost contemporaneous fetal surgery, where practitioners went out of their way to shun publicity and critique.

33. See Seal (1990, 70–73).

34. See Anspach (1989).

35. Klein (1989, 18, 33–34, 109, 192–197).

36. Coalition to Fight Infant Mortality (1989).

37. Ince (1989) and Scutt (1990). In commercial surrogacy, a woman is paid a fee over and above expenses for becoming pregnant with the sperm of the husband of a commissioning couple. Conventional surrogacy refers to surrogacy pregnancies where the egg(s) that result in the pregnancy came from the woman who was carrying the pregnancy. The surrogate is genetically as well as gestationally the mother in conventional surrogacy. In gestational surrogacy, the surrogate gestates the embryo but does not provide the egg(s). See chapter 5.

38. Corea (1989, 133; emphasis in original).

39. Zelizar (1985).

40. Hochschild (1983).

41. Allen (1990).

42. Cannell (1990) and Corea (1989, 137).

43. Feminist Self-Insemination Group (1980).

44. Klein (1989, 245).

45. Baruch et al. (1988), Franklin (1991), Ginsburg and Tsing (1990), and Petchesky (1987).

46. Franklin and Ragoné (1998, introduction).

47. Rosaldo and Lamphere (1974) and Rubin (1975).

48. Chodorow (1978) and Ortner (1974).

49. Gilligan (1982), Griffin (1978), and Treblicot (1984).

50. Daly (1978), McMillan (1982), and Ruddick (1983). As described in chapter 1, some scholars in this period argued that being a woman, like belonging to any other oppressed group, conferred "epistemological privilege" (e.g., Hartsock 1983; Kuhn and Wolpe 1978). Still other feminist theorists rejected both standpoint theory and the maternalist writings mentioned above for their essentialism about women and their denial of differences among women. For examples of these debates in feminist epistemology, see Harding (1986), Harding and Hintikka (1983), Lloyd (1984), and Longino (1990).

51. Sandelowski (1990, 1993).

52. See Butler (1990).

53. Haraway (1989, 1991).

54. Edwards et al. (1993), Franklin (1997), and Strathern (1992).

55. Schneider (1980) and Yanagisako and Delaney (1995).

56. Strathern (1992, 2003). Also see chapter 5 and table 5.1 concerning landmark legal cases.

57. Various feminist scholars, myself included, have expressed this as a coming apart of blood and genes (Cussins 1998b), but this is not quite right. While the gestational mother does feed and hence give form to the baby with her own blood and tissue, usually a blood test is used to definitively prove genetic relationship. Thus, in *Calvert v. Johnson* (see below), Christopher/Matthew had a blood test done that showed that he could not have been genetically related to Anna Johnson.

58. See, from increasingly many, Franklin and Mackinnon (2001), Howell (2005), Lewin (1993), Modell (1994), and Weston (1991).

59. See Epstein (1995) and Rabinow (1992).

60. Pfeffer (1993).

61. May (1995) and Rainwater (1960).

62. Marsh and Ronner (1996).

63. Rapp (1987) and Rothman (1986).

64. Clarke (1998) and Rapp (1996, 1999).

65. Becker (1994, 2000), Cussins (1996, 1998a, 1998b), Franklin (1997), Inhorn (1994, 1996), Layne (1994), McNeil and Franklin (1993), and Sandelowski (1993).

66. See Akrich and Pasveer (1997), Clarke and Montini (1993), Davis-Floyd (1992), Davis-Floyd and Dumit (1998), Hartouni (1997), Oudshoorn (1994), Todd (1998), and Van der Ploeg (1998).

67. Hogden et al. (1994) and National Institutes of Health (1994).

68. See Chung et al. (1995), Collins (1995), Duka and DeCherney (1994), Edwards (1993), Egozcue (1993), Katz (1995), Robertson (1996), Schenker (1996), Seibel et al. (1994), and Shenfield (1994). Also see Farquhar (1996) for comment on this.

69. Pfeffer (1993, 230).

70. Finkel (1999). Prominent U.S. infertility specialists also thought that the practice of transferring large numbers of embryos at once was misguided and cited this as a major reason for the need for nationwide standards of best practice, such as the outlawing of the transfer of more than three embryos at once

by the United Kingdom's Human Fertilisation and Embryology Authority (HFEA). See chapter 7 for the reasons why the U.K. and U.S. situations are significantly different.

71. Haraway (1995, 1997).

72. See chapter 4.

73. Cussins (1998a) and Pfeffer (1993, 216).

74. See chapter 4, Schmidt and Moore (1998), and Becker (2000). See also Waller (2002).

75. Chang and Fourcey (1994), Ladd-Taylor and Umansky (1998), and Layne (2002).

76. Inhorn (1996).

77. Ragoné (1994). See also chapter 5.

78. See table 5.1 for details of *Johnson v. Calvert.*

79. Hartouni (1997, 85–98) and Grayson (1998).

80. The racial logic of the *Johnson v. Calvert* case is actually hard to read as simplistically as black versus white, given that Crispina Calvert and the baby are not white. Grayson, for example, claims that Crispina Calvert was a member of a "model minority" and so counted as an "honorary white" for this case. This seems highly unlikely to me, given among other things that Calvert is not the stereotypical East Asian of model-minority notoriety but from Island Southeast Asia. At the very least, such a claim would need arguing. A systematic comparison between this case and the *In re-Andres* case (see table 5.1) might help sort out these elements (Jaeger 1996, 123–125). See Collins (1990) and Ikemoto (1995).

81. Adams (1994), Cartwright (1992), Casper (1998), Daniels (1993), Duden (1993, unpublished), Gomez (1997), Hartouni (1997), Mitchell (2001), Morgan and Michaels (1999), Oaks (2001), Squier (n.d.), and Taylor (1992).

82. Berlant (1997). See also Solinger (2001).

83. Morantz-Sanchez (1997).

84. Fourteen states offer some kind of insurance coverage for those who are working insured, while none offer contraceptives. Medicaid, the government coverage for the uninsured, does not cover infertility treatment in any state but covers contraceptives in all states. See table 2.2 and King and Harrington Meyer (1997), which was written when ten states offered coverage of some kind.

85. Pfeffer (1993) and Dixon-Mueller (1993).

86. Das (1995, 232).

87. See Foucault (1978) and Sawicki (1991).

88. Inhorn and Van Balen (2002).

89. Dikotter (1998), Greenhalgh (1995), Kaananeh (1997), Lurhmann (1996), Morgan (1997), and Thompson (forthcoming).

90. Bledsoe (1995, 133) and Inhorn (1996, 1–50).

91. See Feldman-Savelsberg (1999), Handwerker (1995), Horn (1994), Kahn (2000), Lurhmann (1997), and Paxson (2004).

92. For example, in 1998 a British woman named Diane Blood traveled to Belgium to be impregnated with her dead husband's sperm, which was not permitted in the United Kingdom.

93. Ginsburg and Rapp (1995). See also, e.g., Ehrenreich and Hochschild (2003).

94. See Davis (1993), Richardson (2003), and Roberts (1999).

95. But see, e.g., Rothman (1998).

96. Rosenberg (1999).

97. Andrews (1999), Brodwin (2000), Cussins (1999), Franklin and MacKinnon (2001), Holland et al. (2001), Nussbaum and Sunstein (1998), Cambrosio et al. (2000), and Squier and Kaplan (1999).

Chapter 3

1. This chapter is drawn primarily from participant observation while I was interning in clinics, and its ethnographic material is based more on field notes than on the interviews on which some other chapters are based. Chapters 7 and 8 also have significant ethnographic and empirical content, but chapter 7 uses mixed methodologies, and chapter 8 ends the book with a manifesto of sorts.

2. See Burawoy (2000) and Charmaz (1991) for examples of ethnography that are alert to the way in which the local involves history, global relations, and long-term psychological horizons.

3. I follow Adele Clarke's (1990) social-worlds-theory solution to selecting a unit of analysis—a meaningful grouping recognized by the actors as such and composed of shared temporal and material commitments. Unlike social-worlds theorists, I do not construct my field of inquiry through actors' perspectives because my political and epistemological interests in the anatomy of normalization and routinization do not (or cannot be presupposed to) coincide with any function of the actors' perspectives. I use the term *socionatural world* rather than *social world* to stress my argument in chapter 1 against the explanatory priority of either the social or the natural.

4. Foucault (1970, preface, xx; 1995, pt. 3; 1994, chap. 7), Goffman (1961, 1969, 1986), Haraway (1991, 480), and Lynch (1985a, 1985b).

5. See Hirschauer (1991).

6. Robert Westman suggested to me that this kind of role normalization is similar to the black-boxing of technical and scientific facts and artifacts.

7. Collins (1994), Fox (1994), Goffman (1961), Katz (1981), Hirschauer (1991), and Lynch (1994).

8. See Lorber (1975).

9. See chapter 7 for a discussion of the history of this distinction, and, e.g., Franklin (1997) and Paris (1994) for two European examples.

10. See tables 5.1 and 7.1 for some information related to gay and lesbian parenting.

11. The standards that are implicitly and explicitly invoked to regulate paying patients and the standards that are invoked to regulate commercial surrogates and donors apply and extend the norms of the wider society. The differential between these standards is mirrored by differences in physicians' comportment toward their "client" patients (the commissioning would-be parents) and toward their "employee" patients (the commercial surrogates and donors). The examples in this chapter are limited to conventional infertility treatments that involve only patients who contribute the gametes that are to be used for their treatment. See chapter 5 for more on this contrast.

12. Whether and how much patients have to pay depends on what country they live in, what the patient's nationality is, whether the procedure is deemed experimental, what their diagnosis dictates, and so on (Inhorn and Van Balen 2002).

13. Thus, day 3 (of the menstrual cycle) estradiol levels and follicle-stimulating hormone levels are routinely measured to predict ovarian response, and sometimes a clomiphene challenge test is administered (Weckstein 2002).

14. See chapter 7 for the history of SART, the SART registry, and associated legislation; also see table 7.2.

15. See Croucher et al. (1998), Evers et al. (1998), Land et al. (1997), Meldrum et al. (1998), Roest et al. (1998), Smith and Buyalos (1996), and Zinaman et al. (1996).

16. Psychologists have had the "caring" function annexed to them. They also move patients onto alternative treatment options, perform an adjudicatory role in patient assessment for access to treatments and in legal contests, and increasingly specialize in infertility medicine. The Psychological Special Interest Group within the American Society for Reproductive Medicine was inaugurated as early as 1988 to recognize the significance of the

psychological impact of treatment and the growing body of research on the topic.

17. Fausto-Sterling (2000).

18. See Leon (2002) for the historical and cultural specificity of the loss framing in the case of adoption.

19. It is perhaps a measure of the success of these externalizing and normalizing moves in the field, of research into outcomes, and of general publicity about the field that the incidence of patients who disclose their infertility and the provenance of their ART children's conceptions has increased significantly since ARTs began to be offered in the United States. See Becker (2002) as well as the sources she cites.

20. Male infertility and ARTs are discussed in chapter 4. For an example, see Kiekens et al. (1996). A man with idiopathic anejaculation was electroejaculated by rectal probe and found to have unexplained poor sperm quality. Despite her male partner's fertility, the female partner underwent two cycles of IVF with ICSI to become pregnant.

21. IC, WHO Scientific Group on Recent Advances in Medically Assisted Conception, Geneva, April 2–6, 1990.

22. "Beautiful philosophical object" is an expression that was used in a lecture by Bruno Latour to describe a soil-collecting device.

23. See Latour (1988) and Lynch (1985b, 1991).

24. This is what Hirschauer (1991) seems to suggest is the case in surgery.

25. This is in the spirit of Michael Lynch's fundamental question "How do graphic properties merge with and come to embody the natural object?" (Lynch 1985b, 43).

26. The characteristic T-shaped uterus of daughters of women who took diethylstilbestrol (DES) while pregnant with them, which can interfere with conception, also shows up.

27. See, e.g., Oreskes's (1997) discussion of Eleanor Lamson. See also Martin's (1991) history of women telephone operators for examples of the ways in which presumptions about gender are built into the very nature of the skills required to perform certain jobs.

28. Hutchins (1995) uses the hierarchical nature of skills as distributed in teams to produce his account of the organic reproduction of overlap of skill—through moving upward in a skill hierarchy, masking other epistemic functions of such role differences, and overcoming the boundaries that most workplaces raise to mobility across the boundaries. Hutchins's navy work on team navigation is arguably at one extreme of the discussion of distributed cognition. The U.S. Navy takes care institutionally of the enforcement and

reproduction of the necessary hierarchies, so Hutchins did not confront them in his own fieldwork.

29. Ragoné (1994) provides a fascinating discussion of hierarchical role overlap in American surrogacy, where the commercial surrogate temporarily experiences class mobility as she is disciplined (through diet, drink, and medical surveillance) and pampered (through friendship with the commissioning parents and lavish gifts) across class lines. At the end of the nine months, however, the relationship is over, and no permanent mobility is achieved. I have found that in my own interviews with surrogates they frequently say that they cannot understand why they were "dropped" by the socioeconomically privileged commissioning parents after the baby's birth (see chapter 5).

30. See Knorr-Cetina (1999).

31. Many writers commented on the problematic interpretation of success rates based on statistics in ARTs before reporting was standardized and in non-U.S. environments (e.g., Crowe 1990, 30; Steinberg 1990, 84; Klein and Rowland 1989; Marcus-Steiff 1991). These accounts focus on how a prospective patient should interpret published statistics: Which cycles get counted? Do multiple births count as more than one success, inflating the figures? Are only live births considered to be successes, or are all pregnancies counted, including perhaps the infamous "biochemical pregnancies," in which no fetus is ever present? Are the figures significantly better than comparable figures for different treatments and figures for no treatment at all?

32. Very local success rates, such as the number of pregnancies that week, can have a surprisingly large effect on morale.

33. This turn of phrase brings out particularly well the fact that the statistics must be produced by the very procedures that they are used to justify.

34. Young (1989).

35. At another clinic where I worked, all the exam rooms were arranged in a square around a central nurses' station. Because none of them was sufficiently private for masturbation, the men's bathroom was used instead. This room was stocked with an open rack of heterosexual pornographic magazines and sterile collecting pots.

36. Increasingly, clinics are offering deals where patients pay for a set number of treatment cycles, say three, and then get any additional cycles almost free. If they get pregnant in fewer than three cycles, though, they do not get refunded, and these deals tie a patient to a given clinic. See table 2.1.

37. Some clinics make do with a single laboratory, and other clinics have a suite of rooms that allows them to separate diagnostic work from therapeutic work and separate andrology from embryology.

38. Sarah Franklin, personal communication.

Chapter 4

1. Butler (1999, 10–11).

2. See Fausto-Sterling (1995, 127–134).

3. Butler (1999, 179).

4. I draw on Erving Goffman's landmark work on stigma (Goffman 1986).

5. The predominance of feminist scholarship is beginning to give way to more mainstream scholarship in the history of the life sciences, medicine, eugenics, and biomedical ethics as genetics becomes a significant part of clinical practice, but these interventions tend to be entirely uncritical of gender formations in biomedicine.

6. Van Der Ploeg (1998). See chapter 3.

7. See Inhorn (1996, 2002) and Kahn (2000).

8. Gay Becker pioneered the systematic study of the experience of infertility for the American male (see, e.g., Becker 2000, 2002).

9. Nachtigall et al. (1992).

10. See Becker (2000, 136–137) for the assessment that the system of patriarchy leads men to identify with their genetic contribution. For the contemporary American obsession with genetics and genetic idioms of relatedness, see, e.g., Nelkin and Lindee (1995). For the historical newness and cultural specificity of the genetic idiom of fatherhood in the United States and an analysis of "believed-in imaginings of fatherhood" with undisclosed artificial insemination, see Morawski (1997).

11. In addition to my own findings in this regard, Becker (2000, 2002) and Inhorn (1996, 2002) report similar findings of supportiveness, Becker for the United States and Inhorn for Egypt.

12. Dumit (2000).

13. See Brinsden et al. (1997).

14. Palermo et al. (1992).

15. See Silber and Johnson (1998) and Vanderzwalmen et al. (1998).

16. Silber and Johnson (1998, 509)

17. Van Steiteghem et al. (1996).

18. ESHRE (1998, 2029).

19. ESHRE (1998).

20. Gutmann (1997). See also, e.g., Kimmel and Messner (1998).

21. For an excellent theoretical discussion as well as ethnography of masculinity, see Peletz (1996, 1–11, 257–308). On masculinity and sexual anxiety, see Gregor (1985). Among well-known ethnographic accounts of masculinity, see Brandes (1981), Herzfeld (1985), and Herdt (1982).

22. Studies of masculinity have pointed to hierarchies within maleness, contingencies about what counts as maleness, and cultural and historical specificities of ways of manifesting as male. In this regard, the irreducible connections between other categories of identity—such as race, ethnonation, class, and sexuality—and masculinity have begun to be revealed. See, e.g., Awkward (1995), Dent (1992), and Ong and Peletz (1995). The role of tropes of (heterosexual) masculinity in defining femininity and in defining lesbian, gay, and bisexual identity has been interestingly explored (see Halberstam 1998).

23. This is sometimes referred to as the "taken-for-granted" nature of masculinity as it figures in most theoretical discussions of gender (see, e.g., Gilmore 1990).

24. Marlon Bailey's work on black queer "worldmaking" in Detroit ballroom culture explores the parodic performance of gender and the ways in which it makes kinship (Bailey 2005; see also Munoz 1999).

25. I am grateful to Irving Leon for introducing me to the expression "womb envy."

26. For these and the following details of electroejaculation and spinal-cord injury, see Brinsden et al. (1997).

27. Rosaldo and Lamphere (1974), Ortner (1974), and Chodorow (1974).

28. See Harding (1991).

29. De Beauvoir (1953, 59).

30. Levi-Strauss (1969) argues in his 1949 book *The Elementary Structures of Kinship* that all human societies are built on a fundamental separation between the domestic unit (for reproducing and socializing) and the society and all its networks. The former is biology, the other, culture. Women are associated with the domestic realm of nature, reproduction, and socialization, while men are associated with the public realm of culture. To be human is to prevent the domestic sphere of nature from becoming a closed and self-reproducing merely biological system. This task is the role of culture and is seen in the mandatory and universal taboos against incest and endogamy, which together, as this scheme is gendered, require men to "exchange" women. Ortner's position grants women the possibility of transcending their biological functions and the obligation to socialize at least boy children into culture. It was actually a more emancipatory statement of the status of women than was current at the time, however unrealistically structurally confining it now seems.

31. For a refutation of the structural functionalist feminist psychosocial relationality argument about women that takes a poststructuralist antiessentialist feminist perspective, see Fraser and Nicholson (1990).

32. See Peletz (1996) and Stack (1974, chap. 7).

33. For example, "The demasculinization of colonized men and the hypermasculinity of European males represent principal assertions of white supremacy" (Stoler 1991, 56).

34. Butler (1993, 15).

35. This is in evidence and contentious in the same-sex marriage debate.

36. Hacking (1995).

37. Butler (1990, 7; 1993, 4–12) and Haraway (1991).

38. Haraway (1991, 148).

Chapter 5

1. Strathern (2003) and Dolgin (1995) for discussions of the question of procreative intent and its legal counterpart of the contract.

2. See Weston (2001) for a discussion of the only sometimes kin-conferring qualities of the transmission of substance.

3. See Ragoné (1994) and Grayson (1998).

4. See Strathern (1992, 2003), Franklin (1997), Franklin and Ragoné (1998), Franklin and MacKinnon (2001), and Edwards et al. (1993).

5. Dolgin (1995, 47–48).

6. See Andrews and Douglass (1991), Andrews (1999), and Dolgin (1993).

7. See American Society for Reproductive Medicine (2002b).

8. See Barkan (1992) for an account of the interwar retreat of scientific racism. See Strong and Winkle (1996) for, as Pauline Strong (2001, 468) expresses it, showing that "the reckoning of identity through 'blood quanta' (that is, percentages or 'degrees' of Indian blood) is at once a 'tragic absurdity' and a 'tragic necessity' for many contemporary Native Americans in the United States." See Wailoo (1997) and Tapper (1999) for the historical intersection between disease concepts and African Americanness. Jennifer Robertson, anthropologist of Japan, is working on the expression of Japanese eugenics through the use by Japanese couples of certain West Coast ART clinics (personal communication).

9. Yanagisako and Delaney (1995).

10. Ginsburg and Rapp (1995).

11. As is explained in chapter 1, I do not imply a lack of belief in the reality of any of these procedures by arguing for their fungibility. I take this not as anti-realism but as attentiveness to the contingent realities of complex phenomena.

12. See Catholic Church, Congregation for the Doctrine of the Faith 1987; see also chapters 2 and 6.

13. Bruno Latour (personal communication) expressed it as showing that "high tech processes can be re-employed to provide powerful fetishes—in the good sense of the word—to produce affiliation and identity."

14. See Farquhar (1996).

15. See chapter 7.

16. See tables I.1, 2.1, 2.2, 5.2, 5.3, and 7.3 for details on the current stratification of the egg, embryo, donor, and surrogate markets in the United States. The hunt for the ideal egg donor with particular characteristics, the high prices (up to $50,000) being offered to chosen egg donors, and the seduction and competition that characterize efforts by elite college students to be selected have become something of a cultural phenomenon in the United States. There is an unspoken etiquette (not always adhered to but pervasive) that parties seek roughly to match the qualities and phenotype of the infertile partner in both donor egg and sperm, but it is nonetheless expected that this will be interpreted leniently enough to permit upward mobility in such areas as physical prowess or IQ.

17. See, e.g., Casper (1998) and Daniels (1993).

18. This woman was what I would call upper middle class and living in an affluent neighborhood (like most infertility patients). I think that her relevant notion of community, based on her explanations of alternative kinds of mothering, was based primarily on a shared African American identity.

19. In the United Kingdom, daughters are not allowed to be donors for their own mothers because the relationship is seen as necessarily involving a coercive element. Frances Price (Edwards et al. 1993, 36–37) recounts the 1986 British test case that resulted in what was then the Voluntary Licensing Authority, which declared that mother and daughter egg donation was illegal.

20. I'm not sure that this relationship has a name.

21. In this usage, I am following the common definition of the word *custody* as "a guarding or keeping safe, care, protection, guardianship."

22. According to the Roman Catholic Church (1987), no part of human reproduction can be made instrumental and alienated. Every part of the process must be relational, tracing a trajectory of kin and presence at all points. In commercial surrogacy, the mother comes close to being merely instrumental at the moment of handover (see the text on this point), but typically even if the surrogacy contract was entered into for economic gain, the infant is not

treated as a means, for a moral discourse intervenes between the transaction and the gift of life (Ragoné 1994).

23. Compare Dolgin (1995) on the law, sperm, and fatherhood and Laqueur (2000) on the lack of motherhood in egg donation.

24. See Seibel and Crockin (1996, 111–216).

Chapter 6

1. I gratefully acknowledge the helpful discussions I had with Adrian Cussins while writing the paper from which this chapter is derived. The notion of trails is his.

2. The continental critical tradition especially has theorized technology as part and parcel of modern industrialization and so has tended to equate the social and personal effects of technology with the class relations that are mediated by the means of production of industrial capitalism. Compare Marx (1982) and Weber (1958, 181–182), where the technological apparatus of bureaucracy is seen as the iron bars out of which the "iron cage" of increased rationalization and control of social life is forged, with recent contributions developing themes of deskilling (e.g., Braverman 1974, Noble 1984, and with speculations that increasing bureaucratization and surveillance are implicated in a loss of personhood, autonomy, and agency (e.g., Foucault 1995; Garson 1988; Zuboff 1988).

3. See Foucault 1994 and 1982 which argue together for the historical specificity of modern medicine in enabling the individual to be "both subject and object of his own knowledge," and Horkheimer and Adorno (1979), who argue for an instrumental approach to objectification.

4. In addition to the works cited in chapter 2 for phase 2, see the individual contributions in Berg and Mol (1998). See also Clarke and Montini (1993), Singleton and Michael (1993), and Traweek (1993).

5. See Franklin (1997) for a review of Squier (1994).

6. See Michele Martin (1991), who reworks Foucault's later ideas on the possibilities for the development of self-identity and resistance around technology (e.g., Foucault 1982). She recounts how a new telephone exchange is produced and how simultaneously gendered and disciplined identities for its feminizing workforce are constructed. See also Cockburn and Furst-Dilic (1994) for European case studies of the coproduction of gendered identities and technology.

7. Katy Charmaz's work (e.g., 1991) on the maintenance of coherent self-narratives over time for the chronically ill is exemplary. See Casper (1994), Franklin (1991), and Layne (2002) for fascinating medical examples from the fields of ARTs and obstetrics. See, e.g., Bell and Kennedy (2000), Cutting Edge (2000), and Downey (1998) for digital examples.

8. See Carritthers et al. (1985), Ferguson et al. (1990), Gergen (1991), Kondo (1990), and Prins (1995).

9. See Shapin (1993, 338) for an excellent discussion of the central role of assigning value in the "realization stories" that are an integral part of our sense of self: "We tell 'realization stories' about ourselves, and those who approve of us help us tell them, as a way of pouring value on those actions. Because such narratives are storehouses of value, their plausibilities are highly protected by a wide range of everyday and academic practices."

10. This debate comes directly from Locke and then Hume, where the self is seen as consisting in a bundle of experiences. Each of the elements of the bundle is constituted independently of the self. This is in contrast to the Kantian view, where all experiences are irreducibly stamped with the first-person act of being experienced and so are not dissociable from the self. See, e.g., part 3 of Parfit (1984) for an important recent statement of the Lockean and Humean tradition of work on personal identity, which covers the exotic thought experiments that are so characteristic of this line of inquiry.

11. See Taylor (1989) and MacIntyre (1981). Taylor criticizes Parfit for continuing in the tradition of conceiving of personal identity as an "aspiration to a disengaged subject of rational control" and therefore being unable to see the irreducibly collective, temporal, and moral dimensions of identity.

12. The tension that results from deconstructing (as opposed to taking for granted or to constructing) the human agent is embraced as a defining contradiction for postmodernism and critical theory: agency is necessary for sustaining viable oppositional or marginal identities but is illicit because of its historical embedding in capitalism and post-Enlightenment phallogocentrism. Haraway (1992, 98) begins a wonderfully provocative exploration of the theme of charismatics "who might trouble our notions—all of them: classical, biblical, scientific, modernist, postmodernist, and feminist—of 'the human,' while making us remember why we cannot not want this problematic universal."

13. For a summary of the aporia in medicine, see Dodier (1993, 325). Unlike Dodier, I argue for agency and subjectivity under some circumstances *within* the objective and public aspect of the body under the medical gaze.

14. The principle of charity as introduced by Quine (e.g., 1960) enjoins the analyst to interpret the utterances of those studied in a way that makes as many of them come out true (according to the analyst's beliefs) as possible. For Davidson and Quine, the analyst is a fixed reference point into whose interpretive schema the utterances of the other are interpreted or translated. Charity as it is used in sociology and anthropology picks up on the Verstehende aspect of the principle in a way that requires the initiation and conceptual movement of the analyst. To this extent, then, the Quinean and ethnographic principles are in conflict; my intended usage is in line with the latter.

15. See Kathy Charmaz's discussion of the self in sufferers of chronic illness (Charmaz 1991) and of the way that such things as "denial" are not simply

failures of understanding; understanding chronicity requires change and management of the multiple selves of illness and ordinary life.

16. These two positions correspond to the sociology that was criticized by Garfinkel and to his own methodology for the breaching experiments, respectively: "In accounting for the persistence and continuity of the features of concerted actions, sociologists commonly select some set of stable features of an organization of activities and ask for the variables that contribute to their stability. An alternative procedure would appear to be more economical: to start with a system with stable features and ask what can be done to make for trouble. The operations that one would have to perform in order to produce and sustain anomic features of perceived environments and disorganized interaction should tell us something about how social structures are ordinarily and routinely maintained" Garfinkel (1963, 187).

17. Jordan and Lynch (1993) showed this: "The social constructivists' black-box analogy places diversity and fragmentation at a preliminary stage of the narrative, whereas we see a persistent dispersion of innovations even within the frame of a highly consensual practice." They dispelled the idea that diversity and contingency operate only at moments of uncertainty and controversy. The socionatural order displays continuous diversity and fragmentation, preserving and disrupting stability across settings and other features of the black box.

18. See figure 6.1, the ASRM press release addressing the allegations against the UC Irvine doctors.

19. For a discussion of the primacy of women as patients, see chapter 2 and Van der Ploeg (1998).

20. See Franklin (1990) on the construction of the "desperateness" of infertility patients. "Achieving" pregnancy is an interesting formulation that is used by practitioners of infertility medicine and counselors when the "most natural thing in the world" (getting pregnant) becomes work.

21. I found no fail-safe connection between experiences of agency and any particular conception of the greater good.

22. For interesting critical uses of testimony, see Arditti et al. (1984) and especially Klein (1989). For a more valorizing account drawing on women's experiences, see Birke et al. (1990). See, e.g., Wynne (1988) for a discourse analysis of patient testimony showing how multiple sclerosis patients ascribe scientific rationality to doctors even when the doctors fail to diagnose correctly or to help the patient in any way.

23. These comments were gathered shortly after the failed or successful procedure (at subsequent treatment, at early prenatal checkups, or outside the clinic setting). As Nicolas Dodier pointed out to me, the testimony probably would change in interesting ways again by a later point in time as an evolving first-person narrative developed.

24. Jake wasn't in the lab when the sperm assay (the Ham test) was performed; only his sperm was. Sperm outside bodies is normally waste, but in this context it continues to refer back to the person. The clinic works only if it is able to maintain the sperm on a trail that marks out potential pregnancy. Under these conditions, body parts can be separated without losing their mereological role or being alienated from the subject. This part separation without alienation therefore applies to male clinic patients as well as female patients.

25. The *Shorter Oxford English Dictionary* defines *synecdochism* as used in ethnology as "belief or practice in which a part of an object or person is taken as equivalent to the whole, so that anything done to, or by means of, the part is held to take effect upon, or have the effect of, the whole."

26. Some of the controversies over success rates for reproductive technologies have focused on the possible disjuncture between fixing the parts and fixing the infertility. Some clinics claim successes when the functioning of parts or stages is restored rather than when a baby is born. See, e.g., Marcus-Steiff (1991) for a thorough look at the discrepancies between success rates reported by clinics that include ectopic and miscarried pregnancies and so are favorable for their bottom line (the higher the pregnancy rates, the more patients and other funding the clinic will draw) and success rates that matter for patients—"take-home baby rate," as it is called. Nonetheless, many patients report a sense of success after any kind of "result," even in such cases as ectopic pregnancies, where the patient's life might be endangered and she will almost certainly undergo additional surgery to remove the conceptus. I have often heard the following comment, or some equivalent, in these circumstances: "At least now I know I can get pregnant."

27. Renate Klein's (1989) important book charts the testimony of several infertility treatment "survivors" for whom the procedures didn't work. Their tales of the rigors of treatment and their objectification at the hands of the medical establishment are moving, compelling, and often appalling. My work has focused on patients who are still in active treatment. The critical stance that is exhibited in Klein's book is a stronger version of the stance that is taken in my data by women who became pregnant and talked about a previous failed cycle or those who experienced a failed procedure at a previous clinic.

28. See Judith Lorber's (1975) discussion of the "good patient" who enacts compliance as a strategy to, among other things, reduce the stigma of illness for herself and her family.

29. A striking example of this was expressed by a non-IVF patient who had artificial insemination using her husband's sperm, Clomid, and Pergonal. She said that she is fertile only on the weekends and that the protocol worked once she persuaded the treating team that they had to do the insemination on a weekend day. The rationale for this was presented in terms of the regularity of her cycle, her ovulation every four weeks and always on a weekend, and a tight cervix that hampered cannulation for insemination on all days except the very day she ovulated.

30. Studies of patient satisfaction with infertility treatment show high levels of approval of doctors and criticism of aspects of treatment that some might think are physician related (see Souter et al. 1998).

31. See Goffman (1986) for a discussion of the benefits that inmates of "total institutions" can receive by behaving with civility despite their subordinate positions. Although an infertility clinic is very different from a total institution, some parallels between being a good inmate and being a good patient—managing the self to attain the privileges of the site—are striking. But where Goffman uses a sociological idiom that aims to recover a coherent social world independently of the technoscientific world, even when he is talking about medicine, my account emphasizes that technical practice helps to create and maintain this coherence. I avoid the epiphenomenal character of patients' behaviors and meanings in Goffman and other sociological constructivist accounts by bringing aspects of the technoscientific world into the interests of patients and practitioners.

32. Hirsutism is associated with excess male hormones and hence with infertility, especially in Caucasian patients.

33. Another class of options that is potentially available to infertile couples in a culture with a broadly dualistic metaphysics is to transform the self-narrative about the desirability of having children. This can be done by addressing the collective (such as how society treats childless women) or by addressing the individual (such as other ways a woman can express her creativity). Part of the sense that using assisted reproductive technologies should be avoided is that ARTs are much more dramatic and invasive than the options for changing a self-narrative. Technologies are indeed dramatic in the miscegenations they bring about among the economy, the body, social institutions, and the soul and in the numbers and heterogeneity of entities that need to be aligned. But each ART stage is mundane. Attempts to change self-narratives without technologies also are dramatic—in the sense of the power that is required to change personal and social narratives without ARTs.

34. See Latour (1999, 24–79).

35. See Cartwright (1992) on x-rays and Rapp (1995b) and Mitchell (2001) on the role that is played by the sonogram in contemporary Western pregnancies.

36. Because laparoscopy is usually at least partially covered by insurance in America and IVF and many of the other "treatments" are not, such surgery is much more routine here than in the United Kingdom, for example. As insurance coverage improves, routine surgery is becoming less frequent, particularly because pregnancy rates after tubal surgery are poor and the costs of tubal surgery are usually higher than the costs of a cycle of IVF (see tables I.1, 2.1, and 2.2).

37. Petchesky (1987) has pointed out that this standard scientific way of representing objects of study can be dangerous in the case of reproductive politics.

Fetuses can be visualized on ultrasound and other screens in vivo, as if in splendid isolation, which has been important fodder for prolife proponents. Hartouni (1991) further documents this link in her analysis of the prolife film *The Silent Scream.* See also Casper (1998) and Mitchell (2001).

38. The art and technicality are discussed—how good the image definition is, what the image size is, and how fast the image can be translated into stills—but these parameters mark the skills of the artisans who are involved and don't question the metaphysics that are enabled by these taken-for-granted representational practices.

39. See Goffman (1986, 29) on the role of putting on institutional clothing in "defacement."

40. Pippin (1996) writes that "one way to allay worries stemming from a rights-based political culture, where human dignity and self-respect are essentially tied to the capacity for self-determination, is simply to integrate such an ethical consideration much more self-consciously and in a much more detailed way, into the transactions between patients and doctors. Thereby the fundamental liberal principle: *volenti non fit iniuria* is preserved. No injury can be done to the willing, or here the well informed health care consumer."

41. The presentism itself is not the problem here. Undue Whiggishness, ideology, and teleology aside, presentism can be a sophisticated appreciation of the irreducible temporal situatedness of the analyst, the analysis, and the function of a temporally extended narrative in creating and maintaining our collective and individual identities. The problem is cutting up time into discrete slices, so that time t is being viewed from a different frame of reference, time t'. This leads inevitably to the kinds of philosophical arguments that are developed against historicism and to the use of actors' categories, all of which are based on the incommensurability or inaccessibility of time t from t'. See, e.g., Kitcher (1993, 100–101 n. 13). Historians rightly tend to react with impatience to such arguments, pointing out that they have long been deploying knowledge of states of affairs at time t from time t' with considerable effect.

42. To transmute Foucault (1994).

Chapter 7

1. Traweek (1988).

2. Jasanoff (1995, 1 and 4).

3. Edwards (1989, xii).

4. The Warnock Committee of Inquiry into Human Fertilisation and Embryology was established in 1982 and was headed by the philosopher Dame Mary Warnock. Based on its recommendations, the Royal College of Obstetricians and Gynaecologists and the Medical Research Council established the

Voluntary Licensing Authority for Human Fertilisation and Embryology, which eventually became the law of the land after the passage in 1990 of the Human Fertilisation and Embryology Act. See Franklin (1997).

5. Winston (1996, 2000).

6. American Society for Reproductive Medicine (2002a).

7. As Monica Casper (1998) points out, this stands in dramatic contrast to the secrecy with which fetal surgery has been developed over roughly the same time period.

8. Tracy (1998, 194).

9. Sunstein (1998, 209–210).

10. As she puts it in unforgettable form, "Banality can also mark the experience of deeply felt emotion, as in the case of 'I love you,' 'Did you come?' or 'Osay, can you see?'" (Berlant 2002, 113). Nonetheless, as she argues, the trauma of difference lies behind and must disrupt the apparent banality of citizenship based on this idea of intimacy.

11. King and Harrington Meyer (1997).

12. Minnow (1990, 269).

13. See Rosen (2001) for the U.S. version of this strong fear and its basis.

14. See McKinnon (1995, 118), who sees little or nothing for women in privacy, which she describes as "the affirmative triumph of the state's abdication of women."

15. As in all the ethnographic data in this book, names and uniquely identifying details in these examples have been changed, but other treatment details and quotations are accurate to the limits of my recording ability.

16. In March 2003, New Jersey Family Court Judge James Farber ruled in a case like this where there was an ongoing pregnancy that both women could be on the birth certificate of the baby, which was due in May 2003. Recent cases of contested parentage in pregnancies that were induced by assisted reproductive technology have found in favor of procreative intent and in favor of genetic or gamete parentage (see chapter 5 and table 5.1). In many states, rules such as the Uniform Parentage Act in California recognize the natural motherhood of birth mothers. These rules were enacted to protect children with absent fathers from the legal repercussions of so-called illegitimacy. Taken together, this hodgepodge of precedent augurs fairly well for this kind of parenting.

17. Bush (1945). See also Kleinman (1995).

18. Proctor (1988) and Lifton (1986).

19. Rothman (1991, 63).

20. Rothman (1990, 1991).

21. Beecher (1966).

22. See Rosenberg (1999) for a compelling call to historicize the bioethical enterprise and for an account of the historiography of American bioethics that starts at Nuremberg and with the individual rights movements of the 1960s. It is my sense that a more "Mertonian" story that links the two through the parallelisms between autonomous science and liberal democracy has yet to be fully told.

23. Pippin (1996).

24. Casper (1998, 139–143).

25. See Skocpol (1992).

26. Jones (1996, 1092). The title of Jones's piece, "The Time Has Come," was taken from John Rock's book of the same name (Rock 1963). Rock is credited with being one of the "fathers" of the birth-control pill, with Gregory Pincus, Celso-Ramon Garcia, and Min-Chueh Chang, and with Miriam Menkin published the first scientific paper that documented human in vitro fertilization (Menkin and Rock 1948).

27. Duka and DeCherney (1994, 187–196).

28. Catholic Church, Congregation for the Doctrine of the Faith (1987).

29. National Institutes of Health (1994).

30. Katz (1995); compare Marcus-Steiff (1991).

31. For details of this regulatory pressure in the United States, see Bonnicksen (1991). For European Union comparisons, see, e.g., Gunning and English (1994) and Gunning (2000).

32. Duka and DeCherney (1994, 180).

33. Steinberg et al. (1998) and Garcia (1998).

34. American Society for Reproductive Medicine (2002a).

35. American Society for Reproductive Medicine (2002).

36. Clarke (1998). For a large range of histories of sex and contraception in twentieth-century America, see, e.g., Fausto-Sterling (2000, 170–194), Gordon (1977), and Reed (1983).

37. The first pregnancies were reported after the administration of human FSH that was extracted from the pituitary gland in 1960, long before the development of IVF (Gemzell et al. 1960).

38. Shoham and Insler (1996).

39. Out et al. (1997).

40. See table 7.4 for more details on Serono and Organon.

41. The comparative details between u-FSH and r-FSH are adapted from Prevost (1998).

42. See Landecker (2000).

43. See Prevost (1998).

44. Porchet et al. (1998).

45. Neuman et al. (1994).

46. Collins (1995).

47. VanderLaan et al. (1998).

48. Cain (1993).

49. Griffin and Panak (1998).

50. VanderLaan et al. (1998, 1205).

51. Hurst et al. (1997).

52. Quote from the American Infertility Association's Web site on ARTs, "Reimbursement of ART Procedures," provided by Brian Allen, financial counselor ⟨http://www.americaninfertility.org/⟩.

53. Garcia (1998).

54. RESOLVE "Fast Facts," ⟨http://www.resolve.org⟩.

55. See Hogle (1996) and Radin (1996).

56. See Epstein (1996) and Bové and Dufour (2001).

57. See Bretell and Sargent (2001), Jetter et al. (1997), Landes (1998), Radcliffe and Westwood (1993), and Yuval-Davis and Werbner (1999).

58. To the best of my knowledge, RESOLVE awaits comprehensive analysis by an academic historian or social movements specialist. It has not received the same kind of attention from feminist historians and sociologists as *Our Bodies, Ourselves,* in large because of its location in these multiple private spheres.

59. Sandelowski (1993, 46–54) and Cussins (1998).

60. Pfeffer (1987) and Franklin (1990).

61. Callon (1998a, 1998b). Monetarism, for example, proposes stabilizing the monetary system at low levels of inflation and consistent exchange rates as the means of ensuring economic growth.

62. For example, such questions as to what extent do such competencies inhere in individuals rather than in networks or social context, or to what extent can agents be calculative in the face of risk or uncertainty.

63. Berlant (1997, 1–54).

64. Compare Galison (1997).

65. Rapp (1987).

66. "People are at home when they can jump across different domains of experience without feeling they have left sense behind" (Strathern 2003, 170).

67. Judith Butler, in a critique of Paul Gilroy's *The Black Atlantic* (Butler 1995, 128).

Chapter 8

1. Rifkin (1999). See also Nelkin and Lindee (1995).

2. Keller (2000, 132).

3. Rabinow (1999).

4. Kleinman, Fox, and Brandt (1999) and Rosenberg (1999, 2002).

5. Rheinberger (2000, 29).

6. Franklin and Lock (2003).

7. Berg and Mol (1998) and Viveiros de Castro (1998).

8. Jasanoff (forthcoming).

9. Beck (1992, 1995).

10. Appadurai (1996), Castells (2000), Harvey (2001), Lash and Urry (1987), Law (1994), Mirowski and Sent (2001), and Poster (1990).

11. Strathern (1992).

12. Haraway (1991, 2003).

13. Verran (2001).

14. Fortun and Mendelsohn (1999).

15. Latour (2002, 34).

16. Rabinow (2000).

17. Rabinow (1999, 12).

18. See Lock et al. (2000) and Rheinberger (1997, 2000).

19. I am using the expression "mode of reproduction" in a manner loosely indebted to the Marxist notion of "mode of production," which is typically defined by social scientists as including productive forces (such things as labor, raw materials, and instrumentation), the relations of production (the structural relations among groups of people—and I would include things—in relation to production), and the forms of life and consciousness that are possible under a given mode of production. The key insight is that the economic system, social order, and consciousness are not independent but in some way reflect one another.

20. In *The Biotech Century,* Jeremy Rifkin (1999, 8–9) describes seven "strands that make up the operational matrix of the Biotech Century." These strands —gene pool as raw resource, gene patenting, rise of a global life-science industry, gene mapping, a new sociobiology, the fusing of computational and genetic technologies, and a post-Darwinian cosmological narrative—are all different from the dimensions that I isolate here.

21. Anecdotes abound of informal methods of artificial insemination (other than simply having sexual intercourse with someone other than one's legal partner) stretching back to the distant past. Historian and political economist Naomi Pfeffer (1993) has shown that medical infertility procedures date back to gynecological surgical procedures in the early twentieth century and that artificial insemination began to be performed regularly by doctors in the United Kingdom in the 1930s.

22. See Casper (1998), Daniels (1993), and Rapp (1999).

23. For a summary of the key debates as well as many of the classic papers on socialist feminism, see Hansen and Philipson (1990).

24. See Thompson (1982).

25. See Lie (2004).

26. See Rabinow (1992), Rapp (1999), Epstein (1995), and Petryna (2002).

27. Knorr Cetina (1999, 153–155); see also Clarke and Fujimura (1992).

28. Haraway (1991, 5) and Franklin (2003).

29. Franklin and Ragoné (1998).

30. Foucault (1990, 140–141).

31. See Callon (1998a, 1998b).

32. Scherer (1990).

33. Folbre (1994) and Michel (1999).

34. Rapp (1978).

35. Colen (1995).

36. Rubin (1990) and McKinnon (2001).

37. Strathern (1992, 31–43).

38. Marx (1972, 1:171) and Engels (1972, 71–72).

39. Zelizer (1985).

40. "Chains of custody" is an expression that I use throughout this conclusion. I use it with the meaning given it by Michael Lynch, who himself draws on Derrida's notion of the archive: "During the [O. J.] Simpson [murder] case, the well-resourced defense team attacked the chain of custody wherever it could, and with notable success. In most other U.S. and U.K. cases in which forensic DNA profiling has been used, such chains have resisted attack and held together well enough to convince juries that the samples traveled intact through their chains and gave definitive evidence of criminal identity" (Lynch 1999, 81; Derrida 1996). This use is augmented by the legal and other meanings of custody that are associated with parentage and with Veena Das's arguments about the role of power differentials and the state in the "networks" and "translations" that decide organ custody in transplant cases (Das 2000).

41. This alienation of kinship does not involve a loss of materiality or embodiment. All these procedures, even when they are contested and including procedures that rely on genetic understandings of parenthood to preserve procreational intent, are essentially embodied at all stages.

42. Scheper-Hughes and Wacquant (2003).

43. See, e.g., Grayson (1998), Laqueur (2000), and Hartouni (1997). Also see tables 5.1 and 5.2. Custody disputes arise in the case of surrogacy and not in the case of egg donations or embryo donations because of an experiential mismatch between procreational intent and procreational actuality in the case of surrogacy. There also are well-established precedents if family law that involve after-birth legitimacy and custody determinations. In other words, surrogacy cases can easily be subsumed to previously existing disputes, so they have been able to be brought to court, thereby contributing to changing the law in this area.

44. See Thompson (2004). In my research on race in the field of ARTs and third-party reproduction, I have found that in fact a few sperm donors do advertise Native ancestry in their own ethnoracial identification.

45. In the United States, commercial egg and sperm donation and gestational surrogacy are all widely available (see table 7.3). In some other countries, gamete donation and gestational surrogacy are illegal, or commercial variants thereof are illegal. See, e.g., Kahn (2000) and Melhuus Howell (2003).

46. Vieira (personal communication), Das (2000), Landecker (2000), Lynch (1999), and Rabinow (1996a, 1996b, 1999).

47. Haraway (1997, 143).

48. See Fortun (2001).

49. Uniquely among tissue transplants in humans, pregnancies usually do not trigger severe rejection responses (Rh-negative mothers with Rh-positive fetuses are an exception), and so the biological criteria are nothing like as dominant as they are for other kinds of tissue donation.

50. Schmidt and Moore (1998).

51. The majority of the well-known high-order (especially quads or greater) births result from infertile women who have taken infertility drugs to induce ovulation and not from ARTs.

52. Yaron et al. (1998).

53. ⟨http://www.asrm.org/Media/Practice/NoEmbryosTransferred.pdf⟩.

54. ⟨http://www.sart.org/⟩.

55. Hoffman et al. (2003).

56. The U.K. Human Fertilisation and Embryology Act (HFEA) came into effect on August 1, 1990. One of its provisions was that no embryos could be frozen in infertility clinics for longer than five years without being used (with the patients' consent, that could be extended). On July 31, 1996, all unclaimed and unwanted human embryos that were frozen when the act became law and that were still frozen—approximately 4,000 embryos from 900 couples—were thawed and discarded.

57. A *New York Times* editorial, "400,000 Embryos and Counting," of May 15, 2003, A34, called for a U.K.-style limit after which frozen embryos would be thawed and discarded, but this is to misunderstand U.S. ARTs.

58. American Infertility Association Web Page, "The Frozen Embryo Dilemma: A Matter of Privacy, Responsibility, and Choice," May 2003. ⟨http://www.americaninfertility.org/⟩.

59. Burawoy (2000).

60. Kleinman (1995).

61. Charmaz (1991) and Sandelowski (1993, 181).

62. See Lock (2000, 2002) and Franklin and Lock (2003).

63. See Yanagisako and Delaney (1995) and Dumit (2000).

64. See Squier (1994).

65. The United Kingdom has endorsed this train of thought, while the United States has banned cloning but is stumbling to implement the bizarre compromise that permits federal funding of public and private embryonic stem-cell research from existing stem-cell lines. The U.S. solution of July 2001 was hailed

as a masterful political compromise even though it ruled out federal funding of the histocompatibility promise of therapeutic cloning.

66. See Butler (2002).

67. See Laqueur (2000).

68. Wilmut, Cambell, and Tudge (2000, 282).

69. See Dumit (2000), Rosenberg (1999), and Finkler (2000) on medicalization.

70. Dolgin (1995).

Glossary

adhesions The bands of fibrous scar tissue that can result from surgery or infection. Adhesions may bind the pelvic organs together and render them nonfunctional.

American Society for Reproductive Medicine (ASRM) A nonprofit organization that was founded in 1944 as the American Fertility Society (AFS) and now includes more than 10,000 gynecologists, obstetricians, urologists, reproductive endocrinologists, and related professionals who specialize in reproductive medicine.

andrology The science of diseases that are peculiar to the male sex, particularly infertility, spermatogenesis, and sexual dysfunction.

anejaculation The absence of regular voluntary ejaculations.

artificial (assisted) insemination by donor (AID) The process of placing sperm from a donor (a man who is not a woman's sexual partner) into a woman's reproductive tract for the purpose of producing a pregnancy. Also called *donor insemination (DI).*

artificial (assisted) insemination by husband (AIH) The process of placing sperm from the woman's partner into a woman's reproductive tract for the purpose of producing a pregnancy.

aspiration The application of light suction to the ovarian follicle to remove the egg.

assisted reproductive technology (ART) A treatment for infertility that involves the laboratory handling of eggs, sperm, or embryos. Some examples of ARTs are **in vitro fertilization (IVF)**, **gamete intrafallopian transfer (GIFT)**, pronuclear stage tubal transfer (PROST), tubal embryo transfer (TET), zygote intrafallopian transfer (ZIFT), and micromanipulation techniques, such as **intracytoplasmic sperm injection (ICSI)**.

azoospermia The absence of sperm in the ejaculate.

capacitation The alteration of sperm during their passage through the female reproductive tract that gives them the capacity to penetrate and fertilize the ovum.

cervix The lower section of the uterus that protrudes into the vagina and dilates during labor to allow the passage of a baby.

cloning The development of an organism bypassing sexual reproduction. It offers the possibility of creating genetically identical individuals. *Reproductive cloning* refers to cloning as a means of having a child. In *human therapeutic cloning,* a cloned embryo is used for the purposes of growing particular tissue types that were genetically identical to the person whose DNA was used and not for the purposes of reproducing an entire individual.

cryopreservation The freezing at a very low temperature, such as in liquid nitrogen (−196°C), of embryos (to keep them viable and store them for future transfer into a uterus) or of sperm (to keep them viable for future insemination or assisted reproductive technology procedures). At present, eggs cannot be cryopreserved, although ovarian tissue containing immature follicles can be cryopreserved for later in vitro maturation.

diethylstilbestrol (DES) A synthetic estrogen that is used occasionally as a morning-after intercourse pill. Formerly thought to prevent miscarriage, it caused cancer and fertility problems in some of the offspring of women who took it during pregnancy.

donor eggs The eggs that are retrieved from the ovaries of one woman and given to an infertile woman to be used in an assisted reproductive technology procedure.

donor insemination (DI) See **artificial (assisted) insemination by donor**.

donor sperm The sperm that are contributed by a donor to an infertile couple to be used in donor insemination.

ectopic pregnancy A pregnancy in which the fertilized egg implants anywhere but in the uterine cavity (usually in the fallopian tube, the ovary, or the abdominal cavity).

egg donation The process of fertilizing eggs from a donor with the infertile woman's partner's sperm in a laboratory dish and transferring the resulting embryos to the infertile woman's uterus for the purpose of producing a pregnancy. The infertile woman will not be genetically related to the child although she will be the birth mother. The male partner will be genetically related to the child.

embryo transfer The introduction of an embryo into a woman's uterus after in vitro fertilization.

endometrium The lining of the uterus that is shed each month during menstruation. During pregnancy, the endometrium provides a site for the implantation of the fertilized egg.

endometriosis The presence of endometrial tissue (the normal uterine lining) in abnormal locations such as the tubes, ovaries, and peritoneal cavity, often causing painful menstruation and infertility.

epididymus A tightly coiled system of tiny tubing where sperm collect after leaving the testis. Sperm continue to mature and gain the ability to move as they are pushed through the epididymus, which covers the top and back side of each testis.

estradiol The main estrogen hormone that is produced by the ovaries. Estradiol is released during the development of the egg.

estrogen A class of hormones that are produced mainly by the ovaries from the onset of puberty and continuing until menopause and that are responsible for the development of secondary sexual characteristics in women.

fallopian tubes A pair of tubes, one attached to each side of the uterus, where sperm and egg meet in normal conception.

follicle The cyst, located just below the surface of the ovary, in which an egg matures and is later released.

follicle-stimulating hormone (FSH) The pituitary hormone that is responsible for stimulating the follicle cells around the egg. Used in **in vitro fertilization (IVF)**. Can be urinary or recombinant in origin. Fertinex, Follistim, Gonal-F, and Repronex are brand names.

follicular phase The first half of the menstrual cycle when ovarian follicle development takes place.

gamete intrafallopian transfer (GIFT) A procedure where clomiphene citrate, **human menopausal gonadotropins (hMG)**, or **luteinizing hormone (LH)** is given to stimulate ovulation. The eggs are removed from the ovaries via laparoscopy or transvaginal ultrasound aspiration and immediately mixed with washed sperm. The sperm-egg mixture is then transferred by **laparoscopy** or **hysteroscopy** into the **fallopian tubes**, where fertilization may then take place.

gestational surrogate or gestational carrier A woman who carries and delivers a baby for an infertile couple and is not genetically related to the child. Eggs are removed from the infertile woman and fertilized with her partner's sperm, and the resulting embryos are placed into the uterus of a gestational surrogate, who will carry the baby to delivery.

gonadotropin-releasing hormone (GnRH) A hormone that is secreted by the hypothalamus and that prompts the pituitary gland to release **follicle-stimulating hormone (FSH)** and **luteinizing hormone (LH)** into the bloodstream. Lupron or Lupron Depot is the brand name.

hamster (Ham) test (sperm-penetration assay) Ah assay where sperm are added to defrosted hamster eggs in a laboratory dish to see if the sperm penetrate the eggs. Thought to be evidence of the sperm's penetrating ability with human eggs.

hirsutism The growth of long, coarse hair on the face, chest, upper arms, and upper legs of women in a pattern similar to that of men.

human chorionic gonadotropin (hCG) A hormone that is secreted by the placenta during pregnancy that prolongs the life of corpus luteum progesterone production and thus preserves the pregnancy. This hormone is what is detected in positive pregnancy tests. It is administered during **in vitro fertilization (IVF)** treatment cycles and is sometimes given to men to treat cryptorchidism (undescended testes). Profasi is the brand name.

human menopausal gonadotropins (hMG) A drug that contains **follicle-stimulating hormone (FSH)** and **luteinizing hormone (LH)**. It is derived from the urine of postmenopausal women. It is used in the treatment of both male and female infertility and to stimulate the development of multiple oocytes in ovulatory patients participating in **in vitro fertilization (IVF)** programs. Pergonal and Humegon are the brand names for preparations that are 50 percent FSH and 50 percent LH.

hyperandrogenism Elevated levels of androgens that are associated in women with polycystic ovarian disease.

hyperstimulation syndrome A syndrome that is caused in some women by a reaction to the high doses of drugs that are given to promote multiple oocyte development. Symptoms may include ovarian enlargement, gastrointestinal symptoms, abdominal distension, and weight gain. Severe cases can be complicated with cardiovascular, pulmonary, and electrolyte disturbances, which can require hospitalization and in some documented cases even lead to death.

hysterosalpingogram An x-ray study in which a contrast dye is injected into the uterus to show the delineation of the body of the uterus and the patency (that they are open) of the fallopian tubes.

hysteroscopy The insertion of a thin, lighted viewing instrument with a telescopic lens into the vagina to examine the inside of the uterus and fallopian tubes.

idiopathic (unexplained) infertility A diagnostic category that is used when no reason has been found by physicians for a couple's infertility.

implantation The embedding of the embryo in the endometrium (lining) of the uterus.

intracytoplasmic sperm injection (ICSI) A micromanipulation technique whereby a single sperm can be injected directly into an egg in an attempt to achieve fertilization.

insemination An office procedure in which sperm are placed into a woman's uterus or cervix via a syringe to achieve pregnancy. Can be **artificial (assisted) insemination by husband (AIH)** or **artificial (assisted) insemination by donor (AID)**.

in vitro fertilization (IVF) A process in which eggs are surgically removed from a woman's ovaries and combined with sperm in a laboratory dish to bring about fertilization. The fertilized eggs are left to divide for about two days in a protected environment. The resulting embryos are transferred to the uterus of the woman being treated for infertility or can be frozen, donated, or discarded. Also called *test-tube baby.*

laparoscopy The insertion of a long, thin lighted, telescope-like instrument called a *laparoscope* into the abdomen through an incision in the navel to visually inspect the organs in the abdominal cavity (which is inflated with carbon dioxide). Other small incisions may also be made, and additional instruments may be inserted to faciliate diagnosis and allow surgical correction of pelvic abnormalities. The laparoscope can be used both to diagnose and operate.

lesbian IVF The use of in vitro fertilization by lesbian couples wishing to coparent and to be related biologically to the resulting child. One woman provides the egg(s), and the other gestates the embryo(s) and gives birth. Also a possible means of making it possible for both women to be listed on the birth certificate.

luteal phase The second half of the ovarian cycle after ovulation when the corpus luteum secretes large amounts of progesterone.

luteinizing hormone (LH) The hormone that triggers ovulation and stimulates progesterone secretion. Present in Humegon and Pergonal, which, with FSH, are sometimes used to stimulate the development of multiple follicles for **in vitro fertilization (IVF)**.

micromanipulation The manipulation of eggs, sperm, and embryos with the aid of a microscope. Micromanipulation techniques to facilitate fertilization include injection of a single sperm into an egg, or **intracytoplasmic sperm injection (ICSI)**; injection of sperm under the membrane surrounding the egg (subzonal injection); or drilling of a small hole in the membrane to allow the sperm easier access to the egg (partial zona dissection or drilling). Embryos are micromanipulated to facilitate pregnancy. The two most common embryo micromanipulation procedures are assisted (embryo) hatching, a procedure in which the **zona pellucida** (outer covering) of the embryo is partially disrupted, usually by injection of acid, to facilitate embryo implantation and pregnancy, and embryo biopsy, in which one or more cells are removed from the embryo for prenatal diagnosis.

morphology (semen sample) The shape of individual sperm as seen under a microscope. At least 35 percent of the sperm in a normal semen sample should have oval heads and slightly curving tails.

motility (semen sample) The percentage of all moving sperm in a semen sample. Normally 50 percent or more are moving rapidly.

necrospermia Semen containing a high proportion of dead sperm.

oligospermia An abnormally low number of sperm in the ejaculate of the male.

oocyte, ovum An egg.

oocyte retrieval A surgical procedure, which can be done in the office with pain killers but is usually done under general anesthesia, to collect the eggs that are contained within the ovarian follicles. A needle is inserted into the follicle, and the fluid and egg are aspirated into the needle and then placed into a culture dish. It is sometimes done transabdominally (the needle is passed through the abdominal wall) or even transurethrally but is usually done transvaginally (the needle is passed through the vaginal wall to reach the ovary).

ooplasm and ooplasm transfer The nonnuclear material in an oocyte and the transfer of nonnuclear material by micromanipulation to another oocyte for the purpose of improving the "quality" (and thus likelihood of fertilization and implantation) of the recipient oocyte.

ovarian reserve An estimate of a woman's remaining fertility potential. A woman's age is a strong predictor of her response to **assisted reproductive technologies (ARTs)**. Other tests, such as antral follicle counts and day 3 **follicle-stimulating hormone (FSH)** levels, are also used to estimate the likelihood that a woman will become pregnant. Antral follicle counts involve counting small follicles that are present on the patient's ovaries. Day 3 FSH levels are used, often in conjunction with estradiol levels, to see if a woman has a hormonal profile that approximates menopause. It is worth noting that the concept of ovarian reserve was developed with the assumption that women have a finite number of eggs of ever diminishing quantity. The assumption of a finite number of eggs has lately been called into question, although these tests of "egg quality" do correlate with a woman's chances of becoming pregnant with ARTs.

ovulation induction and ovarian stimulation The use of hormone therapy—including **luteinizing hormones (LH)**, **follicle-stimulating hormone (FSH)**, **human menopausal gonadotropin (hMG)**, and **human chorionic gonadotropin (hCG)**—to stimulate oocyte development and ovulation (release of the ripened egg in midcycle).

parthenogenesis The development of an egg into a complete organism without fertilization from sperm. It occurs in some species but not yet in humans.

pelvic inflammatory disease (PID) Any inflammatory disease of the pelvis. PID often leaves scar tissue that can impair fertility.

polycystic ovarian disease (PCOD) A condition in which the ovaries contain many cystic follicles that are associated with chronic anovulation and increased secretion of androgens. Symptoms include irregular or absent menstrual periods and usually obesity and excessive growth of central body hair.

postcoital test A diagnostic test for infertility in which vaginal and cervical secretions are obtained following intercourse and then analyzed under a microscope. Normal test results show large numbers of motile sperm.

preimplantation diagnosis The screening of a zygote or embryo by examining its chromosomes after biopsying a single cell. Can be used to screen for sex-linked diseases as well as a growing number of other genetic anomalies that would otherwise be fatal or cause a miscarriage. Only apparently unaffected embryos are transferred to the uterus for implantation.

premature ovarian failure (early menopause) A condition that results when ovaries stop producing eggs and menstruation ceases before the age of forty.

progesterone A female hormone that is secreted by the ovaries during the second half of the menstrual cycle. It prepares the lining of the uterus for implantation of a fertilized egg.

reproductive endocrinologist A gynecologist who has a subspecialty fellowship and is trained in women's infertility.

RESOLVE Inc. A U.S. nationwide organization that was founded in 1974 that provides referrals, support groups, and medical information to people who are coping with infertility and that engages in advocacy to support family building.

Society for Assisted Reproductive Technologies (SART) registry An ongoing collection of IVF results from participating clinics that was developed and is maintained by SART, which is an affiliate of the **American Society for Reproductive Medicine (ASRM)**, in conjunction with the Centers for Disease Control and Prevention (CDC) and RESOLVE. Carries out the success rates and certification mandate of the Wyden law, the Fertility Clinic Success Rate and Certification Act of 1992.

semen analysis The study of fresh ejaculate under the microscope to count the number of sperm and to check the shape, size, and motility of the sperm. A normal count is usually 20 million or more sperm per milliliter.

sperm washing A technique that separates the sperm from the seminal fluid.

surrogacy The carrying of an infertile couple's baby by another woman. There are several types of surrogacy. In *traditional, or conventional, surrogacy*, the woman who is to carry the pregnancy conceives through insemination with the infertile woman's partner's sperm (or donor sperm) and then carries the pregnancy, resulting in a child that is the genetic offspring of the surrogate and the infertile woman's partner (or donor). *Gestational surrogates* conceive using an embryo or embryos from the infertile woman and her partner. These embryos are placed into the surrogate's uterus, and she then carries the baby until delivery. The resulting child is the genetic offspring of the infertile woman and her partner.

testosterone The hormone that in males is produced in the testes and is necessary for sperm production.

ultrasound High-frequency sound waves that produce a picture image of internal organs on a screen. Can be done abdominally or vaginally. An ultrasound image is used to visualize the growth of ovarian follicles during stimulation and is used during surgery for egg retrieval.

urologist A physician who specializes in diseases of the urinary tract in men and women and the genital organs in men.

varicocele A varicose vein of the testicles, sometimes a cause of male infertility.

vas deferens The two muscular tubes that carry sperm from the epididymus to the urethra.

vasectomy Surgery to excise part of the **vas deferens** to sterilize a man.

zona pellucida The outer covering of the ovum that the sperm must penetrate before fertilization can occur.

References

Adams, Alice. 1994. *Reproducing the Womb: Images of Childbirth in Science, Feminist Theory, and Literature.* Ithaca: Cornell University Press.

Akrich, Madeleine, and Bernicke Pasveer. 1995. *De la conception à la naissance: Comparison France/Pays-Bas des réseaux et des pratiques obstétriques.* Paris: École Nationale Supèrieure des Mines.

Allen, A. L. 1990. "Surrogacy, Slavery, and the Ownership of Life." *Harvard Journal of Law and Public Policy* 13: 139–149.

American Society for Reproductive Medicine (ASRM). 2002a. "More Than 35,000 Babies Born in 2000 as a Result of ART Procedures." *ASRM Bulletin* 4, December 20, 47.

American Society for Reproductive Medicine (ASRM). 2002b. "Race Does Not Affect IVF Outcomes." *Highlights from ASRM 2002,* Press Release, October 15.

Andrews, Lori B. 1999. *The Clone Age: Adventures in the New World of Reproductive Technology.* New York: Holt.

Andrews, L., and L. Douglass. 1991. "Alternative Reproduction." *Southern California Law Review* 65: 623.

Anspach, Reneé. 1989. "From Principles to Practice: Life-and-Death Decisions in the Intensive-Care Nursery." In L. Whiteford and M. Poland, eds., *New Approaches to Human Reproduction.* Boulder: Westview Press.

Appadurai, Arjun. 1996. *Modernity at Large.* Minneapolis: University of Minnesota Press.

Arditti, Rita, Renate Duelli Klein, and Shelley Minden, eds. 1989. *Test-Tube Women: What Future for Motherhood?* London: Pandora Press. Originally published in 1984.

Arms, Suzanne. 1975. *Immaculate Deception: A New Look at Women and Childbirth in America.* Boston: Houghton Mifflin.

Awkward, Michael. 1995. *Negotiating Difference: Race, Gender, and the Politics of Positionality.* Chicago: University of Chicago Press.

Baier, Annette. 1985. *Postures of the Mind: Essays on Mind and Morals.* Minneapolis: University of Minnesota Press.

Bailey, Marlon. 2005. "Queering the African Diaspora: On Ballroom Cultural Performance and Black Queer 'World-Making.'" Ph.D. dissertation, University of California, Berkeley.

Barad, Karen. 1999. "Agential Realism: Feminist Interventions in Understanding Scientific Practices." In Biagioli (1999, 1–11). Originally published in 1998.

Barkan, Elazar. 1992. *The Retreat of Scientific Racism: Changing Concepts of Race in Britain and the United States between the World Wars.* Cambridge: Cambridge University Press.

Barnes, Barry. 1971. "Making Out in Industrial Research." *STS* 1: 157–175.

Barnes, Barry. 1985. *About Science.* Oxford: Blackwell.

Barnes, Barry, and David Edge, eds. 1982. *Science in Context: Readings in the Sociology of Science.* Milton Keynes, UK: Open University Press.

Barnes, Barry, and Steven Shapin, eds. 1979. *Natural Order: Historical Studies in Scientific Culture.* London: Sage.

Bartels, Dianne, Reinhard Priester, Dorothy Wawter, and Arthur Caplan, eds. 1990. *Beyond Baby M.: Ethical Issues in New Reproductive Techniques.* Clifton, NJ: Humana Press.

Bartholet, Elizabeth. 1999. *Family Bonds: Adoption, Infertility, and the New World of Child Production.* Boston: Beacon Press.

Baruch, Elaine, Amadeo D'Adamo, and Joni Seager, eds. 1988. *Embryos, Ethics, and Women's Rights: Exploring the New Reproductive Technologies.* New York: Harrington Park Press.

Beck, Ulrich. 1992. *Risk Society: Towards a New Modernity.* Trans. Mark Ritter. London: Sage.

Beck, Ulrich. 1995. *Ecological Politics in the Age of Risk.* Cambridge: Polity Press.

Becker, Gay. 1994. "Metaphors in Disrupted Lives: Infertility and Cultural Constructions of Continuity." *Medical Anthopology Quarterly* 8: 383–410.

Becker, Gay. 2000. *The Elusive Embryo: How Women and Men Approach New Reproductive Technologies.* Berkeley: University of California Press.

Becker, Gay. 2002. "Deciding Whether to Tell Children about Donor Insemination: An Unresolved Question in the United States." In Inhorn and Van Balen (2002, 119–133).

Beecher, Henry. 1966. "Ethics and Clinical Research." *New England Journal of Medicine* 74: 1354–1360.

Bell, David, and Barbara Kennedy, eds. 2000. *The Cybercultures Reader.* New York: Routledge.

Benhabib, Seyla, Judith Butler, Drucilla Cornell, and Nancy Fraser. 1995. *Feminist Contentions: A Philosophical Exchange.* New York: Routledge.

Berg, Marc. 1997. *Rationalizing Medical Work.* Cambridge, MA: MIT Press.

Berg, Marc, and Annemarie Mol, eds. 1998. *Differences in Medicine: Unraveling Practices, Techniques and Bodies.* London: Libbey.

Berlant, Lauren. 1997. *The Queen of America Goes to Washington City: Essays on Sex and Citizenship.* Durham: Duke University Press.

Berlant, Lauren. 2002. "The Subject of True Feeling: Pain, Privacy, and Politics." In Brown and Halley (2002, 105–133).

Bernard, Jessie. 1974. *The Future of Motherhood.* New York: Dial Press.

Biagioli, Mario. 1990. "Galileo the Emblem Maker." *ISIS* 81.

Biagioli, Mario, ed. 1999. *The Science Studies Reader.* New York: Routledge.

Bijker, Wiebe, Thomas Hughes, and Trevor Pinch. 1987. *The Social Construction of Technological Systems: New Directions in the Sociology and History of Technology.* Cambridge, MA: MIT Press.

Birke, Lynda, Susan Himmelweit, and Gail Vines. 1990. *Tomorrow's Child: Reproductive Technologies in the 1990s.* London: Virago Press.

Bledsoe, Caroline. 1995. "Marginal Members: Children of Previous Unions in Mende Households in Sierra Leone." In Greenhalgh (1995, 130–153).

Bleier, Ruth. 1984. *Science and Gender: A Critique of Biology and Its Theories on Women.* New York: Pergammon Press.

Bloor, David. 1991. *Knowledge and Social Imagery* (2nd ed.). Chicago: Chicago University Press. Originally published in 1976.

Bonnicksen, Andrea. 1991. *In Vitro Fertilization: Building Policy from Laboratories to Legislatures.* New York: Columbia University Press.

Bordo, Susan. 1986. "The Cartesian Masculinization of Thought." *Signs* 11: 439–456.

Bové, José, and François Dufour. 2001. *The Word Is Not for Sale: Farmers against Junk Food.* London: Verso.

Braidotti, Rosi. 1994. *Nomadic Subjects: Embodiment and Sexual Difference in Contemporary Feminist Theory.* New York: Columbia University Press.

Braidotti, Rosi, and Nina Lykke, eds. 1996. *Between Monsters, Goddesses and Cyborgs: Feminist Confrontations with Science, Medicine and Cyberspace.* London: Zed Books.

Brandt, Allan. 1985. *No Magic Bullet: A Social History of Venereal Disease in the United States since 1880.* Oxford: Oxford University Press.

Braverman, Harry. 1974. *Labor and Monopoly Capitalism.* New York: Monthly Review.

Bretell, Caroline, and Carolyn Sargent, eds. 2001. *Gender in Cross-Cultural Perspective.* Upper Saddle River, NJ: Prentice-Hall.

Brinsden, Peter, Susan Avery, Samuel Marcus, and Michael Macnamee. 1997. "Transrectal Electroejaculation Combined with In-Vitro Fertilization: Effective Treatment of Anejaculatory Infertility Due to Spinal Cord Injury." *Human Reproduction* 12: 2687–2692.

Brodwin, Paul, ed. 2000. *Biotechnology and Culture: Bodies, Anxieties, Ethics.* Bloomington: Indiana University Press.

Brown, Wendy, and Janet Halley, eds. 2002. *Left Legalism, Left Critique.* Durham: Duke University Press.

Burawoy, Michael, ed. 2000. *Global Ethnography: Forces, Connections, and Imaginations in a Postmodern World.* Berkeley: University of California Press.

Bush, Vannevar. 1945. *Science, the Endless Frontier: A Report to the President by Vannevar Bush Director of the Office of Scientific Research and Development.* Washington, DC: U.S. Government Printing Office.

Butler, Judith. 1990. *Gender Trouble: Feminism and the Subversion of Identity.* New York: Routledge.

Butler, Judith. 1993. *Bodies That Matter: On the Discursive Limits of "Sex."* New York: Routledge.

Butler, Judith. 1995. "For a Careful Reading." In Benhabib et al. (1995, 127–143).

Butler, Judith. 1999. *Gender Trouble: Feminism and the Subversion of Identity.* New York: Routledge.

Butler, Judith. 2002. "Is Kinship Always Already Heterosexual?" *Differences* 13(1): 1–18.

Butler, Judith, and Joan Scott, eds. 1992. *Feminists Theorize the Political.* New York: Routledge.

Cain, J. M. 1993. "Is Deception for Reimbursement for Obstetrics and Gynecology Justified?" *Obstetrics and Gynecology* 82: 475–478.

Callon, Michel. 1986. "Some Elements of a Sociology of Translation: Domestication of the Scallops and Fishermen of St. Brieuc Bay." In Law (1986, 196–233).

Callon, Michel. 1998a. "An Essay on Framing and Overflowing: Economic Externalities Revisited by Sociology." In Callon (1998b, 244–269).

Callon, Michel, ed. 1998b. *The Laws of the Market.* Oxford: Blackwell.

Cambrosio, Alberto, Margaret Lock, Allan Young, eds. 2000. *Living and Working with the New Medical Technologies.* Cambridge: Cambridge University Press.

Canguilhem, Georges. 1991. *The Normal and the Pathological* (115–179). New York: Zone Books. Originally published in 1966.

Cannell, Fenella. 1990. "Concepts of Parenthood: The Warnock Report, the Gillick Debate, and Modern Myths." *American Ethnologist* 17: 667–688.

Carp, Wayne, ed. 2003. *Adoption in America: Historical Perspectives.* Ann Arbor: University of Michigan Press.

Carrithers, M., S. Collins, and S. Lukes, eds. 1985. *The Category of the Person: Anthropology, Philosophy, History.* Cambridge: Cambridge University Press.

Cartwright, Lisa. 1992. "Women, X-rays, and the Public Culture of Prophylactic Imaging." *Camera Obscura* 29: 19–56.

Cartwright, Nancy. 1983. *How the Laws of Physics Lie.* Oxford: Clarendon.

Casper, Monica. 1994. "Reframing and Grounding 'Non-human' Agency: What Makes a Fetus an Agent?" *American Behavioral Scientist* 37(6): 839–857.

Casper, Monica. 1998. *The Making of the Unborn Patient: A Social Anatomy of Fetal Surgery.* New Brunswick, NJ: Rutgers University Press.

Castells, Manuel. 2000. *The Rise of the Network Society.* Oxford: Blackwell.

Catholic Church, Congregation for the Doctrine of the Faith. 1987. "Instruction on Respect for Human Life in Its Origin and on the Dignity of Procreation." In Pellegrino, Harvey, and Langan (1990, 1–41).

Chang, Grace, and Linda Rennie Forcey, eds. 1994. *Mothering: Ideology, Experience, and Agency.* New York: Routledge.

Charmaz, Kathy. 1991. *Good Days, Bad Days: The Self in Chronic Illness and Time.* New Jersey: Rutgers University Press.

Chodorow, Nancy. 1974. "Family Structure and Feminine Personality." In Rosaldo and Lamphere (1974, 43–66).

Chodorow, Nancy. 1978. *The Reproduction of Mothering: Psychoanalysis and the Sociology of Gender.* Berkeley: University of California Press.

Chung, P., T. Yeko, J. Meyer, E. Sanford, and G. Maroulis. 1995. "Assisted Fertility Using Electroejaculation in Men with Spinal Cord Injury: A Review of Literature." *Fertility and Sterility* 64: 1–9.

Clarke, Adele. 1990. "A Social Worlds Research Adventure: The Case of Reproductive Science." In Cozzens and Gieryn (1990, 15–42).

Clarke, Adele. 1991. "Social Worlds/Arenas Theory as Organizational Theory." In D. R. Maines, ed., *Social Organization and Social Process: Essays in Honor of Anselm Strauss.* New York: de Gruyter.

Clarke, Adele. 1998. *Disciplining Reproduction: Modernity, American Life Sciences, and "the Problems of Sex."* Berkeley: University of California Press.

Clarke, Adele, and Joan Fujimura, eds. 1992. *The Right Tools for the Job: At Work in Twentieth-Century Life Sciences.* Princeton, NJ: Princeton University Press.

Clarke, Adele, and Theresa Montini. 1993. "The Many Faces of RU 486: Tales of Situated Knowledges and Technological Contestations." *Science, Technology and Human Values* 18: 42–78.

Clarke, Adele, Janet Shim, L. Mamo, J. R. Fosket, and J. R. Fishman. 2003. "Biomedicalization: Theorizing Technoscientific Transformations of Health, Illness, and U.S. Biomedicine." *American Sociological Review* 68: 161–194.

Coalition to Fight Infant Mortality. 1989. "Equal Opportunity for Babies? Not in Oakland!" In Arditti, Klein, and Minden (1989, 391–396).

Cockburn, Cynthia. 1982. *Brothers: Male Dominance and the New Technology.* London: Pluto Press.

Cockburn, C., and R. Furst-Dilic. 1994. *Bringing Technology Home: Gender and Technology in a Changing Europe.* Milton Keynes, UK: Open University Press.

Cohen, Sherill, and Nadine Taub, eds. 1989. *Reproductive Laws for the 1990s.* Clifton, NJ: Humana Press.

Colen, Shellee. 1995. "'Like a Mother to Them': Stratified Reproduction and West Indian Childcare Workers and Employers in New York." In Ginsburg and Rapp (1995, 78–102).

Collins, Harry. 1985. *Changing Order: Replication and Induction in Scientific Practice.* London: Sage.

Collins, Harry. 1994. "Dissecting Surgery: Forms of Life Depersonalized." *Social Studies of Science* 24: 311–333.

Collins, Harry, and Trevor Pinch. 1998. *The Golem: What Everyone Should Know about Science* (2nd ed.). Cambridge: Cambridge University Press.

Collins, Harry, and Trevor Pinch. 1999. *The Golem at Large: What Everyone Should Know about Technology.* Cambridge: Cambridge University Press.

Collins, J. 1995. "An Estimate of the Cost of In Vitro Fertilization Services in the United States in 1995." *Fertility and Sterility* 64: 538–545.

Collins, Patricia Hill. 1990. *Black Feminist Thought: Knowledge, Consciousness, and the Politics of Empowerment.* Boston: Unwin Hyman.

Corea, Gena. 1985. *The Mother Machine: Reproductive Technologies from Artificial Insemination to Artificial Wombs.* New York: Harper & Row.

Corea, Gena. 1989. "Surrogacy: Making the Links." In Klein (1989, 133–166).

Corson, Stephen, and Andrea Mechanick-Braverman. 1998. "Why We Believe There Should Be a Gamete Registry." *Fertility and Sterility* 69: 809–811.

Cowan, Ruth Schwartz. 1983. *More Work for Mother: The Ironies of Household Technology from the Open Hearth to the Microwave.* New York: Basic.

Cozzens, Susan, and Thomas Gieryn, eds. 1990. *Theories of Science in Society.* Bloomington: Indiana University Press.

Crary, Jonathan, and Sanford Kwinter, eds. 1992. *Incorporations.* New York: Zone Books.

Croucher, Carolyn, Amir Lass, Raul Margara, and Robert Winston. 1998. "Predictive Value of the Results of a First In-vitro Fertilization Cycle on the Outcome of Subsequent Cycles." *Human Reproduction* 13: 403–408.

Crowe, Christine. 1985. "'Women Want It': In Vitro Fertilization and Women's Motivations for Participation." *Women's Studies International Forum* 8: 547–552.

Crowe, Christine. 1990. "Whose Mind over Whose Matter? Women, In Vitro Fertilisation and the Development of Scientific Knowledge." In McNeil (1990, 27–57).

Cussins, Adrian. 1992. "Content, Embodiment and Objectivity: The Theory of Cognitive Trails." *MIND* 101: 651–688.

Cussins, Charis. 1996. "Ontological Choreography: Agency through Objectification in Infertility Clinics." *Social Studies of Science* 26: 575–610.

Cussins, Charis. 1998a. "Producing Reproduction: Techniques of Normalization and Naturalization in an Infertility Clinic." In Franklin and Ragoné (1998, 66–101).

Cussins, Charis. 1998b. "Quit Snivelling Cryo-Baby, We'll Work Out Which One's Your Mama: Kinship in an Infertility Clinic." In Davis-Floyd and Dumit (1998, 40–66).

Cussins, Charis Thompson. 1999. "Confessions of a Bioterrorist: Subject Position and the Valuing of Reproductions." In Squier and Kaplan (1999, 189–219).

Cutting Edge, ed. 2000. *Digital Desires: Language, Identity and New Technologies.* London: Taurus.

Daly, Mary. 1978. *Gyn/ecology: The Metaethics of Radical Feminism.* Boston: Beacon Press.

Daniels, Cynthia R. 1993. *Women's Expense: State Power and the Politics of Fetal Rights.* Cambridge, MA: Harvard University Press.

Das, Veena. 1995. "National Honor and Practical Kinship: Unwanted Women and Children." In Ginsburg and Rapp (1995, 212–233).

Das, Veena. 2000. "The Practice of Organ Transplants: Networks, Documents, Translations." In Lock, Young, and Cambrosio (2000, 263–287).

Davis, Angela. 1983. *Women, Race, and Class.* London: Women's Press.

Davis, Angela. 1993. "Outcast Mothers and Surrogates: Racism and Reproductive Health in the Nineties." In Linda Kauffman, ed., *American Feminist Thought at Century's End: A Reader.* Cambridge, MA: Blackwell.

Davis, Angela. 1999. *Blues Legacies and Black Feminism: Gertrude Ma Rainey, Bessie Smith, and Billie Holiday.* New York: Vintage Books.

Davis-Floyd, Robbie. 1992. *Birth as an American Rite of Passage.* Berkeley: University of California Press.

Davis-Floyd, Robbie, and Joseph Dumit, eds. 1998. *Cyborg Babies: From Techno Tots to Techno Toys.* New York: Routledge.

Dear, Peter. 1985. Totius in Verba: Rhetoric and Authority in the Early Royal Society." *ISIS* 76: 145–161.

De Beauvoir, Simone. 1953. *The Second Sex.* Trans. H. M. Parshley. New York: Knopf. Originally published in 1949.

Dennis, Michael. 1994. "'Our First Line of Defense': Two University Laboratories in the Post-War American State." *ISIS* 85: 427–456.

Dent, Gina, ed. 1992. *Black Popular Culture.* Seattle: Bay Press.

De Rivera, Joseph, and Theodore Sarbin, eds. 1997. *Believed-In Imaginings: The Narrative Construction of Reality.* Washington, DC: American Psychological Association.

Dikotter, Frank. 1998. *Imperfect Conceptions: Medical Knowledge, Birth Defects, and Eugenics in China.* New York: Columbia University Press.

Di Leonardo, M. ed. 1991. *Gender at the Crossroads of Knowledge: Feminist Anthropology in the Postmodern Era.* Berkeley: University of California Press.

Dixon-Mueller, Ruth. 1993. *Population Policy and Women's Rights: Transforming Reproductive Choice.* Westport, CT: Praeger.

Dodier, Nicolas. 1993. *L'Expertise médicale: Essai de sociologie sur l'exercise du jugement.* Paris: Métailié.

Dolgin, Janet. 1993. "Just a Gene: Judicial Assumptions about Parenthood." *UCLA Law Review* 40(3): 637–694.

Dolgin, Janet. 1995. "Family Law and the Facts of Family." In Yanagisako and Delaney (1995, 47–67).

Donnison, Jean. 1977. *Midwives and Medical Men: A History of Interprofessional Rivalry and Women's Rights.* London: Heinemann.

Douglas, Mary. 1966. *Purity and Danger.* New York: Praeger.

Downey, Gary Lee. 1998. *The Machine in Me: An Anthropologist Sits among Computer Engineers.* New York: Routledge.

Downey, Gary, Joe Dumit, and Sharon Traweek, eds. 1995. *Cyborgs and Citadels.* Seattle: University of Washington Press.

Duden, Barbara. 1993. *Disembodying Women: Perspectives on Pregnancy and the Unborn.* Cambridge, MA: Harvard University Press.

Duden, Barbara. 1997. "The History of Security in the Knowledge of Pregnancy." Paper prepared at the Max Planck Institute for the History of Science while a Research Fellow.

Duka, Walter E., and Alan H. DeCherney. 1994. *From the Beginning: A History of the American Fertility Society.* Birmingham, AL: American Fertility Society.

Dumit, Joseph. 2000. "When Explanations Rest: 'Good-enough' Brain Science and the New Socio-medical Disorders." In Lock, Young, and Cambrosio (2000, 209–232).

Dupré, John. 1993. *The Disorder of Things: Metaphysical Foundations of the Disunity of Science.* Cambridge: Harvard University Press.

Dworkin, Andrea. 1981. *Pornography: Men Possessing Women.* New York: Perigree.

Easlea, Brian. 1981. *Science and Sexual Oppression: Patriarchy's Confrontation with Women and Nature.* London: Weidenfeld and Nicholson.

Easlea, Brian. 1983. *Fathering the Unthinkable: Masculinity, Scientists and the Nuclear Arms Race.* London: Pluto Press.

Edwards, Jeanette, Sarah Franklin, Eric Hirsch, Frances Price, and Marilyn Strathern. 1993. *Technologies of Procreation: Kinship in the Age of Assisted Conception.* Manchester: Manchester University Press.

Edwards, Robert. 1989. *Life before Birth: Reflections on the Embryo Debate.* London: Hutchinson.

Edwards, Robert. 1993. "Pregnancies Are Acceptable in Post-Menopausal Women." *Human Reproduction* 8: 1542–1544.

Egozcue, J. 1993. "Sex Selection: Why Not?" *Human Reproduction* 8: 1777.

Ehrenreich, Barbara, and Deirdre English. 1973. *Complaints and Disorders: The Sexual Politics of Sickness.* Old Westbury, NY: Feminist Press.

Ehrenreich, Barbara, and Deirdre English. 1978. *For Her Own Good: One Hundred Fifty Years of the Experts' Advice to Women.* New York: Doubleday.

Ehrenreich, Barbara, and Arlie Hochschild, eds. 2003. *Global Woman: Nannies, Maids, and Sex Workers in the New Economy.* New York: Metropolitan Books.

Engels, Frederick. 1972. *The Origin of the Family, Private Property, and the State.* New York: International.

Epstein, Steven. 1995. "The Construction of Lay Expertise: AIDS Activism and the Forging of Credibility in the Reform of Clinical Trials." *Science, Technology and Human Values* 20: 408–437.

Epstein, Steven. 1996. *Impure Science: AIDS Activism, and the Politics of Knowledge.* Berkeley, CA: University of California Press.

Ethics Committee of the American Society for Reproductive Medicine. 2000. "Financial Incentives in Recruitment of Oocyte Donors." *Fertility and Sterility* 74(2): 216–220.

European Society for Human Reproduction and Embryology (ESHRE). 1998. "Male Infertility Update." *Human Reproduction* 13: 2025–2032.

Evers, J. L. H., H. W. de Haas, J. A. Land, J. C. M. Dumoulin, and G. A. J. Dunselman. 1998. "Treatment-Independent Pregnancy Rate in Patients with Severe Reproductive Disorders." *Human Reproduction* 13: 1206–1209.

Farquhar, Dion. 1996. *The Other Machine: Discourse and Reproductive Technologies.* New York: Routledge.

Fausto-Sterling, Anne. 1985. *Myths of Gender: Biological Theories about Women and Men.* New York: Basic Books.

Fausto-Sterling, Anne. 1995. "Gender, Race, and Nation: The Comparative Anatomy of 'Hottentot' Women in Europe, 1815–1817." In Terry and Urla (1995, 19–48).

Fausto-Sterling, Anne. 2000. *Sexing the Body: Gender Politics and the Construction of Sexuality.* New York: Basic Books.

Fee, Elizabeth. 1979. "Nineteenth-Century Craniology: The Study of the Female Skull." *Bulletin of the History of the Medicine* 53: 415–433.

Feldman-Savelsberg, Pamela. 1999. *Plundered Kitchens, Empty Wombs: Threatened Reproduction and Identity in the Cameroon Grassfields.* Michigan: University of Michigan Press.

Feminist Self-Insemination Group. 1980. *Self-Insemination.* P.O. Box 3, 190 Upper Street, London N1.

Ferguson, R., M. Gever, Trin Minh-ha, and Cornel West. 1990. *Out There: Marginalization and Contemporary Cultures.* Cambridge, MA: MIT Press.

Findlen, Paula. 1991. "The Economy of Scientific Exchange in Early Modern Italy." In B. T. Moran, ed., *Patronage and Institutions: Science, Technology and Medicine at the European Court, 1500–1750* (5–24). Rochester: Boydell Press.

Finkel, David. 1999. "A Human Toll in the Battle on Infertility." *Washington Post,* March 21, A1.

Finkler, Kaja. 2000. *Experiencing the New Genetics: Family and Kinship on the Medical Frontier.* Philadelphia: University of Pennsylvania Press.

Firestone, Shulamith. 1970. *The Dialectic of Sex: The Case for Feminist Revolution.* New York: Morrow.

Fleck, Ludwig. 1935. *Genesis and Development of a Scientific Fact.* Chicago: University of Chicago Press.

Folbre, Nancy. 1994. *Who Pays for the Kids? Gender and the Structures of Constraint.* New York: Routledge.

Fortun, Michael. 2001. "Mediated Speculations in the Genomics Futures Markets." *New Genetics and Society* 20: 139–156.

Fortun, Michael, and Everett Mendelsohn, eds. 1999. *The Practices of Human Genetics.* Boston: Kluwer.

Foucault, Michel. 1970. *The Order of Things.* New York: Random House.

Foucault, Michel. 1982. "Subject of Power." *Critical Inquiry* 8: 777–795.

Foucault, Michel. 1990. *The History of Sexuality* (Vol. 1). Trans. Robert Hurley. New York: Vintage Books. Originally published in 1978.

Foucault, Michel. 1994. *The Birth of the Clinic: An Archaeology of Medical Perception* (3–21, 22–37). New York: Vintage. Originally published in 1963.

Foucault, Michel. 1995. *Discipline and Punish.* Trans. A. Sheridan. New York: Vintage Books. Originally published in 1975.

Fox, Nicholas. 1994. "Fabricating Surgery: A Response to Collins." *Social Studies of Science* 24: 347–354.

Franklin, Sarah. 1990. "Deconstructing 'Desperateness': The Social Construction of Infertility in Popular Representations of New Reproductive Technologies." In McNeil, Maureen, Ian Varcoe, and Steven Yearley, eds., *The New Reproductive Technologies* (200–229). London: Macmillan.

Franklin, Sarah. 1991. "Fetal Fascinations: New Medical Constructions of Fetal Personhood." In Sarah Franklin, Celia Lury, and Jackie Stacey, eds., *Off-Centre: Feminism and Cultural Studies* (190–206). New York: Harper Collins Academic.

Franklin, Sarah. 1997. *Embodied Progress: A Cultural Account of Assisted Conception.* London: Routledge.

Franklin, Sarah, forthcoming. 2003. "Ethical Bio-capital: New Strategies of Cell Culture." In Franklin and Lock (2003, 97–128).

Franklin, Sarah, and Margaret Lock. 2003. "Introduction: Animation and Cessation: The Remaking of Life and Death." In Franklin and Lock (2003, 3–22).

Franklin, Sarah, and Margaret Lock, eds. 2003. *Remaking Life and Death: Toward an Anthropology of the Biosciences.* Santa Fe: School of American Research Press.

Franklin, Sarah, Celia Lury, and Jackie Stacey. 2000. *Global Nature, Global Culture.* London: Sage.

Franklin, Sarah, and Susan McKinnon, eds. 2001. *Relative Values: Reconfiguring Kinship Studies.* Durham: Duke University Press.

Franklin, Sarah, and Maureen McNeil. 1988. "Reproductive Futures: Recent Feminist Debate of New Reproductive Technologies." *Feminist Studies* 14(3): 545–574.

Franklin, Sarah, and Helena Ragone, eds. 1998. *Reproducing Reproduction.* Philadelphia: University of Pennsylvania Press.

Fraser, Nancy, and Linda J. Nicholson. 1990. "Social Criticism without Philosophy: An Encounter between Feminism and Postmodernism." In Nicholson (1990, 19–38).

Galison, Peter. 1997. *Image and Logic: A Material Culture of Microphysics.* Chicago: University of Chicago Press.

Galison, Peter, and David Stump, eds. 1995. *The Disunity of Science.* Stanford: Stanford University Press.

Garcia, Jairo. 1998. "Profiling Assisted Reproductive Technology: The Society for Reproductive Technology Registry and the Rising Costs of Assisted Reproductive Technology." *Fertility and Sterility* 69: 624–626.

Garfinkel, Harold. 1963. "A Conception of, and Experiments with, 'Trust' as a Condition of Stable Concerted Actions." In O. Harvey, ed., *Motivation and Social Interaction* (187–238). New York: Ronald Press.

Garfinkel, Harold. 1967. *Studies in Ethnomethodology.* Englewood Cliffs, NJ: Prentice-Hall.

Garfinkel, Harold, Michael Lynch, and Eric Livingston. 1981. "The Work of a Discovering Science Construed with Materials from the Optically Discovered Pulsar." *Philosophy of the Social Sciences* 11: 131–158.

Garson, Barbara. 1988. *The Electronic Sweatshop.* New York: Simon and Schuster.

Gemzell, C. A., E. Dicfaluzy, and K. G. Tillinger. 1960. "Human Pituitary Follicle-Stimulating Hormone I: Clinical Effect of a Partly Purified Preparation." *Ciba Foundation Colloquia on Endocrinology* 13: 191–200.

Gergen, Kenneth. 1991. *The Saturated Self: Dilemmas of Identity in Contemporary Life.* New York: Basic Books.

Giddens, Anthony, and David Held, eds. 1982. *Classes, Power and Conflict: Classical and Contemporary Debates.* Berkeley: University of California Press.

Gieryn, Thomas. 1983. "Boundary Work and the Demarcation of Science from Non-Science: Strains and Interests in Professional Ideologies of Scientists." *American Sociological Review* 48: 781–795.

Gieryn, Thomas, and Anne Figert. 1990. "Ingredients for the Theory of Science in Society: O-Rings, Ice Water, C-Clamp, Richard Feynman, and the Press." In Cozzens and Gieryn (1990, 67–97).

Gill, Brian Paul. 2002. "Adoption Agencies and the Search for the Ideal Family." In Carp (2002, 160–180).

Gilligan, Carol. 1982. *In a Different Voice: Psychological Theory and Women's Development.* Cambridge, MA: Harvard University Press.

Gilmore, David. 1990. *Manhood in the Making: Cultural Concepts of Masculinity.* New Haven: Yale University Press.

Ginsburg, Faye. 1989. *Contested Lives: The Abortion Debate in an American Community.* Berkeley: University of California Press.

Ginsburg, Faye, and Rayna Rapp, eds. 1995. *Conceiving the New World Order: The Global Politics of Reproduction.* Berkeley: University of California Press.

Ginsburg, Faye, and Anna Lowenhaupt Tsing, eds. 1990. *Uncertain Terms: Negotiating Gender in American Culture.* Boston: Beacon Press.

Goffman, Erving. 1961. *Encounters: Two Studies in the Sociology of Interaction.* Harmondsworth: Penguin.

Goffman, Erving. 1969. *The Presentation of Self in Everyday Life.* Harmondsworth, UK: Penguin. Originally published in 1959.

Goffman, Erving. 1986. *Asylums: Essays on the Social Situation of Mental Patients and Other Inmates.* Harmondsworth: Pelican. Originally published in 1961.

Goffman, Erving. 1986. *Stigma: Notes on the Management of Spoiled Identity.* New York: Touchstone. Originally published in 1963.

Gomez, Laura. 1997. *Misconceiving Mothers: Legislators, Prosecutors, and the Politics of Prenatal Drug Exposure.* Philadelphia: Temple University Press.

Goodman, Nelson. 1978. *Ways of Worldmaking.* Indianapolis: Hackett.

Gordon, Linda. 1977. *Woman's Body, Woman's Rights: A Social History of Birth Control in America.* New York: Penguin.

Grayson, Deborah. 1998. "Mediating Intimacy: Black Surrogate Mothers and the Law." *Critical Inquiry* 24: 525–546.

Gregor, Thomas. 1985. *Anxious Pleasures: The Sexual Lives of an Amazonian People.* Chicago: University of Chicago Press.

Greenhalgh, Susan, ed. 1995. *Situating Fertility: Anthropology and Demographic Inquiry.* Cambridge: Cambridge University Press.

Griffin, Martha, and William Panak. 1998. "The Economic Cost of Infertility-Related Services: An Examination of the Massachusetts Infertility Insurance Mandate." *Fertility and Sterility* 70: 22–29.

Griffin, Susan. 1978. *Woman and Nature: The Roaring inside Her.* New York: Harper and Row.

Gross, Paul, and Norman Levitt. 1994. *Higher Superstition: The Academic Left and Its Quarrels with Science.* Baltimore: Johns Hopkins University Press.

Grosz, Elizabeth, and Elspeth Probyn, eds. 1995. *Sexy Bodies: The Strange Carnalities of Feminism.* London: Routledge.

Gunning, Jennifer, ed. 2000. *Assisted Conception: Research, Ethics, and Law.* Burlington, VT: Ashgate/Dartmouth.

Gunning, Jennifer, and Veronica English. 1994. *Human In Vitro Fertilization: A Case Study in the Regulation of Medical Innovation.* Brookfield, VT: Dartmouth.

Guttman, Matthew. 1997. "Trafficking in Men: The Anthropology of Masculinity." *Annual Review of Anthropology* 26: 385–409.

Hacking, Ian. 1983. *Representing and Intervening: Introductory Topics in the Philosophy of Natural Science.* Cambridge: Cambridge University Press.

Hacking, Ian. 1995. "The Looping Effects of Human Kinds." In Sperber, Premack, and Premack (1995, 351–383).

Hacking, Ian. 2000. *The Social Construction of What?* Cambridge, MA: Harvard University Press.

Halberstam, Judith. 1998. *Female Masculinity.* Durham, NC: Duke University Press.

Hammonds, Evelynn. 1995. "Missing Persons: African American Women, AIDS, and the History of Disease." In Beverly Guy Sheftall, ed., *Words of Fire An Anthology of African American Feminist Thought* 434–449.

Handwerker, Lisa. 1995. "The Hen That Can't Lay an Egg (Bu Xia Dan de Mu Ji): Conceptions of Female Infertility in China." In Terry and Urla (1995, 358–386).

Hansen, Karen, and Ilene Philipson, eds. 1990. *Women, Class, and the Feminist Imagination: A Socialist-Feminist Reader.* Philadelphia: Temple University Press.

Haraway, Donna. 1989. *Primate Visions: Gender, Race and Nature in the World of Modern Science.* New York: Routledge.

Haraway, Donna. 1991. *Simians, Cyborgs and Women: The Reinvention of Nature.* New York: Routledge.

Haraway, Donna. 1992. "Ecce Homo, Ain't (Ar'n't) I a Woman, and Inappropriate/d Others: The Human in a Post-Humanist Landscape." In Butler and Scott (1992, 86–100).

Haraway, Donna. 1995. "Universal Donors in a Vampire Culture: It's All in the Family: Biological Kinship Categories in the Twentieth-Century United

States." In W. Cronon, ed., *Uncommon Ground: The Reinvention of Nature.* New York: Norton.

Haraway, Donna. 1997. *Modest Witness@Second Millenium—FemaleMan© Inc. Meets Oncomouse™: Feminism and Technoscience.* New York: Routledge.

Haraway, Donna. 2003. *The Companion Species Manifesto: Dogs, People, and Significant Otherness.* Chicago: Prickly Paradigm Press.

Harding, Sandra. 1986. *The Science Question in Feminism.* Ithaca, NY: Cornell University Press.

Harding, Sandra. 1991. *Whose Science? Whose Knowledge? Thinking from Woman's Lives.* Milton Keynes, UK: Open University Press.

Harding, Sandra, and M. Hintikka, eds. 1983. *Discovering Reality: Feminist Perspectives on Epistemology, Metaphysics, Methodology, and Philosophy of Science.* Dordrecht: Reidel.

Hartouni, Valerie. 1997. *Cultural Conceptions: On Reproductive Technologies and the Remaking of Life.* Minneapolis: University of Minnesota Press.

Hartsock, Nancy. 1983. "The Feminist Standpoint: Developing the Ground for a Specifically Feminist Historical Materialism." In Harding and Hintikka (1983, 283–310).

Harvey, David. 2001. *Spaces of Capital: Towards a Critical Geography.* Edinburgh: Edinburgh University Press.

Haste, Helen. 1994. *The Sexual Metaphor: Men, Women, and the Thinking That Makes the Difference.* Cambridge, MA: Harvard University Press.

Herdt, Gilbert, ed. 1982. *Rituals of Manhood: Male Initiation in Papua New Guinea.* Berkeley: University of California Press.

Hernstein-Smith, Barbara. 1997. "The Microdynamics of Incommensurability: Philosophy of Science Meets STS." In B. Hernstein-Smith, and A. Plotnisky, eds., *Mathematics, Science, and Post-Classical Theory* (243–266). Durham, NC: Duke University Press.

Herzfeld, Michael. 1985. *The Poetics of Manhood: Contest and Identity in a Cretan Mountain Village.* Princeton: Princeton University Press.

Hirschauer, Stefan. 1991. "The Manufacture of Bodies in Surgery." *Social Studies of Science* 21: 279–319.

Hochschild, Arlie Russell. 1983. *The Managed Heart: Commercialization of Human Feeling.* Berkeley: University of California Press.

Hoffman, David, Gail Zellman, et al. 2003. "Cryopreserved Embryos in the United States and Their Availability for Research." *Fertility and Sterility* 79: 1063–1069.

Hogden, G., E. Wallach, and L. Mastroianni. 1994. "Ethical Considerations of Assisted Reproductive Technologies." Paper prepared on behalf of the Ethics Committee of the American Fertility Society. Birmingham, AL: American Fertility Society.

Hogle, Linda. 1996. "Transforming 'Body Parts' into Therapeutic Tools: A Report from Germany." *Medical Anthropology Quarterly* 10(4): 675–682.

Holland, Suzanne, Karen Lebacqz, and Laurie Zoloth, eds. 2001. *The Human Embryonic Stem Cell Debate: Science, Ethics, and Public Policy.* Cambridge, MA: MIT Press.

Holmes, Helen Bequaert, ed. 1992. *Issues in Reproductive Technology: An Anthology.* New York: Garland.

Holmes, Helen, Betty Hoskins, and Michael Gross, eds. 1981. *The Custom-Made Child? Woman-Centered Perspectives.* Clifton, NJ: Humana Press.

Homans, Hilary. 1986. *The Sexual Politics of Reproduction.* Aldershot, UK: Gower.

Horkheimer, M., and T. Adorno. 1979. *Dialectic of Enlightenment.* New York: Seabury Press.

Horn, David. 1994. *Social Bodies: Science, Reproduction, and Italian Modernity.* Princeton, NJ: Princeton University Press.

Howell, Signe. 2005. "'Kinning': The Creation of Life Trajectories in Transnational Adoptive Families." *Journal of the Royal Anthropological Institute.*

Hubbard, Ruth. 1982. *Biological Woman, the Convenient Myth.* Cambridge, MA: Schenkman.

Hubbard, R., M. S. Henifin, and B. Fried, eds. 1979. *Women Look at Biology Looking at Women: A Collection of Feminist Critiques.* Cambridge, MA: Schenkman.

Hughes, Thomas. 1983. *Networks of Power: Electrification in Western Society 1880–1930.* Baltimore: Johns Hopkins University Press.

Hurst, Bradley, Kathleen Tucker, Caleb Awoniyi, and William Schlaff. 1997. "Preprogrammed, Unmonitored Ovarian Stimulation Reduces Expense without Compromising the Outcome of Assisted Reproduction." *Fertility and Sterility* 68: 282–286.

Hutchins, Edwin. 1995. *Cognition in the Wild.* Cambridge, MA: MIT Press.

Ikemoto, Lisa. 1995. "The Code of Perfect Pregnancy: At the Intersection of the Ideology of Motherhood, the Practice of Defaulting to Science, and the Interventionist Mindset of the Law." In Richard Delgado ed., *Critical Race Theory: The Cutting Edge* (478–497). Philadelphia: Temple University Press.

Ince, Susan. 1989. "Inside the Surrogate Industry." In Arditti, Klein, and Minden (1989, 99–116). Originally published in 1984.

Inhorn, Marcia. 1994. *Quest for Conception: Gender, Infertility, and Egyptian Medical Traditions.* Philadelphia: University of Pennsylvania Press.

Inhorn, Marcia. 1996. *Infertility and Patriarchy: The Cultural Politics of Gender and Family Life in Egypt.* Philadelphia: University of Pennsylvania Press.

Inhorn, Marcia, and Frank Van Balen, eds. 2002. *Infertility around the Globe: New Thinking on Childlessness, Gender, and Reproductive Technologies.* Berkeley: University of California Press.

Irwin, Alan, and Brian Wynne, eds. 1996. *Misunderstanding Science? The Public Reconstruction of Science and Technology.* Cambridge: Cambridge University Press.

Jaeger, Ami. 1996. "Laws Surrounding Reproductive Technologies." In M. Seibel and S. Crockin, eds., *Family Building through Egg and Sperm Donation* (113–130). London: Jones and Bartlett.

Jasanoff, Sheila. 1990. *The Fifth Branch: Science Advisors as Policymakers.* Cambridge, MA: Harvard University Press.

Jasanoff, Sheila, ed. 1994. *Learning from Disaster: Risk Management after Bhopal.* Philadelphia: University of Pennsylvania Press.

Jasanoff, Sheila. 1995. *Science at the Bar.* Cambridge, MA: Harvard University Press.

Jasanoff, Sheila. 1996a. "Beyond Epistemology: Relativism and Engagement in the Politics of Science." *Social Studies of Science* 26: 396–418.

Jasanoff, Sheila. 1996b. "Is Science Socially Constructed—and Can It Still Inform Public Policy?" *Science and Engineering Ethics* 2: 263–276.

Jasanoff, Sheila. 1996c. "Science and Norms in Global Environmental Regimes." In F. O. Hampson and J. Reppy, eds., *Earthly Goods: Environmental Change and Social Justice.* Ithaca: Cornell.

Jasanoff, Sheila, ed. 2004. *States of Knowledge.* London: Routledge.

Jasanoff, Sheila, Gerald Markle, James Petersen, and Trevor Pinch, eds. 1995. *Handbook of STS.* London: Sage.

Jetter, Alexis, Annelise Orleck, and Diana Taylor, eds. 1997. *The Politics of Motherhood: Activist Voice from Left to Right.* Hanover: University Press of New England.

Jones, Howard. 1996. "The Time Has Come." *Fertility and Sterility* 65: 1090–1092.

Jordan, Kathleen, and Michael Lynch. 1993. "The Sociology of a Genetic Engineering Technique: Ritual and Rationality in the Performance of the 'Plasmid Prep.'" In Clarke and Fujimura (1992, 77–114).

Jordanova, Ludmilla. 1989. *Sexual Visions: Images of Gender in Science and Medicine between the Eighteenth and Twentieth Centuries.* London: Routledge.

Kahn, Susan. 2000. *Reproducing Jews: A Cultural Account of Assisted Conception in Israel.* Durham; NC: Duke University Press.

Kanaaneh, Rhoda A. 1997. "Conceiving Difference: Birthing the Palestinian Nation in the Galilee." *Critical Public Health* 3&4: 64–79.

Katz, M. 1995. "Federal Trade Commission Staff Concerns with Assisted Reproductive Technology Advertising." *Fertility and Sterility* 64: 10–12.

Katz, Pearl. 1981. "Ritual in the Operating Room." *Ethnology* 20: 335–350.

Kaupen-Hass, Heidrun. 1988. "Experimental Obstetrics and National Socialism: The Conceptual Basis of Reproductive Technology." *Reproductive and Genetic Engineering: Journal of International Feminist Analysis* 1: 127–132.

Keller, Evelyn Fox. 1985. *Reflections on Gender and Science.* New Haven, CT: Yale University Press.

Keller, Evelyn Fox. 2000. *The Century of the Gene.* Cambridge, MA: Harvard University Press.

Kevles, Daniel. 1998. *The Baltimore Case: A Trial of Politics, Science and Character.* New York: Norton.

Kiekens, Carlotte, Carl Spiessens, Francis Duyck, Danielle Vandenweghe, Willy Coucke, and Dirk Vanderschureren. 1996. "Pregnancy after Electroejaculation in Combination with Intracytoplasmic Sperm Injection in a Patient with Idiopathic Anejaculation." *Fertility and Sterility* 66: 834–836.

Kimmel, Michael, and Michael Messner, eds. 1998. *Men's Lives* (4th ed.). Boston: Allyn and Bacon.

King, Leslie, and Madonna Harrington Meyer. 1997. "The Politics of Reproductive Benefits: U.S. Insurance Coverage of Contraceptive and Infertility Treatments." *Gender and Society* 11: 8–30.

Kirejczyk, Marta. 1994. "Cassandra's Warnings: Feminist Discourse, Gender and Social Entrenchment of In Vitro Fertilization in the Netherlands." *European Journal of Women's Studies* 1: 151–164.

Kitcher, Philip. 1993. *The Advancement of Science: Science without Legend, Objectivity without Illusions.* Oxford: Oxford University Press.

Kitzinger, Sheila. 1978. *Women as Mothers: How They See Themselves in Different Cultures.* New York: Vintage Books.

Klein, Renate, ed. 1989. *Infertility: Women Speak Out about Their Experiences of Reproductive Technologies.* London: Pandora Press.

Klein, Renate, and Robin Rowland. 1989. "Hormone Cocktails: Women as Test-Sites for Fertility Drugs." *Women's Studies International Forum* 12: 333–348.

Kleinman, Arthur, Renée Fox, and Allan Brandt, eds. 1999. "Bioethics and Beyond: Special Issue." *Dædalus* 129(4).

Kleinman, Daniel. 1995. *Politics on the Endless Frontier: Postwar Research Policy.* Durham: Duke University Press.

Knorr-Cetina, Karin. 1999. *Epistemic Cultures: How the Sciences Make Knowledge.* Cambridge, MA: Harvard University Press.

Knorr-Cetina, Karin, and Michael Mulkay, eds. 1983. *Science Observed: Perspectives on the Social Study of Science.* Beverly Hills: Sage.

Kondo, Dorinne. 1990. *Crafting Selves: Power, Gender and Discourses of Identity in a Japanese Workplace.* Chicago: University of Chicago Press.

Kuhn, Annette, and AnnMarie Wolpe, eds. 1978. *Feminism and Materialism: Women and Modes of Production.* Boston: Routledge & Kegan Paul.

Kuhn, Thomas. 1970. *The Structure of Scientific Revolutions.* Chicago: Unviersity of Chicago Press. Originally published in 1962.

Kuhn, Thomas. 1977. *The Essential Tension: Selected Studies in Scientific Tradition and Change.* Chicago: Chicago University Press.

Ladd-Taylor, Molly, and Lauri Umansky, eds. 1998. *"Bad Mothers": The Politics of Blame in Twentieth-Century America.* New York: New York University Press.

Land, Jolande, Dorette Courtar, and Johannes Evers. 1997. "Patient Dropout in an Assisted Reproductive Technology Program: Implications for Pregnancy Rates." *Fertility and Sterility* 68: 278–281.

Landecker, Hannah. 2000. "Immortality, In Vitro: A History of the HeLa Cell Line." In Brodwin (2000, 53–72).

Landes, Joan, ed. 1998. *Feminism, the Public, and the Private.* New York: Oxford University Press.

Laqueur, Thomas. 2000. "'From Generation to Generation': Imagining Connectedness in the Age of Reproductive Technologies." In Brodwin (2000, 75–98).

Lash, Scott, and John Urry. 1987. *The End of Organized Capital.* Cambridge: Polity Press.

Lasker, Judith, and Susan Borg. 1989. *In Search of Parenthood: Coping with Infertility and High Tech Conception.* London: Pandora.

Laslett, Barbara, Sally Gregory Kohlstedt, Helen Longino, and Evelyn Hammonds, eds. 1996. *Gender and Scientific Authority.* Chicago: Chicago University Press.

Latour, Bruno. 1987. *Science in Action.* Cambridge, MA: Harvard University Press.

Latour, Bruno. 1988. *The Pasteurization of France, Followed by Irreductions: A Politico-Scientific Essay.* Cambridge, MA: Harvard University Press.

Latour, Bruno. 1988. "Drawing Things Together." In Lynch and Woolgar (1988, 19–68).

Latour, Bruno. 1993. *We Have Never Been Modern.* Trans. Catherine Porter. Cambridge, MA: Harvard University Press.

Latour, Bruno. 1996. *Aramis, or the Love of Technology.* Harvard: Harvard University Press.

Latour, Bruno. 1999. *Pandora's Hope: Essays on the Reality of Science Studies.* Cambridge, MA: Harvard University Press.

Latour, Bruno. 2002. *The War of Worlds: What about Peace?* Chicago: Prickly Paradigm Press.

Latour, Bruno, and Steve Woolgar. 1986. *Laboratory Life: The Construction of Scientific Facts.* Princeton: Princeton University Press. Originally published in 1979.

Law, John, ed. 1986. *Power, Action and Belief: A New Sociology of Knowledge?* London: Routledge and Kegan Paul.

Law, John, ed. 1991. *A Sociology of Monsters: Essays on Power, Technology and Domination.* London: Routledge.

Law, John. 1994. *Organizing Modernity.* Oxford: Blackwell.

Law, John, and Annemarie Mol, eds. 2002. *Complexities: Social Studies of Knowledge Practices.* Durham: Duke University Press.

Layne, Linda. 2002. *Motherhood Lost: A Feminist Account of Pregnancy Loss in America.* New York: Routledge.

Leon, Irving. 2002. "Adoption Losses: Naturally Occurring or Socially Constructed?" *Child Development* 73: 652–664.

Lévi-Strauss, Claude. 1969. *The Elementary Structures of Kinship.* Boston: Beacon Press. Originally published in 1949.

Lewin, Ellen. 1993. *Lesbian Mothers: Accounts of Gender in American Culture.* Ithaca: Cornell University Press.

Lie, John. 2004. *Modern Peoplehood.* Cambridge, MA: Harvard University Press.

Lifton, Robert Jay. 1986. *The Nazi Doctors: Medical Killing and the Psychology of Genodice.* New York: Basic Books.

Lindberg, David, and Robert Westman, eds. 1990. *Reappraisals of the Scientific Revolution.* Cambridge: Cambridge University Press.

Lloyd, Genevieve. 1984. *The Man of Reason: "Male" and "Female" in Western Philosophy.* Minneapolis: University of Minnesota Press.

Lock, Margaret. 2000. "On Dying Twice: Culture, Technology and the Determination of Death." In Lock, Young, and Cambrosio (2000, 233–262).

Lock, Margaret. 2002. *Twice Dead: Organ Transplants and the Reinvention of Death.* Berkeley: University of California Press.

Lock, Margaret, Allan Young, and Alberto Cambrosio, eds. 2000. *Living and Working with the New Medical Technologies.* Cambridge: Cambridge University Press.

Lofland, J. and Lyn Lofland. 1984. *Analyzing Social Settings: A Guide to Qualitative Observation and Analysis* (2nd ed.). Belmont, CA: Wadsworth. Originally published in 1971.

Longino, Helen. 1990. *Science as Social Knowledge: Values and Objectivity in Scientific Inquiry.* Princeton, NJ: Princeton University Press.

Longino, Helen, and R. Doel. 1983. "Body, Bias, and Behavior: A Comparative Analysis of Reasoning in Two Areas of Biological Science." *Signs* 9: 207–227.

Lorber, Judith. 1975. "Good Patients, Bad Patients." *Journal of Health and Social Behavior* 16: 213–225.

Luker, Kristen. 1975. *Taking Chances: Abortion and the Decision Not to Contracept.* Berkeley: University of California Press.

Luker, Kristen. 1984. *Abortion and the Politics of Motherhood.* Berkeley: University of California Press.

Lurhmann, Tanya. 1996. *The Good Parsi: The Fate of a Colonial Elite in a Postcolonial Society.* Cambridge, MA: Harvard University Press.

Lynch, Michael. 1985a. *Art and Artifact in Laboratory Science: A Study of Shop Work and Shop Talk in a Research Laboratory.* London: Routledge & Kegan Paul.

Lynch, Michael. 1985b. "Discipline and the Material Form of Images: An Analysis of Scientific Visibility." *Social Studies of Science* 15: 37–66.

Lynch, Michael. 1991. "Laboratory Space and the Technological Complex: An Investigation of Topical Contextures." *Science in Context* 4: 51–78.

Lynch, Michael. 1993. *Scientific Practice and Ordinary Action: Ethnomethodological and Social Studies of Science.* Cambridge: Cambridge University Press.

Lynch, Michael. 1994. "Collins, Hirschauer, and Winch: Ethnography, Exoticism, Surgery, Antisepsis, and Dehorsification." *Social Studies of Science* 24: 354–369.

Lynch, Michael. 1999. "Archives in Formation: Privileged Spaces, Popular Archives, and Paper Trails." *History of the Human Sciences* 12(2): 65–87.

Lynch, Michael, and Steven Woolgar, eds. 1988. *Representation in Scientific Practice.* Cambridge, MA: MIT Press.

MacIntyre, Alistair. 1981. *After Virtue.* London: Duckworth.

Mackenzie, Donald. 1990. *Inventing Accuracy: A Historical Sociology of Nuclear Missile Guidance.* Cambridge, MA: MIT Press.

Mackenzie, Donald, and Judy Wajcman, eds. 1999. *The Social Shaping of Technology* (2nd ed.). Milton Keynes, UK: Open University Press. Originally published in 1985.

MacKinnon, Catherine. 1987. *Feminism Unmodified: Discourses on Life and Law.* Cambridge, MA: Harvard University Press.

MacKinnon, Catherine. 1995. "Reflections on Law in the Everyday Life of Women." In Sarat and Kearns (1993, 109–122).

Mannheim, Karl. 1936. *Ideology and Utopia: An Introduction to the Sociology of Knowledge.* Trans. L. Wirth and E. Shils. London: Kegan Paul, Trench, Trubner. Originally published in 1929.

Marcus-Steiff, Joachim. 1991. "Les taux de 'succès' de FIV—Fausses transparences et vrais mensonges." *La Recherche* 20: 25.

Marsh, Margaret, and Wanda Ronner. 1996. *The Empty Cradle: Infertility in America from Colonial Times to the Present.* Baltimore: Johns Hopkins University Press.

Martin, Emily. 1987. *The Woman in the Body: A Cultural Analysis of Reproduction.* Boston: Beacon Press.

Martin, Emily. 1991. "The Egg and the Sperm: How Science Has Constructed a Romance Based on Stereotypical Male-Female Roles." *Signs* 16: 485–501.

Martin, Michele. 1991. *"Hello Central?" Gender, Technology and Culture in the Formation of Telephone Systems.* Montreal: Queen's University Press.

Marx, Karl. 1972. *Capital.* New York: International Publishers.

Marx, Karl. 1982. "The Economic and Philosophical Manuscripts." In Anthony Giddens and David Held, eds., *Classes, Power and Conflict: Classical and Contemporary Debates.* (1982, 12–39).

May, Elaine. 1995. *Barren in the Promised Land: Childless Americans and the Pursuit of Happiness.* New York: Basic Books.

McKinnon, Susan. 2001. "The Economies in Kinship and the Paternity of Culture: Origin Stories in Kinship Theory." In Franklin and McKinnon (2001, 277–301).

McMillan, Carol. 1982. *Women, Reason, and Nature.* Princeton, NJ: Princeton University Press.

McNeil, Maureen, and Sarah Franklin, eds. 1993. "Procreation Stories." *Science as Culture* (special issue) 3(4).

McNeil, Maureen, Ian Varcoe, and Steven Yearley, eds. 1990. *The New Reproductive Technologies.* London: Macmillan.

Meldrum, David, Kaylen Silverberg, Maria Bustillo, and Lynn Stokes. 1998. "Success Rate with Repeated Cycles of In Vitro Fertilization—Embryo Transfer." *Fertility and Sterility* 69: 1005–1009.

Melhuus, Marit, and SigneHowell. 2003. "Kinship Beyond Biology: NRT and Transnational Adoption." Paper presented at Anthropology of Science Association Decennial Conference, Manchester, UK, July 2003.

Menkin, Miriam, and John Rock. 1948. "In-Vitro Fertilization and Cleavage of Human Ovarian Eggs." *American Journal of Obstetrics and Gynecology* 55: 440–452.

Merchant, Caroline. 1980. *The Death of Nature: Women, Ecology, and the Scientific Revolution.* San Francisco: Harper and Row.

Merton, Robert K. 1942. "Science and Technology in a Democratic Order." *Journal of Legal and Political Sociology* 1: 15–26.

Merton, Robert K. 1970. *Science, Technology and Society in Seventeenth-Century England.* New York: Harper and Row. Originally published in 1938.

Merton, Robert K. 1973a. "The Normative Structure of Science." In Storer (1973, 267–278). Originally published in 1942.

Merton, Robert K. 1973b. "Science and the Social Order." In Storer (1973, 254–266). Originally published in 1938.

Michel, Sonya. 1999. *Children's Interests/Mothers' Rights: The Shaping of America's Child Care Policy.* New Haven: Yale University Press.

Mies, Maria. 1987. "Sexist and Racist Implications of New Reproductive Technologies." *Alternatives* 12: 323–342.

Minnow, Martha. 1990. *Making All the Difference: Inclusion, Exclusion, and American Law.* Ithaca: Cornell University Press.

Mirowski, P., and E. Sent, eds. 2001. *Science Bought and Sold: The New Economies of Science.* Chicago: University of Chicago Press.

Mitchell, Lisa. 2001. *Baby's First Picture: Ultrasound and the Politics of Fetal Subjects.* Toronto: University of Toronto Press.

Modell, Judith. 1989. "Last Chance Babies: Interpretations of Parenthood in an IVF Program." *Medical Anthropology Quarterly* 3: 124–138.

Modell, Judith. 1994. *Kinship with Strangers: Adoption and Interpretations of Kinship in American Culture.* Berkeley: University of California Press.

Mol, Annemarie. 2002. *The Body Multiple: Ontology in Medical Practice.* Durham: Duke University Press.

Morantz-Sanchez, Regina. 1997. "Coming to Grips with the Limitations of Science: Infertility and Heredity in American History." *Reviews in American History* 25: 207–212.

Morawski, Jill. 1997. "Imaginings of Parenthood: Artificial Insemination, Experts, Gender Relations, and Paternity." In de Rivera and Sarbin (1997, 229–246).

Morgan, Lynn. 1997. "Ambiguities Lost: Fashioning the Fetus into a Child in Ecuador and the United States." In Carolyn Sargent and Nancy Scheper-Hughes, eds., *The Cultural Politics of Child Survival.* Berkeley: University of California Press.

Morgan, Lynn, and Meredith Michaels, eds. 1999. *Fetal Subjects, Feminist Positions.* Philadelphia: University of Pennsylvania Press.

Moscucci, Ornella. 1990. *The Science of Woman. Gynaecology and Gender in England, 1800–1929.* Cambridge: Cambridge University Press.

Mukerji, Chandra. 1989. *A Fragile Power: Scientists and the State.* Princeton: Princeton University Press.

Mulkay, Michael. 1976. "Norms and Ideology in Science." *Social Science Information* 15: 637–656.

Mulkay, Michael. 1979. *Science and Sociology of Knowledge.* London: Allen and Unwin.

Munoz, Jose. 1999. *Disidentifications: Queers of Color and the Performance of Politics.* Minneapolis: University of Minnesota Press.

Nachtigall, R. D., G. Becker, and M. Wozny. 1992. "Effects of Gender-Specific Diagnosis on Men's and Women's Response to Infertility." *Fertility and Sterility* 57: 113–121.

National Institutes of Health. 1994. *Final Report of the Human Embryo Research Panel (Muller panel).* Bethesda, MD: National Institutes of Health.

Needham, Joseph. 1988. *Science and Civilization in China* (Vol. 1). Cambridge: Cambridge University Press. Originally published in 1954.

Nelkin, Dorothy, and M. Susan Lindee. 1995. *The DNA Mystique: The Gene as a Cultural Icon.* New York: Freeman.

Nelson, Jennifer. 2003. *Women of Color and the Reproductive Rights Movement.* New York: New York University Press.

Neuman, P. J., S. D. Gharib, and M. C. Weinstein. 1994. "The Cost of Successful Delivery with In Vitro Fertilization." *New England Journal of Medicine* 331: 239–243.

Newman, Karen. 1996. *Fetal Positions: Individualism, Science, Visuality.* Stanford, CA: Stanford University Press.

Nicholson, Linda, ed. 1990. *Feminism/Postmodernism.* New York: Routledge.

Noble, David. 1984. *Forces of Production: A Social History of Industrial Automation.* New York: Knopf.

Nussbaum, Martha, and Cass R. Sunstein, eds. 1998. *Clones and Clones: Facts and Fantasies about Human Cloning.* New York: Norton.

Oakley, Ann. 1984. *The Captured Womb: A History of the Medical Care of Pregnant Women.* Oxford: Blackwell.

Oaks, Laurie. 2001. *Smoking and Pregnancy: The Politics of Fetal Protection.* New Brunswick, NJ: Rutgers University Press.

O'Brien, Mary. 1981. *The Politics of Reproduction.* London: Routledge and Kegan Paul.

Ong, Aiwa, and Michael Peletz, eds. 1995. *Bewitching Women, Pious Men: Gender and Body Politics in Southeast Asia.* Berkeley: University of California Press.

Oreskes, Naomi. 1997. "Objectivity or Heroism? On the Invisibility of Women in Science." *Osiris* 11: 87–113.

Ortner, Sherry. 1974. "Is Female to Nature as Male Is to Culture?" In Rosaldo and Lamphere (1974, 67–96).

Oudshoorn, Nelly. 1994. *Beyond the Natural Body: An Archaeology of Sex Hormones.* New York: Routledge.

Out, Henk, Stefan Driessen, Bernadette Mannaerts, and Herhan Coelingh Bennink. 1997. "Recombinant Follicle-Stimulating Hormone (follitropin beta, Puregon®) Yields Higher Pregnancy Rates in In Vitro Fertilization Than Urinary Gonadotropins." *Fertility and Sterility* 68: 138–142.

Palermo, G., H. Joris, P. Devroey, and A. Van Steirteghem. 1992. "Pregnancies after Intracytoplasmic Injection of Single Spermatozoon into an Oocyte." *Lancet* 340: 17–18.

Parfit, Derek. 1984. *Reasons and Persons.* Oxford: Oxford University Press.

Paris, Gilles. 1994. "Les sénateurs réservent l'assistance à la procréation aux couples formés depuis deux ans au moins." *Le Monde,* January 20, 11.

Patel, Vibhuti. 1987. "Eliminate Inequality, Not Women." *Connexions* 25: 2–3.

Paxson, Heather. 2004. *Making Modern Mothers: Ethics and Family Planning in Urban Greece.* Berkeley: University of California Press.

Peletz, Michael. 1996. *Reason and Passion: Representations of Gender in a Malay Society.* Berkeley: University of California Press.

Pellegrino, Edmund, John Collins Harvey, and John Langan, eds. 1990. *Gift of Life: Catholic Scholars Respond to the Vatican Instruction.* Washington, DC: Georgetown University Press.

Petchesky, Roaslind Pollack. 1990. *Abortion and Women's Choice: The State, Sexuality, and Reproductive Freedom.* Boston: Northeastern University Press. Originally published in 1984.

Petchesky, Rosalind Pollack. 1987. "Foetal Images: The Power of Visual Culture in the Politics of Reproduction." In Stanworth (1987, 57–80).

Petryna, Adriana. 2002. *Life Exposed: Biological Citizens after Chernobyl.* Princeton, NJ: Princeton University Press.

Pfeffer, Naomi. 1987. "Artificial Insemination, In Vitro Fertilization and the Stigma of Infertility." In Stanworth (1987, 81–97).

Pfeffer, Naomi. 1993. *The Stork and the Syringe: A Political History of Reproductive Medicine.* Cambridge: Polity Press.

Pfeffer, Naomi, and Anne Woollett. 1983. *The Experience of Infertility.* London: Virago.

Pickering, Andrew. 1984. *Constructing Quarks: A Sociological History of Particle Physics.* Edinburgh: Edinburgh University Press.

Pickering, Andrew, ed. 1992. *Science as Practice and Culture.* Chicago: Chicago University Press.

Pickering, Andrew. 1995. *The Mangle of Practice: Time, Agency, and Science.* Chicago: Chicago University Press.

Pinch, Trevor. 1985. "Towards an Analysis of Scientific Observation: The Externality and Evidential Significance of Observation Reports in Physics." *Social Studies of Science* 15: 167–187.

Pinch, Trevor, and Wiebe Bijker. 1984. "The Social Construction of Facts and Artifacts, or How the Sociology of Science and the Sociology of Technology Might Benefit Each Other." *Social Studies of Science* 14: 399–441.

Pippin, Robert. 1996. "Medical Practice and Social Authority." *Journal of Medicine and Philosophy* 21: 357–373.

Porchet, H. C., J. Y. Le Cotonnec, S. Canali, and G. Zanolo. 1998. "Pharmacokinetics of Recombinant Human Follicle Stimulating Hormone after Intravenous, Intramuscular, and Subcutaneous Administration in Monkeys, and Comparison with Intravenous Administration of Urinary Follicle Stimulating Hormone." *PMID*: 8095209, PubMed indexed for Medline. Ares Serono, Geneva, Switzerland.

Poster, Mark. 1990. *The Mode of Information: Poststructuralism and Social Context.* Oxford: Blackwell.

Prevost, R. R. 1998. "Recombinant Follicle-Stimulating Hormone: New Biotechnology for Infertility." *Pharmacotherapy* 18(5): 1001–1010.

Prins, Baukje. 1995. "The Ethics of Hybrid Subjects: Feminist Constructivism according to Donna Haraway." *Science, Technology, and Human Values* 20: 352–367.

Probyn, Elspeth. 1995. "Queer Belongings: The Politics of Departure." In Grosz and Probyn (1995, 1–18).

Proctor, Robert. 1988. *Racial Hygiene: Medicine under the Nazis.* Cambridge, MA: Harvard University Press.

Proctor, Robert. 2000. *The Nazi War on Cancer.* Princeton, NJ: Princeton University Press.

Purdy, Laura, ed. 1989. *Hypatia* (special issue on ethics and reproduction) 4(3).

Quine, W. V. O., ed. 1953. *From a Logical Point of View.* Cambridge: Cambridge University Press.

Quine, W. V. O. 1960. *Word and Object.* Cambridge, MA: MIT Press.

Quine, W. V. O. 1969. *Ontological Relativity and Other Essays.* New York: Columbia University Press.

Rabinow, Paul. 1992. "Artificiality and Enlightenment: From Sociobiology to Biosociality." In Crary and Kwinter (1992, 234–252).

Rabinow, Paul. 1996a. *Essays on the Anthropology of Reason.* Princeton, NJ: Princeton University Press.

Rabinow, Paul. 1996b. *Making PCR: A Story of Biotechnology.* Chicago and London: University of Chicago Press.

Rabinow, Paul. 1999. *French DNA: Trouble in Purgatory.* Chicago: University of Chicago Press.

Rabinow, Paul. 2000. "Epochs, Presents, Events." In Lock, Young, and Cambrosio (2000, 31–46).

Radcliffe, Sarah, and Sallie Westwood, eds. 1993. *Viva: Women and Popular Protest in Latin America.* New York: Routledge.

Radin, Margaret. 1996. *Contested Commodities: The Trouble with Trade in Sex, Children, Body Parts, and Other Things.* Cambridge, MA: Harvard University Press.

Ragoné, Helena. 1994. *Surrogate Motherhood: Conception in the Heart.* Boulder, CO: Westview Press.

Rainwater, Lee. 1960. *And the Poor Get Children: Sex, Contraception and Family Planning in the Working Class.* Assisted by Karol Kane Weinstein. Chicago: Quadrangle Books.

Rapp, Rayna. 1978. "Family and Class in Contemporary America." *Science and Society* 42: 278–300.

Rapp, Rayna. 1987. "Moral Pioneers: Women, Men and Fetuses on a Frontier of Reproductive Technology." *Women and Health* 13: 101–116.

Rapp, Rayna. 1995a. "Heredity, or: Revising the Facts of Life." In Yanagisako and Delaney (1995, 69–86).

Rapp, Rayna. 1995b. "Real Time Fetus: The Role of the Sonogram in the Age of Monitored Reproduction." In Downey, Dumit, and Traweek (1995, 31–48).

Rapp, Rayna. 1996. "Extra Chromosomes: Medico-Familial Interpretations." Paper prepared for SSRC Conference on "Cultures of Biomedicine." Cambridge, UK, July 3–6.

Rapp, Rayna. 1999. *Testing Women, Testing the Fetus: The Social Impact of Amniocentesis in America.* London: Routledge.

Raymond, Janice. 1993. *Women as Wombs: Reproductive Technologies and the Battle over Women's Freedom.* San Francisco: Harper.

Reed, James. 1983. *The Birth Control Movement and American Society: From Private Vice to Public Virtue.* Princeton, NJ: Princeton University Press.

Reynolds et al. 2003. "Does Insurance Coverage Decrease the Risk for Multiple Births Associated with Assisted Reproductive Technology?" *Fertility and Sterility* 80: 1.

Rheinberger, Hans-Jorg. 1997. *Toward a History of Epistemic Things: Synthesizing Proteins in the Test Tube.* Stanford: Stanford University Press.

Rheinberger, Hans-Jorg. 2000. "Beyond Nature and Culture: Modes of Reasoning in the Age of Molecular Biology and Medicine." In Lock, Young, and Cambrosio (2000, 19–30).

Rich, Adrienne. 1976. *Of Woman Born: Motherhood as Experience and as Institution.* New York: Norton.

Richardson, Mattie Udora. 2003. "No More Secrets, No More Lies: African American History and Compulsory Heterosexuality." *Journal of Women's History* 15: 63–76.

Rifkin, Jeremy. 1999. *The Biotech Century: Harnessing the Gene and Remaking the World.* New York: Tarcher and Putnam.

Roberts, Dorothy. 1999. *Killing the Black Body: Race, Reproduction, and the Meaning of Liberty.* New York: Vintage Books.

Robertson, John. 1996. "Legal Troublespots in Assisted Reproduction." *Fertility and Sterility* 65: 11–12.

Rock, John. 1963. *The Time Has Come: A Catholic Doctor's Proposals to End the Battle over Birth Control.* New York: Knopf.

Rodin, Judith, and Aila Collins, eds. 1991. *Women and New Reproductive Technologies: Medical, Psychosocial, Legal and Ethical Dilemmas.* Hillside, NJ: Erlbaum.

Roest, J., A. M. van Heusden, G. H. Zeilmaker, and A. Verhoeff. 1998. "Cumulative Pregnancy Rates and Selective Drop-out of Patients in In-Vitro Fertilization Treatment." *Human Reproduction* 13: 339–341.

Roggencamp, Viola. 1989. "Abortion of a Special Kind: Male Sex Selection in India." In Arditti, Klein, and Minden (1989, 266–277). Originally published in 1984.

Rosaldo, Michelle, and Louise Lamphere, eds. 1974. *Women, Culture and Society.* Stanford, CA: Stanford University Press.

Rose, Steven, ed. 1982. *Towards a Liberatory Biology: The Dialectics of Biology Group.* London: Allison & Busby.

Rosen, Jeffrey. 2001. *The Unwanted Gaze: The Destruction of Privacy in America.* New York: Vintage Books.

Rosenberg, Charles. 1999. "Meanings, Policies, and Medicine: On the Bioethical Enterprise and History." *Daedalus* 128(4): 27–46.

Rosenberg, Charles. 2002. "What Is Disease? In Memory of Owsei Temkin." Paper prepared for the Oswei Temkin Memorial Symposium, Johns Hopkins School of Medicine, October 5.

Rothman, Barbara Katz. 1982. *In Labor: Women and Power in the Birthplace.* New York: Norton.

Rothman, Barbara Katz. 1986. *The Tentative Pregnancy: Prenatal Diagnosis and the Future of Motherhood.* New York: Viking Press.

Rothman, Barbara Katz. 1989. *Recreating Motherhood: Ideology and Technology in a Patriarchal Society.* New York: Norton.

Rothman, Barbara Katz. 1998. *Genetic Maps and Human Imaginations: The Limits of Science in Understanding Who We Are.* New York: Norton.

Rothman, David. 1990. "Human Experimentation and the Origins of Bioethics in the United States." In George Weisz, ed., *Social Science Perspectives on Medical Ethics* (185–200). Philadelphia: University of Pennsylvania Press.

Rothman, David. 1991. *Strangers at the Bedside: A History of How Law and Bioethics Transformed Medical Decision Making.* New York: Basic Books.

Rowe, Patrick, Frank Comhaire, Timothy Hargreave, and Heather Mellows. 1993. *WHO Manual for the Standardized Investigation and Diagnosis of the Infertile Couple.* Cambridge: Cambridge University Press, for the World Health Organization.

Rowland, Robyn. 1992. *Living Laboratories: Women and Reproductive Technologies.* Bloomington: Indiana University Press.

Rubin, Gayle. 1975. "The Traffic in Women: Notes on the 'Political Economy' of Sex." In R. Reiter, ed., *Toward an Anthropology of Women* (157–210). New York: Monthly Review Press.

Ruddick, Sara. 1983. "Maternal Thinking." In Joyce Treblicot, ed., *Mothering: Essays in Feminist Theory* (213–230). Totowa, NJ: Rowman and Allenheld.

Rudwick, Martin. 1985. *The Great Devonian Controversy: The Shaping of Scientific Knowledge among Gentlemanly Specialists.* Chicago: University of Chicago Press.

Sandelowski, Margarete. 1990. "Failures of Volition: Female Agency and Infertility in Historical Perspective." *Signs* 15: 475–499.

Sandelowski, Margarete. 1991. "Compelled to Try: The Never Enough Quality of Conceptive Technology." *Medical Anthropology Quarterly* 5: 29–40.

Sandelowski, Margarete. 1993. *With Child in Mind: Studies of the Personal Encounter with Infertility.* Philadelphia: University of Pennsylvania Press.

Sarat, Austin, and Thomas Kearns, eds. 1993. *Law in Everyday Life.* Ann Arbor: University of Michigan Press.

Sawicki, Jan. 1991. *Disciplining Foucault: Feminism, Power, and the Body.* London: Routledge.

Saxton, Marsha. 1989. "Born and Unborn: The Implications of Reproductive Technologies for People with Disabilities." In Arditti, Klein, and Minden (1989, 298–312). Originally published in 1984.

Schaffer, Simon. 1992. "Late Victorian Metrology and Its Instrumentation: A Manufactory of Ohms." In Robert Bud and Susan Cozzens, eds., *Invisible Connections: Instruments, Institutions, and Science* (23–56). Bellingham, WA: SPIE Optical Engineering Press.

Schenker, Joseph. 1996. "Religious Views regarding Gamete Donation." In M. Seiber and S. Crockin, eds., *Family Building through Egg and Sperm Donation* (238–250). Sudbury: Jones and Bartlett.

Scheper-Hughes, Nancy, and Loic Wacquant, eds. 2003. *Commodifying Bodies.* London: Sage.

Scherer, Donald, ed. 1990. *Upstream/Downstream: Issues in Environmental Ethics.* Philadelphia: Temple University Press.

Schiebinger, Londa. 1989. *The Mind Has No Sex? Women in the Origins of Modern Science.* Cambridge, MA: Harvard University Press.

Schmidt, Matthew, and Lisa Jean Moore. 1998. "Constructing a 'Good Catch,' Picking a Winner: The Development of Technosemen and the Deconstruction of the Monolithic Male." In Davis-Floyd and Dumit (1998, 21–39).

Schneider, David M. 1980. *American Kinship: A Cultural Account* (2nd ed.). Chicago: University of Chicago Press. Originally published in 1968.

Schutz, Alfred. 1970. *On Phenomenology and Social Relations: Selected Writings*. Ed. Helmut Wagner. Chicago: Chicago University Press.

Scutt, Jocelynne, ed. 1990. *The Baby Machine: Reproductive Technology and the Commercialization of Motherhood*. London: Green Print.

Seal, Vivian. 1990. *Whose Choice? Working Class Women and the Control of Fertility*. London: Fortress Books.

Seibel, Machelle, and Susan Crockin, eds. 1996. *Family Building through Egg and Sperm Donation: Medical, Legal, and Ethical Issues*. Sudbury, MA: Jones and Bartlett.

Seibel, M. M., S. Glazier Seibel, and M. Zilberstein. 1994. "Gender Distribution—Not Sex Selection." *Human Reproduction* 9: 569–570.

Shapin, Steven. 1982. "History of Science and Its Sociological Reconstructions." *History of Science* 20: 157–211.

Shapin, Steven. 1984. "Pump and Circumstance: Robert Boyle's Literary Technology." *Social Studies of Science* 14: 481–520.

Shapin, Steven. 1992. "Discipline and Bounding: The History and Sociology of Science as Seen through the Externalism/Internalism Debate." *History of Science* 30: 333–369.

Shapin, Steven. 1993. "Essay Review: Personal Development and Intellectual Biography: The Case of Robert Boyle." *British Journal for the History of Science* 26: 335–345.

Shapin, Steven. 1994. *A Social History of Truth: Civility and Science in Seventeenth-Century England*. Chicago: University of Chicago Press.

Shapin, Steven. 1999. "Scientific Antlers." *London Review of Books* 21(5): 27–28.

Shapin, Steven, and Simon Schaffer. 1985. *Leviathan and the Air-Pump: Hobbes, Boyle, and the Experimental Life*. Princeton: Princeton University Press.

Shenfield, F. 1994. "Sex Selection: A Matter for 'Fancy' or the Ethical Debate?" *Human Reproduction* 9: 569.

Shoham, J., and V. Insler. 1996. "Recombinant Technique and Gonadotropins: New Era in Reproductive Medicine." *Fertility & Sterility* 66(2): 187–201.

Silber, Sherman, and Larry Johnson. 1998. "Are Spermatid Injections of Any Clinical Value?" *Human Reproduction* 13: 509–515.

Singer, Peter, and Deanne Wells. 1985. *Making Babies: The New Science and Ethics of Conception*. New York: Scribner's.

Singleton, Vicky, and Mike Michael. 1993. "Actor Networks and Ambivalence: General Practitioners in the U.K. Cervical Screening Programme." *Social Studies of Science* 23: 227–264.

Skocpol, Theda. 1992. *Protecting Soldiers and Mothers: The Political Origins of Social Policy in the United States.* Cambridge, MA: Harvard University Press.

Smith, Brian Cantwell. 1996. *On the Origins of Objects.* Cambridge, MA: MIT Press.

Smith, Kristen, and Richard Buyalos. 1996. "The Profound Impact of Patient Age on Pregnancy Outcome after Early Detection of Fetal Cardiac Activity." *Fertility and Sterility* 65: 35–40.

Sokal, Alan. 1996a. "Transgressing the Boundaries: Toward Transformative Hermeneutics of Quantum Gravity." *Social Text* (Spring-Summer): 217–252.

Sokal, Alan. 1996b. "A Physicist Experiments with Cultural Studies." *Lingua Franca* (May–June): 61–64.

Solinger, Rickie. 2001. *Beggars and Choosers: How the Politics of Choice Shapes Adoption, Abortion, and Welfare in the United States.* New York: Hill and Wang.

Solomon, Alison. 1989. "Sometimes Perganol Kills." In Klein (1989, 46–50).

Souter, Vivienne, G. Penney, J. Hopton, and A. Templeton. 1998. "Patient Satisfaction with the Management of Infertility." *Human Reproduction* 13: 1831–1836.

Spallone, Patricia. 1989. *Beyond Conception: The New Politics of Reproduction.* London: Macmillan.

Spallone, Patricia. 1992. *Generation Games: Genetic Engineering and the Future for Our Lives.* London: Women's Press.

Spallone, Patricia, and Deborah Steinberg, eds. 1987. *Made to Order: The Myth of Reproductive and Genetic Progress.* London: Pergamon Press.

Sperber, Dan, David Premack, and Ann Premack, eds. 1995. *Causal Cognition: An Interdisciplinary Approach.* Oxford: Oxford University Press.

Squier, Susan Merrill. 1994. *Babies in Bottles: Twentieth-Century Visions of Reproductive Technology.* New Brunswick, NJ: Rutgers University Press.

Squier, Susan and E. Ann Kaplan, eds. 1999. *Playing Dolly: Technocultural Formations, Fantasies, and Fictions of Assisted Reproduction.* New Brunswick, NJ: Rutgers University Press.

Squier, Susan. N.d. "Fetal Subjects and Maternal Objects: Reproductive Technology and the New Fetal/Maternal Relation."

Stacey, Margaret, ed. 1992. *Changing Human Reproduction: Social Science Perspectives.* London: Sage.

Stack, Carol. 1974. *All Our Kin.* New York: Basic Books.

Stanworth, Michelle, ed. 1987. *Reproductive Technologies: Gender, Motherhood, and Medicine.* Cambridge: Polity Press.

Star, Susan Leigh. 1990. "Layered Space, Formal Representations, and Long Distance Control: The Politics of Information." *Fundamenta Scientiae* 10: 125–155.

Star, Susan Leigh, and James Griesemer. 1989. "Institutional Ecology, 'Translations,' and Boundary Objects: Amateurs and Professionals in Berkeley's Museum of Vertebrate Zoology, 1907–1939." *Social Studies of Science* 19: 387–420.

Steinberg, Earl, Patrice Holtz, Erin Sullivan, and Christina Villar. 1998. "Profiling Assisted Reproductive Technology: Outcomes and Quality of Infertility Management." *Fertility and Sterility* 69: 617–623.

Stephen, Elizabeth Hervey, and Anjani Chandra. 1998. "Updated Projections of Infertility in the United States: 1995–2025." *Fertility and Sterility* 70: 30–34.

Stern, J. E., C. P. Cramer, A. Garrod, and R. M. Green. 2001. "Access to Services at Assisted Reproductive Technology Clinics: A Survey of Policies and Practices." *American Journal of Obstetrics and Gynecology* 184: 591–597.

Stolcke, Verena. 1986. "New Reproductive Technologies: Same Old Fatherhood." *Critique of Anthropology* 6: 5–31.

Stoler, A. L. 1991. "Carnal Knowledge and Imperial Power: Gender, Race, and Morality in Colonial Asia." In di Leonardo (1991, 51–101).

Storer, N., ed. 1973. *The Sociology of Science: Theoretical and Empirical Investigations.* Chicago: University of Chicago Press.

Strathern, Marilyn. 1992. *Reproducing the Future: Anthropology, Kinship, and the New Reproductive Technologies.* Manchester: Manchester University Press.

Strathern, Marilyn. 2003. "Emergent Relations." In Mario Biagioli and Peter Galison, eds., *Scientific Authorship: Credit and Intellectual Property in Science* (165–194). New York: Routledge.

Strauss, Anselm. 1991. *Creating Sociological Awareness: Collective Images and Symbolic Representations.* New Brunswick, NJ: Transaction.

Strong, Pauline Turner. 2001. "To Forget Their Tongue, Their Name, and Their Whole Relation: Captivity, Extra-Tribal Adoption, and the Indian Child Welfare Act." In Franklin and McKinnon (2001, 468–493).

Strong, Pauline Turner, and Barrik Van Winkle. 1996. "'Indian Blood': Reflections on the Reckoning and Refiguring of Native North American Identity." *Cultural Anthropology* 11(4): 547–576.

Strum, Shirley, and Linda Fedigan, eds. 2000. *Primate Encounters: Models of Science, Gender, and Society.* Chicago: University of Chicago Press.

Sunstein, Cass. 1998. "The Constitution and the Clone." In Nussbaum and R. Sunstein (1998, 207–220).

Tapper, Melbourne. 1999. *In the Blood: Sickle Cell Anemia and the Politics of Race.* Philadelphia: University of Pennsylvania Press.

Taylor, Charles. 1989. *Sources of the Self: The Making of the Modern Identity.* Cambridge, MA: Harvard University Press.

Taylor, Janelle. 1992. "The Public Fetus and the Family Car." *Public Culture* 4: 67–80.

Terry, Jennifer. 1989. "The Body Invaded: Medical Surveillance of Women as Reproducers." *Socialist Review* 19: 13–43.

Terry, Jennifer, and Jacqueline Urla, eds. 1995. *Deviant Bodies: Critical Perspectives on Difference in Science and Popular Culture.* Bloomington: Indiana University Press.

Thompson, Charis. 2004. "Biological Race Dead and Alive: The Case of Third Party Reproduction." Keynote address, Conference "New Feminist Perspectives on Biotechnology and Bioethics." UC Berkeley.

Thompson, Charis. in preparation. *Charismatic Megafauna and Miracle Babies: Essays in Selective Pronatalism.* Manuscript.

Thompson, E. P. 1982. *Exterminism and Cold War.* New York: Schocken Books.

Todd, Alexandra Dundas. 1998. *Intimate Adversaries: Cultural Conflict between Doctors and Women Patients.* Philadelphia: University of Pennsylvania Press.

Tracy, David. 1998. "Human Cloning and the Public Realm: A Defense of Intuitions of the Good." In Nussbaum and Sunstein (1998, 190–203).

Traweek, Sharon. 1988. *Beamtimes and Lifetimes: The World of High Energy Physicists.* Cambridge: Harvard University Press.

Traweek, Sharon. 1993. "An Introduction to Cultural, Gender, and Social Studies of Science and Technologies." *Culture, Medicine, and Psychiatry* 17: 3–25.

Treblicot, Joyce. 1984. *Mothering: Essays in Feminist Theory.* Totowa, NJ: Rowman & Allanheld.

U.S. Department of Health & Human Services. 2004. "50 States Summary of Legislation Related to Insurance Coverage for Infertility Therapy." ⟨http://www.ncsl.org/programs/health/50infert_htm⟩.

Van Balen, Frank, Jacqueline Verdurmen, and Evert Ketting. 1997. "Choices and Motivations of Infertile Couples." *Patient Education and Counseling* 31: 19–27.

VanderLaan, Burton, Vishvanath Karande, Carol Krohm, Randy Morris, Donna Pratt, and Nrobert Gleicher. 1998. "Cost Considerations with Infertility Therapy: Outcome and Cost Comparison between Health Maintenance Organization and Preferred Provider Organization Care Based on Physician and Facility Cost." *Human Reproduction* 13: 1200–1205.

Van der Ploeg, Irma. 1998. *Prosthetic Bodies: Female Embodiment in Reproductive Technologies.* Maastricht: Maastricht University Press.

Vanderzwalmen, P., M. Nijs, R. Schoysman, G. Bertin, B. Lejeune, B. Vandamme, S. Kahraman, and H. Zech. 1998. "The Problems of Spermatid Microinjection in the Human: The Need for an Accurate Morphological Approach and Selective Methods for Viable and Normal Cells." *Human Reproduction* 13: 515–523.

Van Dyck, J. 1995. *Manufacturing Babies and Public Consent.* London: Macmillan.

Van Steirteghem, A., P. Nagy, and H. Joris et al. 1996. "The Development of Intracytoplasmic Sperm Insertion." *Human Reproduction* 11 (suppl. 1): 59–72.

Verran, Helen. 2001. *Science and an African Logic.* Chicago: University of Chicago Press.

Viveiros de Castro, Eduardo. 1998. "Les pronoms cosmologiques et le perspectivisme amérindien." In ed. E. Alliez, *Gille Deleuze. Une vie philosophique* (429–462). Paris: Les Empêcheurs de penser en rond.

Wacjman, Judith. 1991. *Feminism Confronts Technology.* Cambridge: Polity Press.

Wailoo, Keith. 1997. *Drawing Blood: Technology and Disease Identity in Twentieth-Century America.* Baltimore: Johns Hopkins University Press.

Waller, Maureen Rosamond. 2002. *My Baby's Father: Unmarried Parents and Paternal Responsibility.* Ithaca: Cornell University Press.

Warnock, Mary. 1985. *A Question of Life: The Warnock Report on Human Fertilization and Embryology.* London: Basil Blackwell.

Watson-Verran, Helen, and David Turnbull. 1995. "Science and Other Indigenous Knowledge Systems." In Jasanoff, Markle, Petersen, and Pinch (1995, 115–139).

Weber, Max. 1948. "Science as a Vocation." In H. Gerth and C. Wright Mills, eds., *From Max Weber.* New York: Oxford University Press. Originally published in 1918.

Weber, Max. 1958. *The Protestant Ethic and the Spirit of Capitalism.* Trans. Talcott Parsons. New York: Scribner.

Weckstein, Louis. 2002. *The Value of FSH Levels in Guiding Infertility Treatment.* San Francisco: Reproductive Science Center of the San Francisco Bay Area.

Westman, Robert S. 1990. "Proof, Poetics and Patronage: Copernicus' Preface to *De revolutionibus.*" In Lindberg and Westman (1990, 167–206).

Weston, Kath. 1991. *Families We Choose: Lesbians, Gays, Kinship.* New York: Columbia University Press.

Weston, Kath. 2001. "Kinship, Controversy, and the Sharing of Substance: The Race/Class Politics of Blood Transfusion." In Franklin and McKinnon (2001, 147–174).

Whiteford, Linda, and Marilyn Poland, eds. 1989. *New Approaches to Human Reproduction: Social and Ethical Dimensions.* Boulder: Westview Press.

Williamson, Nancy. 1976. *Sons of Daughters: A Cross-Cultural Survey of Parental Preferences.* London: Sage.

Wilmut, Ian, Keith Campbell, and Colin Tudge. 2000. *The Second Creation: Dolly and the Age of Biological Control.* New York: Farrar, Strauss and Giroux.

Winch, Peter. 1958. *The Idea of a Social Science.* London: Routledge & Kegan Paul.

Winston, Robert. 1996. *Making Babies: A Personal View of IVF.* London: BBC Books.

Winston, Robert. 2000. *The IVF Revolution: The Definitive Guide to Assisted Reproductive Technologies.* London: Vermillion.

Wittgenstein, Ludwig. 1953. *Philosophical Investigations.* Oxford: Blackwell.

Woolgar, Steve, ed. 1988. *Knowledge and Reflexivity: New Frontiers in the Sociology of Knowledge.* London: Sage.

Wylie, Alison. 1992. "Reasoning about Ourselves: Feminist Methodology in the Social Sciences." In E. D. Harvey and K. Okruhlik, eds., *Women and Reason.* Ann Arbor: University of Michigan Press.

Wymelenberg, Suzanne. 1990. *Science and Babies: Private Decisions, Public Dilemmas.* Washington, DC: National Academic Press.

Wynne, A. 1988. "Accounting for Accounts of the Diagnosis of Multiple Sclerosis." In Woolgar (1988, 101–122).

Yanagisako, Sylvia, and Carol Delaney, eds. 1995. *Naturalizing Power: Essays in Feminist Cultural Analysis.* New York: Routledge.

Yaron, Yuval, Karen Johnson, Peter Bryant-Greenwood, Ralph Kramer, Mark Johnson, and Mark Evans. 1998. "Selective Termination and Elective Reduction in Twin Pregnancies: Ten Years of Experience at a Single Centre." *Human Reproduction* 13: 2301–2304.

Young, Robert. 1986. *Darwin's Metaphor.* Cambridge: Cambridge University Press.

Yuval-Davis, Nira, and Pnina Werbner, eds. 1999. *Women, Citizenship, and Difference.* New York: Zed Books.

Zelizar, Viviana. 1985. *Pricing the Priceless Child.* New York: Basic Books.

Zilsel, Edgar. 1942. "The Sociological Roots of Modern Science." *American Journal of Sociology* 47: 245–279.

Zinaman, Michael, Eric Clegg, Charles Brown, John O'Connor, and Sherry Selevan. 1996. "Estimates of Human Fertility and Pregnancy Loss." *Fertility and Sterility* 65: 503–509.

Zuboff, Shoshana. 1988. *In the Age of the Smart Machine.* New York: Basic Books.

Index

Page numbers in *italics* refer to tables.

Inside Technology

edited by Wiebe E. Bijker, W. Bernard Carlson, and Trevor Pinch

Janet Abbate, *Inventing the Internet*

Charles Bazerman, *The Languages of Edison's Light*

Marc Berg, *Rationalizing Medical Work: Decision Support Techniques and Medical Practices*

Wiebe E. Bijker, *Of Bicycles, Bakelites, and Bulbs: Toward a Theory of Sociotechnical Change*

Wiebe E. Bijker and John Law, editors, *Shaping Technology/Building Society: Studies in Sociotechnical Change*

Stuart S. Blume, *Insight and Industry: On the Dynamics of Technological Change in Medicine*

Pablo J. Boczkowski, *Emerging Media: Innovation in Online Newspapers*

Geoffrey C. Bowker, *Science on the Run: Information Management and Industrial Geophysics at Schlumberger, 1920–1940*

Geoffrey C. Bowker and Susan Leigh Star, *Sorting Things Out: Classification and Its Consequences*

Louis L. Bucciarelli, *Designing Engineers*

H. M. Collins, *Artificial Experts: Social Knowledge and Intelligent Machines*

Paul N. Edwards, *The Closed World: Computers and the Politics of Discourse in Cold War America*

Herbert Gottweis, *Governing Molecules: The Discursive Politics of Genetic Engineering in Europe and the United States*

Gabrielle Hecht, *The Radiance of France: Nuclear Power and National Identity after World War II*

Kathryn Henderson, *On Line and on Paper: Visual Representations, Visual Culture, and Computer Graphics in Design Engineering*

David Kaiser, editor, *Pedagogy and the Practice of Science: Historical and Contemporary Perspectives*

Peter Keating and Alberto Cambrosio, *Biomedical Platforms: Reproducing the Normal and the Pathological in Late Twentieth-Century Medicine*

Eda Kranakis, *Constructing a Bridge: An Exploration of Engineering Culture, Design, and Research in Nineteenth-Century France and America*

Pamela E. Mack, *Viewing the Earth: The Social Construction of the Landsat Satellite System*

Donald MacKenzie, *Inventing Accuracy: A Historical Sociology of Nuclear Missile Guidance*

Donald MacKenzie, *Knowing Machines: Essays on Technical Change*

Donald MacKenzie, *Mechanizing Proof: Computing, Risk, and Trust*

Maggie Mort, *Building the Trident Network: A Study of the Enrolment of People, Knowledge, and Machines*

Nelly Oudshoorn and Trevor Pinch, editors, *How Users Matter: The Co-Construction of Users and Technologies*

Paul Rosen, *Framing Production: Technology, Culture, and Change in the British Bicycle Industry*

Susanne K. Schmidt and Raymund Werle, *Coordinating Technology: Studies in the International Standardization of Telecommunications*

Charis Thompson, *Making Parents: The Ontological Choreography of Reproductive Technologies*

Dominque Vinck, editor, *Everyday Engineering: An Ethnography of Design and Innovation*

CPSIA information can be obtained
at www.ICGtesting.com
Printed in the USA
FSHW021913160119
55067FS